W9-CRB-519

CHEMISTRY

Library of Congress Cataloging-in-Publication Data

Knorre, D. G. (Dmitri Georgievich)
  Affinity modification of biopolymers.

  Bibliography: p.
  Includes index.
  1. Affinity labeling. 2. Biopolymers—Affinity
labeling. 3. Proteins—Analysis. 4. Nucleic acids—
Analysis. I. Vlassov, Valentin Viktorovich. II. Title.
QP519.9.A37K66   1988   574.19'285   88-2883
ISBN 0-8493-6925-8

International Standard Book Number 0-8493-6925-8

Library of Congress Number 88-2883
Printed in the United States

PREFACE

The fundamental property of proteins and nucleic acids is the ability to bind definite low molecular weight ligands or other biopolymers. This phenomenon is usually called recognition. It is the basis of a number of the most important biological events. Thus, self-assembly of complicated structures such as ribosomes, chromatins, and viruses proceeds due to mutual recognition of nucleic acids and proteins constituting these structures. The first step of all numerous enzymatic reactions is the specific binding of substrates to the active center of the respective enzyme. Hormones and neurotransmitters interact with specific receptors of cell membranes giving rise to a definite biological response of the target cell. Recognition of antigens by antibodies or receptors of T lymphocytes is one of the crucial steps of the action of the immune system. The specific interaction of nucleic acid strands with complementary sequences of nucleotides is the foundation of the mechanism of storage, multiplication, and expression of the genetic information. This incomplete list of phenomena concerning the formation of biological structures, cell metabolism, regulation, nerve impulse transmission, heredity, and immune response clearly demonstrates the extreme significance of recognition processes in biological systems.

Recognition is not usually followed by any changes of the chemical structure of a biopolymer. Thus, enzymatic conversion of substrates is not accompanied by any alteration of the chemical structure of the enzyme if transient covalent binding of certain fragments of substrates to enzymes in the course of formation of labile intermediates is not taken into account. This permits the repeated use of the same biopolymer molecule. At the same time, the absence of chemical consequences of recognition makes it difficult to determine the area of biopolymer participating in the specific complex formation.

Modern biochemistry demonstrates that it is usually possible to introduce changes in the ligand structure which do not severely damage the ability of the structure to be recognized by the respective biopolymer. This means that in the modified ligand the points are touched which are not critically essential for the specific complex formation. In particular, some reactive groups may be inserted into the ligand. In this case, the specific complex formation may result in the chemical reaction between this reactive group and biopolymer residue adjacent to the recognition area. This reaction should strongly predominate over reactions of the same compound with similar residues of biopolymer due to close contact between reaction partners. Therefore, selective chemical modification of a definite area of biopolymer close to the region of specific interaction may be performed. As the selectivity is achieved due to affinity of the reactive ligand, the process may be called affinity modification of biopolymers.

As a general approach to a specific attack of enzymes and other functional proteins, affinity modification was first realized by B. R. Baker.[1] As the region where recognition and subsequent catalytic transformation of a substrate proceeds is usually referred to as active site (active center), the approach as well as his book were called *Design of Active Site Directed Irreversible Inhibitors*.

About the same time, N. I. Grineva started to elaborate the similar approach to selective modification of nucleic acids. It was proposed to attach reactive groups to oligonucleotides with sequences complementary to a definite region of the target nucleic acid. The method was called complementary addressed modification.[2,3] Oligonucleotide moiety of such reagents may be considered as an address directing the reactive group to the desired fragment of nucleic acid subjected to modification.

In the 2 following decades affinity modification was widely used to study enzymes, receptors, immunoglobulins, nucleoproteins, and other biopolymers and complexes. To more easily localize the residues introduced into biopolymers by means of affinity modification, reagents were usually supplied with radioactive labels. Therefore, the method is referred to as affinity labeling. It has become one of the powerful tools of investigating structure and function of proteins, nucleic acids, and complexes of proteins and nucleic acids. Pharmacological applications of the approach were discussed in a set of papers. Different aspects of affinity labeling were considered in some reviews. A special volume of *Methods in Enzymology*[4] was devoted to this method. In 1981, the Federation of the European Biochemical Society (FEBS) organized special practical courses for affinity labeling in the Institute of Organic Chemistry in Novosibirsk, U.S.S.R. The lectures given by leading scientists in the field were edited as a separate book.[5]

The goal of this book is to give a systematic description of the main principles of affinity modification and applications, consideration of possibilities, and restrictions of the method. Modification within specific complexes is a special case of chemical modification which is widely used in the nonaddressed version in biochemistry and related areas. Therefore, we have included in the first introductory chapter of the book general considerations of chemical modifications of biopolymers and the application of biopolymers.

The main principles and possibilities of affinity modification are illustrated by a great number of experimental works covering the fields of biochemistry, molecular biology, pharmacology, and immunochemistry. Therefore, the primary knowledge of the main concepts of these disciplines by the readers is suggested. However, short descriptions of bio- and immunochemical systems considered in the experimental examples are given in the text.

Among numerous types of specific interactions, the recognition of nucleic acids and the components by proteins attracts the greatest attention of scientists. Several hundred enzymes deal with nucleotides, nucleosidedi-, and triphosphates, and nucleotide coenzymes as substrates and effectors. Almost all proteins and nucleoproteides participating in the storage, multiplication, and expression of genetic information interact specifically with nucleic acids. Among them are DNA and RNA polymerases, ribosomes, translation and transcription factors, enzymes of the reparation and recombination systems, different DNA unwinding proteins, and aminoacyl-tRNA-synthetases. Therefore, some preference is given in this book to affinity labeling of biopolymers with reactive derivatives of nucleic acids and components. A special section is devoted to this group of compounds and the results obtained by affinity modification of polymerases of nucleic acids, ribisomes, and aminoacyl-tRNA-synthetases are given in more detail as compared with the numerous number of other systems discussed in this book.

It is impossible to represent the complete survey of the several thousand scientific papers dealing with affinity modification. At the same time, we tried to present all the reviews concerning the problems and achievements of the last few years.

THE AUTHORS

Dmitri G. Knorre, Professor, is the Director of the Novosibirsk Institute of Bioorganic Chemistry of the Siberian Division of the U.S.S.R. Academy of Sciences. He was a student of Mendeleev Institute of Technology in Moscow. As a postgraduate student, he worked at the Institute of Chemical Physics, and in 1951 he was awarded the doctor's degree in chemistry. In 1968, he became a correspondent member and, in 1981, full member of the U.S.S.R. Academy of Sciences.

Professor Knorre's interest in biochemistry dates back to 1956 when, still in Moscow, he started the first exercises in bioorganic chemistry of peptides and then in nucleic acids chemistry. In 1961, he moved to the Institute of Organic Chemistry in Novosibirsk, and, until 1984, headed the Department of Biochemistry in this institute. Since 1984, Professor Knorre has been the Director of a newly formed Institute of Bioorganic Chemistry in Novosibirsk.

His scientific interests were concentrated around chemical modification of biopolymers, mainly affinity modification of nucleic acids with reactive derivatives of complementary oligonucleotides (complementary addressed modification of nucleic acids). His main achievements are the elucidation of a mechanism of phosphorylation including some steps of oligonucleotide synthesis, design of a series of new nucleoside triphosphate derivatives, identification of ribosomal proteins involved in interaction with transfer and messenger RNAs by means of affinity modification with reactive derivatives of oligo- and polynucleotides.

Being the Head of the Chair of Molecular Biology at the Novosibirsk State University, Professor Knorre holds lectures in physical chemistry, molecular biology, bioorganic chemistry, and biochemistry.

Valentin V. Vlassov is the Deputy Director of the Novosibirsk Institute of Bioorganic Chemistry. Being a student of the Novosibirsk State University, he started experimental work in Prof. D. G. Knorre's laboratory in the Institute of Organic Chemistry in 1967. In 1972, he received a doctor's degree. Since 1982, Dr. Vlassov has been a head of the Laboratory of Biochemistry of Nucleic Acids, studying biochemical and biological applications of oligonucleotide derivatives.

A basic interest of Dr. Vlassov concerns the chemical modification of biopolymers with special regard to specific modifications of nucleic acids. His main contributions to this field have been to develop new approaches for the investigation of tertiary structure and molecular interactions of RNAs by means of chemical modification with alkylating reagents, nitrogen mustard, and ethyl nitrosourea, to design new derivatives of oligo- and polynucleotides, and to study the possibility of using reactive oligonucleotide derivatives for sequence-specific modification of DNA, specific arrest of mRNA translation, and suppression of viruses multiplication. He holds lectures on chemistry of biopolymers at the Novosibirsk State University.

TABLE OF CONTENTS

Chapter 1

MOLECULAR RECOGNITION AND CHEMICAL MODIFICATION OF BIOPOLYMERS — TWO MAIN COMPONENTS OF AFFINITY MODIFICATION

I. MOLECULAR RECOGNITION

A. Noncovalent Interactions and the Role in Molecular Recognition

Recognition results in the formation of a specific complex of biopolymers and ligands. If the stability of a complex with a definite ligand significantly exceeds that of complexes with similar compounds we refer to the phenomenon as recognition of this ligand by the polymer under consideration.

Complexes are formed by noncovalent interactions of some groups of biopolymers with respective groups of ligands. Three main types of interactions participate in recognition: electrostatic, hydrogen bonding, and hydrophobic interactions.

Electrostatic interactions operate between groups bearing opposite charges. Amino acids bear a positive charge at the amino groups in a protonated state and a negative charge at the carboxylic group in an ionized state. However, the formation of a peptide bond results in the disappearance of these charges and the main polypeptide chain of protiens is electro-neutral with N-terminal NH_3^+ and C-terminal COO^- residues.

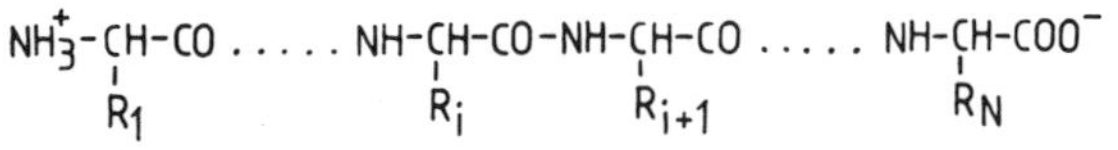

$$\overset{+}{N}H_3-\underset{\underset{R_1}{|}}{C}H-CO \ldots \ldots NH-\underset{\underset{R_i}{|}}{C}H-CO-NH-\underset{\underset{R_{i+1}}{|}}{C}H-CO \ldots \ldots NH-\underset{\underset{R_N}{|}}{C}H-COO^- \qquad (I)$$

Therefore, the main carriers of electric charges in proteins are side-chain groups. These groups are presented in Table 1.

On the contrary, the main chain of nucleic acids

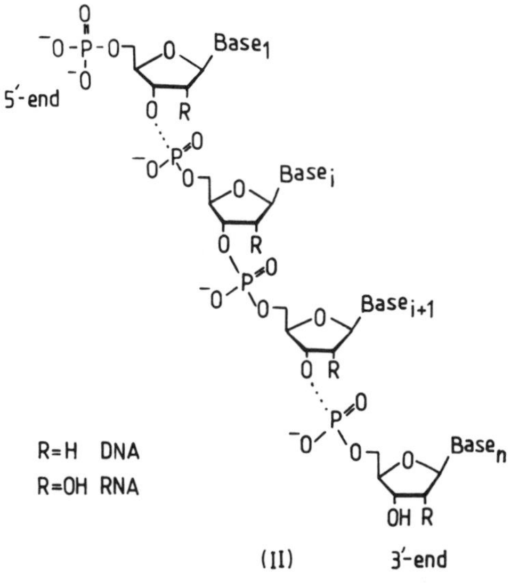

R=H DNA
R=OH RNA

(II)   3′-end

is a carrier of negative charge and, therefore, all nucleic acids are polyanions.

Hydrogen bonding occurs between the polar X-H group (mainly X = N or O) and electron pairs of nonbonding orbitals of F, O, or N atoms. In hydrogen bonding, one of the bonded atoms is the donor, and the other is an acceptor of the proton. The peptide bond contains donor N-H and acceptor C=O groups. Many amino acids (Table 1) bear hydrophilic residues

Table 1
NONCOVALENT INTERACTIONS INHERENT TO SIDE-CHAIN GROUPS OF PROTEINS (AT NEUTRAL pH)

Amino acid	Abbreviation		Structure	Types of interactions
Glycine	Gly	G	$-NH-CH_2-CO-$	—
Alanine	Ala	A	$-NH-\underset{\underset{CH_3}{\mid}}{CH}-CO-$	Weak hydrophobic
Valine	Val	V	$-NH-\underset{\underset{CH(CH_3)_2}{\mid}}{CH}-CO-$	Hydrophobic
Leucine	Leu	L	$-NH-\underset{\underset{CH_2CH(CH_3)_2}{\mid}}{CH}-CO-$	Hydrophobic
Isoleucine	Ile	I	$-NH-\underset{\underset{CH_3-CH-C_2H_5}{\mid}}{CH}-CO-$	Hydrophobic
Proline	Pro	P	$-N-CH-CO-$ (with CH_2, CH_2, CH_2 ring)	Hydrophobic
Phenylalanine	Phe	F	$-NH-\underset{\underset{CH_2-C_6H_5}{\mid}}{CH}-CO-$	Hydrophobic
Tyrosine	Tyr	Y	$-NH-\underset{\underset{CH_2-C_6H_4-OH}{\mid}}{CH}-CO-$	Hydrophobic hydrogen bonding
Tryptophane	Trp	W	$-NH-\underset{\underset{CH_2-\text{(indole)}}{\mid}}{CH}-CO-$	Hydrophobic hydrogen bonding
Serine	Ser	S	$-NH-\underset{\underset{CH_2OH}{\mid}}{CH}-CO-$	Hydrogen bonding
Threonine	Thr	T	$-NH-\underset{\underset{HOCH-CH_3}{\mid}}{CH}-CO-$	Hydrogen bonding
Lysine	Lys	K	$-NH-\underset{\underset{(CH_2)_4-NH_3^+}{\mid}}{CH}-CO-$	Electrostatic hydrogen bonding
Arginine	Arg	R	$-NH-\underset{\underset{(CH_2)_3-NH-C_{\overset{+}{\underset{NH_2}{\parallel}}}NH_2}{\mid}}{CH}-CO-$	Electrostatic hydrogen bonding

Table 1 (continued)
NONCOVALENT INTERACTIONS INHERENT TO SIDE-CHAIN
GROUPS OF PROTEINS (AT NEUTRAL pH)

Amino acid	Abbreviation		Structure	Types of interactions
Histidine	His	H	$-NH-CH-CO-$ CH_2 imidazole (HN, N)	Electrostatic hydrogen bonding
Aspartate	Asp	D	$-NH-CH-CO-$ CH_2-COO^-	Electrostatic hydrogen bonding
Glutamate	Glu	E	$-NH-CH-CO-$ $CH_2-CH_2-COO^-$	Electrostatic hydrogen bonding
Asparagine	Asn	N	$-NH-CH-CO-$ CH_2-CONH_2	Hydrogen bonding
Glutamine	Gln	Q	$-NH-CH-CO-$ $CH_2-CH_2-CONH_2$	Hydrogen bonding
Cysteine	Cys	C	$-NH-CH-CO-$ CH_2-SH	Hydrogen bonding
Methionine	Met	M	$-NH-CH-CO-$ $CH_2-CH_2-SCH_3$	Hydrophobic

able to form hydrogen bonds. Ribose and deoxyribose moieties of nucleotide fragments contain groups capable of participating in hydrogen bonding. Numerous sites for the hydrogen bond formation exist in nucleic acid bases (Table 2).

Hydrophobic interactions reflect the tendency of hydrophobic groups to escape contacts with water molecules in an aqueous solution. Among the groups represented in biopolymer molecules, the main carriers of hydrophobicity are hydrocarbon fragments. The hydrophobic groups of proteins are also given in Table 1. The interaction of these groups takes place due to Van der Waals forces which strongly depend on the distance between interacting atoms (decreasing as the seventh power of the distance). Therefore, an especially strong interaction exists between planar structures providing the best possibilities of contacts between parallel interacting fragments. This type of interaction is called stacking. The ability to stack is inherent to aromatic amino acids and to nucleic acid bases.

Specific complex formation may be observed at the level of simple organic molecules. For example, carboxylic acids are known to be dimerized due to hydrogen bonding.

$$CH_3-C \begin{matrix} O \cdots H-O \\ O-H \cdots O \end{matrix} C-CH_3$$

Ionic species with opposite charges readily form ion pairs. However, this recognition is very low-specific. The specificity is enhanced with the increase of the number of interacting points. It is essential that not only the combination of the types of interaction but to a much

Table 2
HYDROGEN BONDING SITES OF THE NUCLEIC ACID BASES

Base	Abbreviation	Structure	Donor atoms	Acceptor atoms
Uracil	Ura	(structure)	N3	N1, 02, and 04
Thymine	Thy	(structure)	N3	N1, 02, and 04
Cytosine	Cyt	(structure)	N4	N1, 02, and N3
Adenine	Ade	(structure)	N6	N1, N3, N7, and N9
Guanine	Gua	(structure)	N1 and N2	N3, 06, N7, and N9

greater extent the mutual spatial arrangement of the interacting atoms within each of the partners contribute to the diversity of interacting systems.

High aptitude for specific interactions is especially inherent to biopolymers. This is due to a high number of different groups which may serve as interaction sites and sufficient rigidity of the native three-dimensional structure of biopolymers permitting definite mutual orientation of these sites.

As an example of specific recognition in Figure 1, the active center of the enzyme carboxypeptidase A is presented. This enzyme catalyzes the hydrolysis of a peptide bond formed by the C-terminal hydrophobic amino acid. The structure of the enzyme was determined by X-ray crystallography in the absence of a substrate and in the presence of the weak substrate glycyl tyrosine.[6] Beside one glutamate and two histidine residues by which Zn^{2+} is coordinated, an active center contains positively charged arginine residue and adjacent tyrosine and isoleucine residues. These three latter amino acids participate in the recognition of a specific moiety of substrate. It is seen that arginine attracts the negatively charged C-terminal carboxylate hydrophobic part of tyrosine, the side-chain radical of isoleucine interacts with the hydrophobic radical of the C-terminal amino acid, and the OH group of tyrosine may form the hydrogen bond with the amide group to be cleaved.

One of the classic examples of recognition is the specific interaction of complementary sequences of nucleotides in double-stranded nucleic acids. In Figure 2 the complementary base pairs adenine-thymine and guanine-cytosine are represented. In the case of ribonucleic acids (RNAs) uracil is present instead of thymine. Two strands of nucleic acid with nucleotide

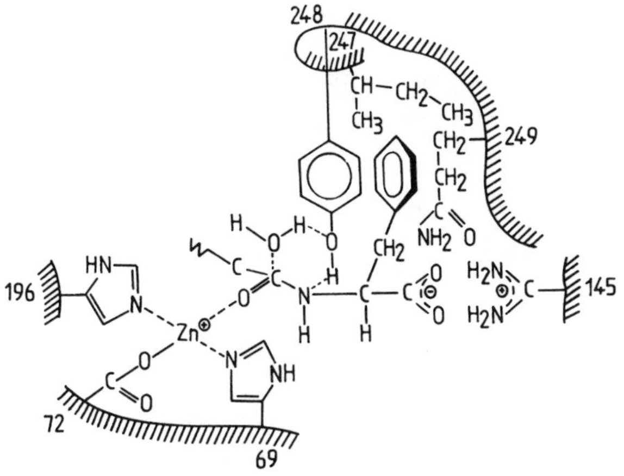

FIGURE 1. Active center of carboxypeptidase A occupied by C-terminal phenylalanine residue of a substrate with a preceding peptide group to be hydrolyzed by a water molecule. The numbers indicate positions of aminoacyl fragments of the enzyme forming the active center in the polypeptide chain: His-69, Glu-72, Arg-145, His-196, Ile-247, Tyr-248, and Gln-249.

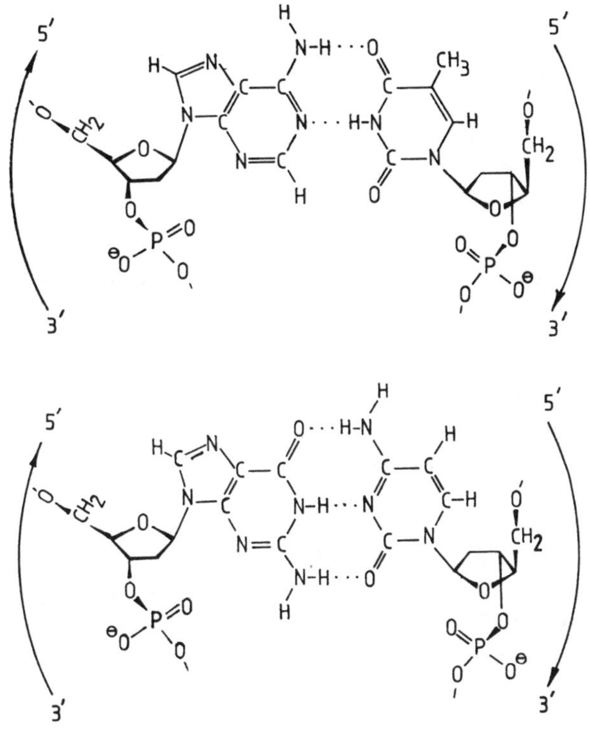

FIGURE 2. Watson-Crick base pairs adenine-thymine (top) and guanine-cytosine (bottom) in a DNA double helix. Arrows indicate the directions of polynucleotide strands.

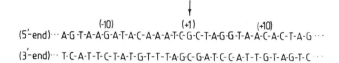

FIGURE 3. Primary structure of the A2 promoter region of double-stranded DNA of bacteriophage T7. The arrow indicates the start point of transcription. The numbers indicate the positions of nucleotide of the upper strand.

sequences with complementary bases along the strands arranged in opposite direction recognize each other and form a specific helical structure.

As an example, the primary structure of the fragment of double-stranded DNA of bacteriophage T7, namely, the region of A2 promoter,[7] is presented in Figure 3. The simultaneous mutual interaction of several different sites of biopolymer and ligand requires a definite conformational state of the ligand. However, for the low molecular weight compounds, different conformations may be considered as being presented in equilibrium concentrations. Complex formation with a ligand in favorable conformation is followed by a displacement of equilibrium towards this state. Therefore, conformation of the ligand within the specific complex of the biopolymer does not necessarily coincide with the conformation predominating in the solution. For example, conformation of deoxythymidine-5'-triphosphate (dTTP) in the complex with *Escherichia coli* DNA polymerase I was elucidated by calculating the distances of a set of protons (CH_3, H_6, H_1', H_2', and H_4') and α, β, and γ phosphates coordinated by the enzyme Mn^{2+} using the frequency dependence of the longitudinal relaxation rate of phosphorus and proton nuclear magnetic resonances. It was found that this conformation is significantly closer to that of dTMP residue in DNA in the double-helical B conformation as compared with dTTP in solution. Thus, the enzyme appears to adjust the conformation of dTTP to Watson-Crick interactions necessary for the selection of a complementary nucleotide in the course of replication.[8]

Certainly biopolymer structure is not absolutely rigid and an interacting area is presented by a number of related conformations. Interaction with a recognized ligand may also displace equilibrium towards a biopolymer conformation more favorable for recognition. Thus, some conformational change of biopolymer may follow the specific complex formation.

B. Intramolecular Recognition in Biopolymers and the Role in the Formation of the Secondary and Tertiary Structure

The ability of biopolymers in particular proteins to function as biologically active molecules is firmly connected with the ability to form definite spatial structures. The most detailed knowledge of coordinates of all atoms of biopolymers is based on the results of X-ray crystallography. One of the most general results of these data is the elucidation of two main levels of spatial structure of biopolymers. The first, more primitive level is periodical secondary structure. For proteins, the main regular elements of the structure are α-helixes and β sheets which are presented in Figures 4 and 5. The formation of these structures is an inherent property of the polypeptide chain. Side-chain radicals distributed along the chain may favor or hinder the formation of these structures. Therefore, the primary structure of a polypeptide (the sequence of amino acid residues) predetermines the preferential conformation of any fragment of polypeptide chain as well as the breaks of these periodic elements. The interaction between side-chain radicals forming α helixes or β sheets results in the formation of a unique three-dimensional tertiary structure with a definite topology of the periodical fragments and joining sequences.

In the single-stranded nucleic acids, a periodicity of the spatial arrangement of the sugar-

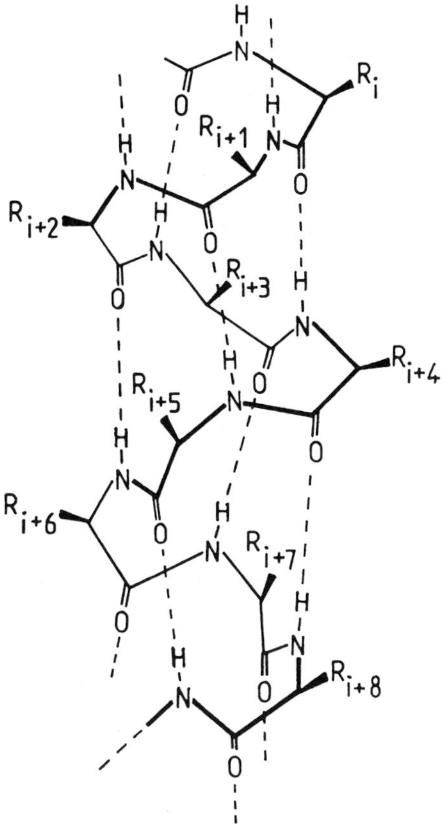

FIGURE 4. Fragment of α helix. R_i, R_{i+1}... side chain radicals of i-th, (i + 1)-th... amino acid residues.

phosphate backbone is met preferentially if some complementary sequences are present at the distance favorable for the double-helix formation. These elements of secondary structure are typical for RNAs which, contrary to DNA, exist in the living cells mainly in a single-stranded form. We shall consider more thoroughly secondary and tertiary structures of transfer ribonucleic acids (tRNAs) as these molecules will often appear in further presentations.

tRNAs are essential components of protein biosynthesis. The building of new polypeptide chains in the living cells proceeds at special nucleoproteins called ribosomes. Details of the structure of ribosomes will be discussed in the next section. The genetic information necessary to combine amino acids into definite sequences is delivered to ribosomes by molecules of messenger ribonucleic acids (mRNAs). In the main region of mRNA coding for a protein (i.e., for the sequence of amino acids) each trinucleotide (codon) corresponds to a definite amino acid in accordance with the genetic code almost universal for all living organisms. Examples are the trinucleotide UUU and UUC codes for phenylalanine, trinucleotide AUG codes for methionine, trinucleotide AAA and AAG codes for lysine, etc. However, codons cannot recognize respective amino acids. To be selected by codons amino acids should be covalently attached to the respective tRNAs. These tRNAs containing special fragments complementary or nearly complementary to the respective codons are recognized by the latter with the participation of ribosome.

The first tRNA which was successfully crystallized in parallel by two groups of investigators and subjected to X-ray diffraction analysis was yeast tRNA specific for phenylal-

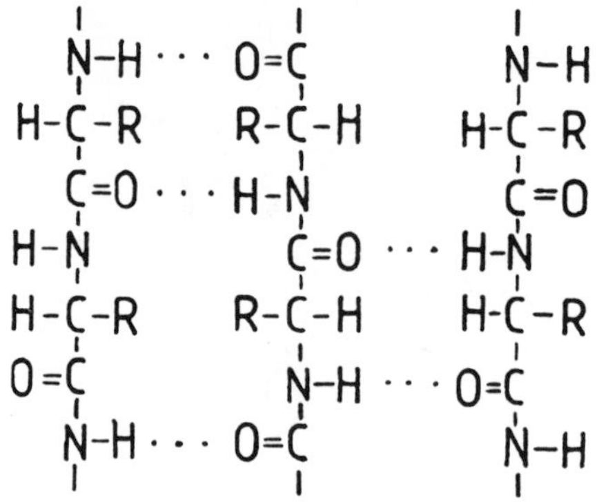

A

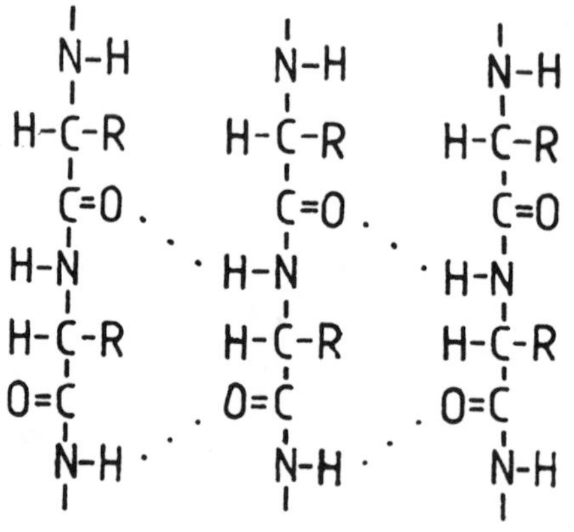

B

FIGURE 5. Antiparallel (A) and parallel (B) β sheets in proteins. Hydrogen bonds between different fragments of polypeptide chain are indicated.

anine, tRNA^{Phe}.[9,10] This tRNA is catalytically acylated with phenylalanine with the help of the special enzyme phenylalnyl-tRNA-synthetase (EC 6.1.1.20) and transfers phenylalanine residue to ribosomes where it is recognized when phenylalanine codons UUU or UUC operate.

Already the knowledge of primary structure of a set of tRNAs of different biological origin and specificity demonstrated that they contain complementary sequences in definite positions. These sequences, separated by a few noncomplementary bases, permit the formation of hairpin structures containing a double-helical stem and a loop. Most tRNAs may form three such hairpins and one stem between the 3′ and 5′ end parts of the molecule. It is traditional to present tRNA nucleotide sequences with these elements of potential secondary structure in the form usually called cloverleaf structure. This form for the tRNA^{Phe} from yeast is presented in Figure 6.[11]

The three-dimensional structure of yeast tRNA^{Phe} is presented in Figure 7. It is seen that tRNA^{Phe} has an L-shaped form. However, all elements of secondary structure present in the clover-leaf form retain the L-shaped structure. The main feature of this molecule is that the T and acceptor stems form a unique double helix as well as do the D and anticodon stems. The molecular reasons for the formation of this structure are analyzed in detail by Rich and

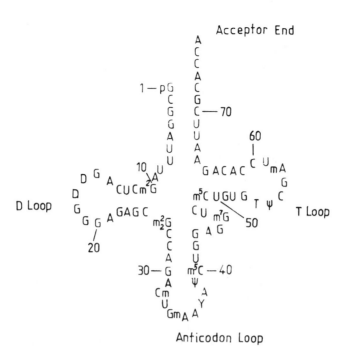

FIGURE 6. The cloverleaf structure of tRNA[Phe] from yeast. Numbers indicate positions of nucleotide residues. Hyphens between nucleoside symbols are omitted. Minor nucleoside abbreviations: m²G, N^2-methylguanosine; D, 5,6-dihydrouridine; m₂²G, N^2,N^2-dimethylguanosine; Cm, 2'-O-methylcytidine; Gm, 2'-O-methylguanosine; Y, wyosine; Ψ, pseudouridine; m⁵C, 5-methylcytidine; M⁷G, 7-methylguanosine; and T, ribothymidine. Sequence GmAA in the anticodon loop represents anticodon, the element recognized by phenylalanine codons UUU and UUC of mRNA.

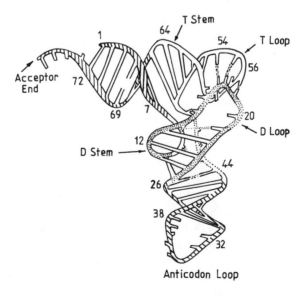

FIGURE 7. Schematic drawing of L-shaped yeast tRNA[Phe]. The main double line indicates the course of ribose-phosphate backbone. Hydrogen bonds are indicated as beams binding different regions of the main chain.

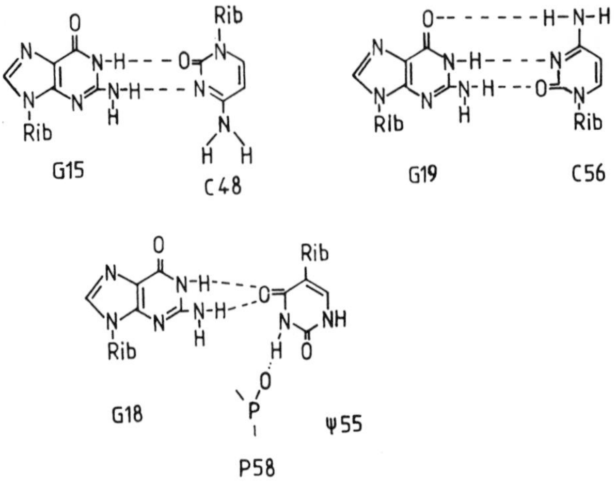

FIGURE 8.   Hydrogen bonds between bases of yeast tRNA[Phe] responsible for joining of D and T loops.

Rajbhandary.[12] There is no need to go into all the details. To illustrate the reasons for the formation of a tertiary structure we shall consider the interactions leading to the joining of T and D loops which are crucial for the transformation of a cloverleaf molecule to an L-shaped one. These are the interactions between G_{15} and C_{48}, G_{18} and Ψ_{55}, and G_{19} and C_{56} (Figure 8). These nucleotide residues do not participate in the formation of secondary structures. Among these three pairs, only G_{19} and C_{56} form the traditional Watson-Crick structure. G_{15} and C_{48} form only two hydrogen bonds instead of three typical of G-C pairs. G_{18} and Ψ_{55} form two hydrogen bonds with the participation of two nonbonding orbitals of the same O atom of pseudouridine.

Thus, as in the previously mentioned case, we see that intramolecular recognition between two parts of a molecule — nucleotides of the D loop and T stem — results in the specific folding of a molecule which is absolutely necessary for further interactions and functioning.

C. Recognition in the Multisubunit Structure Formation: Multipeptide Complexes, Double-Stranded Nucleic Acids, and Nucleoproteins

In most cases proteins and nucleic acids operate in the living organism not as single molecules but as supermolecular complexes. This is typical even of biopolymers with rather simple functions. Thus, it is well known that hemoglobin which functions as a carrier of dioxygen in higher organisms consists of four polypeptide chains bound together by noncovalent forces and each contains one heme residue responsible for the reversible coordination of O_2 molecules. Moreover, these subunits are not identical. There exist pairs of so-called α and β subunits differing in the peptide structure. This is more typical of the enzymes with more complicated functions. Thus, the above-mentioned enzyme phenylalanyl-tRNA-synthetase (EC 6.1.1.20) catalyzes the acylation of tRNA with phenylalanine and uses an additonal adenosine-5'-triphosphate (ATP) molecule as an energy source for the ester bond formation. Thus, this enzyme deals with three substrates (Phe, tRNA[Phe], and ATP). It consists of two α and two β subunits ($\alpha_2\beta_2$ enzyme) and is catalytically active only in this form. Certainly, formation of such structures requires mutual recognition of the definite parts of the polypeptides participating in the complex.

Biochemical conversions of definite physiological significance usually proceed as a number of enzymatic reactions leading to some desired result. As a classic example, we may mention glycolysis. This process starts from the formation of glucose-6-phosphate either by phosphorylation of glucose or by phosphorolysis of glycogen (see Section I.F) to glucose-1-

phosphate with subsequent isomerization. The main goal of glycolysis is the production of energy in the form of

by the phosphorylation of adenosine-5′-diphosphate (ADP). In the course of glycolysis, glucose-6-phosphate converts to pyruvate which in anaerobic conditions is reduced either to lactate or alcohol. In aerobic conditions, pyruvate is further oxidized to CO_2 and H_2O via an intermediate formation of acetyl residue attached to coenzyme A (acetylcoenzyme A)

(III)

and subsequent transfer of this residue to oxaloacetate with formation of citric acid. The latter is further oxidized to oxaloacetate with complete combustion of the acetate fragment to CO_2 and H_2O in the reaction chain known as the tricarbonic acids cycle.

All the oxidation steps in glycolysis, the transformation of pyruvate to $CoASCOCH_3$, and the steps in the tricarbonic acid cycle may be accompanied by phosphorylation of ADP (see Section I.F). As the enzymes participating in glycolysis are extensively studied by affinity labeling, the set of respective reactions are presented in Figure 9.

According to Kurganov et al.,[13] all these enzymes except alcoholdehydrogenase (at least in myofibrils of muscle fiber) are arranged in a single complex containing twelve different enzymes, most of which are present twice. This means that all enzymes recognize their neighbors in the complex and should possess respective recognition areas. Double-stranded DNA contains two different DNA chains and should be formally regarded as a complex with a quaternary structure. However, it is rarely considered this way.

Very complicated structures are met among nucleoproteins. Viruses, chromatin, and ribosomes should be mentioned first. In viruses the information is stored in the nucleic acid which may be represented by double- or single-stranded DNA as well as by double- or single-stranded RNA. In the most simple cases, viruses are built from one nucleic acid molecule and some coat protein. Bacteriophage MS2 and tobacco mosaic virus may be mentioned as such examples. However, even in this case, a great number of the identical coat protein molecules is present in one virus particle.

In many cases the structure of viruses is significantly more complicated. For example, influenza viruses contain eight different RNA molecules coding for different genes and a set of proteins: nucleocapsid protein NP, membrane protein M, hemagglutinin HA, and neuraminidase NA, enzymes PB1, PB2, and PA responsible for the multiplication of a virus.[14]

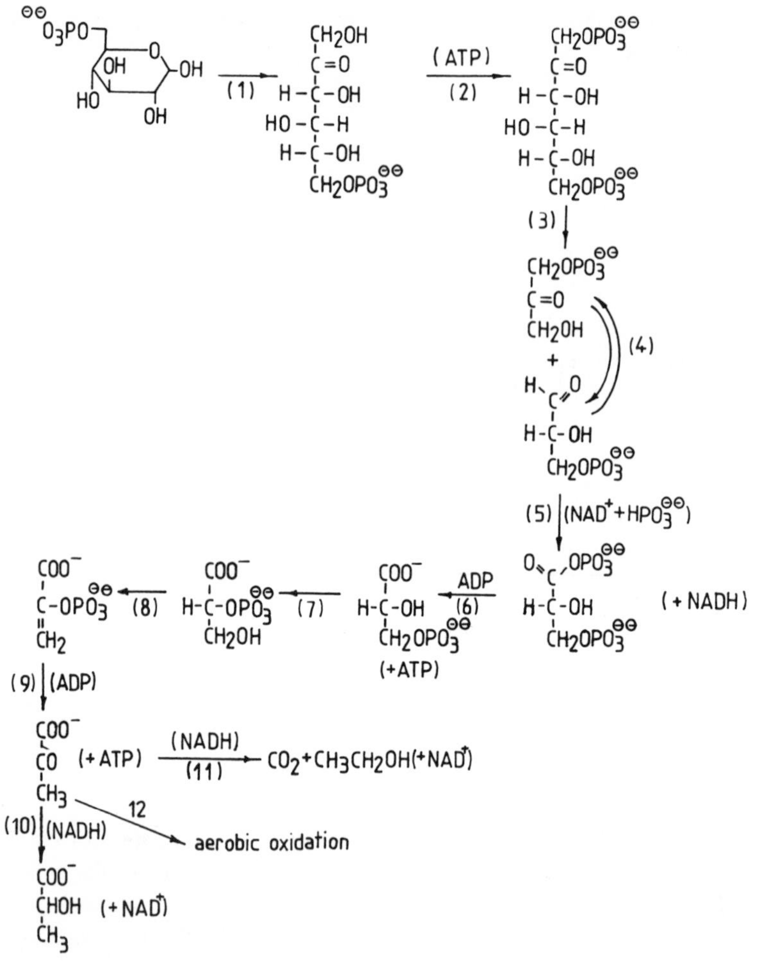

FIGURE 9.   The scheme of glycolysis. Enzymes catalyzed the reactions: (1) glucoso-phosphate isomerase (EC 5.3.1.9); (2) phosphofructokinase (EC 2.7.1.11); (3) aldolase (EC 4.1.2.13); (4) triosophosphate isomerase (EC 5.3.1.1); (5) glyceraldehyde-phosphate dehydrogenase (EC 1.2.1.12); (6) phosphoglycerate kinase (EC 2.7.2.3); (7) phosphog-lycerate mutase (EC 5.4.2.1); (8) enolase [2-phosphoglycerate hydrolase] (EC 4.2.1.11); (9) pyruvate kinase (EC 2.7.1.40); (10) lactate dehydrogenase (EC 1.1.1.27); (11) py-ruvate decarboxylase (EC 4.1.1.1) and alcohol dehydrogenase (EC 1.1.1.1.); and (12) pyruvate dehydrogenase, tricarbonic acid cycle, and the electron transport chain.

The conformity between RNAs and coded proteins as well as information concerning the function of proteins are given in Table 3. The striking feature of this virus is that 5′-terminal sequences for all eight RNAs are identical and represented by the oligonucleotide A-G-U-A-G-A-A-A-C-A-A-G-C-U-C. At the 3′ end all eight RNAs contain the same sequence G-A-C-C-U-G-C-U-U-U-C-G-C-U.[15]

It is essential that such a complicated structure be formed inside the infected cell from all components by self-assembly. It is obvious that this self-assembly suggests the mutual recognition of the constituents.

Ribosomes are another important example. As ribosomes are intensively studied by an affinity labeling technique we shall mention some aspects of the structure. The functions of ribosomes will be discussed in Chapter 5, Section II.D, devoted to affinity labeling of ribosomes.

All ribosomes consist of two unequal subunits, large and small, easily separable at low

Table 3
PROTEINS AND RESPECTIVE CODING RNA SEGMENTS OF INFLUENZA VIRUS

RNA number	Protein	Function
1	PB2	Components for RNA transcribing
2	PB1	system
3	PA	
4	Hemagglutinin HA	Protein responsible for virus adsorption at the cell surface; stimulates agglutination of erythrocytes
5	Nucleocapsid protein NP	Protein covers viral RNAs
6	Neuraminidase (EC 3.2.1.18)	Enzyme which hydrolytically cleaves the glycosidic bond joining sialic acid to D-galactose; function probably is related to the ability of the virus particles to free themselves from neuraminic acid-containing structures including the separation of newly formed virus particles from host cell surface
7	Membrane protein	The most abundant virion protein forming a shell beneath the lipid bilayer surrounding virion
8	Nonstructural proteins NS_1 and NS_2	Proteins which are found in infected cells but are absent from the virion particles; function is not quite clear

Mg^{2+} concentrations. They are usually designated in accordance with the approximate values of the sedimentation constants. Thus, for prokaryotic ribosomes, the whole particles are referred to as 70S ribosomes and the subunits as 50S (large) and 30S (small) subunit. For eukaryotic ribosomes (which are larger than prokaryotic ones), whole ribosomes and subunits of whole ribosomes are referred to as 80S, 60S, and 40S, respectively.

A large subunit of prokaryotic ribosome consists of 2 RNA molecules, 23S and 5S, designated as ribosomes in accordance with the approximate values of sedimentation constants and 32 different proteins designated L1, L2, L3, etc. A small subunit contains 1 molecule of RNA, 16S ribosomal ribonucleic acid (rRNA), and 21 proteins, S1, S2...S21. The numbers of ribosomal proteins bear no biological significance and only reflect the position in the standard electrophoresis system. With some exceptions, the smaller is the number of protein the greater is the molecular mass. A large subunit of eukaryotic ribosomes contains three RNA molecules, 28S, 5.8S, and 5S rRNA, and a number of proteins exceeding that for the prokaryotic system. A small subunit contains one RNA, 18S rRNA, and a set of proteins.

Again, the particles were built in the cell only by self-assembly via exact mutual recognition between all particles. The self-assembly of *Escherichia coli* ribosomal subunits was carried out artificially and it was found that a set of consecutive processes proceeds. The scheme of the in vitro self-assembly may be found in the paper by Nierhaus.[16] There is no need to go into details of this very complicated process. We will only mention that it starts from the recognition of some proteins by rRNA. Thus, definite parts of 16S RNA interact specifically and directly with proteins S15, S17, S4, S8, and S7. Protein S20 is recognized by S4 and S8 bound to 16S RNA. S4 and S20 form the area for recognition of S16, S17 and S16 recognize S12, and so on.

This as well as the preceding section clearly demonstrate the fundamental significance of the biological recognition in the formation of supramolecular structures able to fulfill numerous biological functions. In the next sections we shall consider the role of recognition in the most essential biological phenomena.

D. Active Centers of Enzymes: Recognition of Substrates, Effectors, and Templates

Enzymes are highly efficient catalysts. The rates of reactions catalyzed by enzymes are many orders of magnitude greater than those with similar organic or inorganic catalysts. For example, a zinc ion may serve as a catalyst of the alkaline hydrolysis of a peptide bond. However, in the conditions (pH and temperature) in which this ion operates, being coordinated in the active site of a carboxypeptidase (Figure 1), the reaction catalyzed by an inorganic ion is immeasurably slow.

However, the most striking feature of the enzymes is specificity. Thus, carboxypeptidase A catalyzes the hydrolysis of a C-terminal peptide bond and does not touch the internal bonds. Only the C-terminal L-amino acid is removed. The enzyme strongly prefers C-terminal residues with a hydrophobic side chain. All these features are due to the fact that the substrate should be recognized prior to being hydrolyzed. The recognition of a COO^- group by an arginine residue of the enzyme Arg-145 and of the side chain by hydrohobic Ile-247 and Tyr-248 directs the C-terminal peptide bond to Zn^{2+} which operates as the electrophilic catalyst. In the case of C-terminal D-amino acid, recognition of COO^- group and of the side chain by the same aminoacyl residues of the enzyme should direct to the catalytic center the hydrogen atom bound to assymetrical C_α instead of the peptide bond; no catalysis occurs. Thus, recognition is a foundation of the specificity of enzymes towards substrates including the stereospecificity.

The same is true for the action of effectors (both inhibitors and activators) which influence the reaction rate of enzymatic processes. The sensitivity of enzymatic activity to definite compounds is a feature of many enzymes. We shall restrict ourselves to one example.

One of the key enzymes of glycolysis, 6-phosphofructokinase (see Figure 9), is influenced by a number of compounds, both activators (adenosine-5'-mono- and diphosphates [AMP and ADP]) and inhibitors (ATP and citrate). This means that the enzyme recognizes these compounds and, therefore, possesses special active centers specific for these compounds. The influence on the enzymatic activity is probably due to some conformational changes induced by the interaction with these compounds (see Section I.A). A conformational change induced by AMP and ADP favors catalytic activity whereas the specific binding of ATP or citrate induces some other change distorting the catalytic center of the enzyme.

The biological significance of these interactions is rather obvious. As the main goal of the glycolysis chain is the production of energy in the form of ATP molecules, there is no need to perform glycolysis when ATP predominates among adenylic nucleotides and when citric acid sources are sufficient to produce the necessary amount of energy. In these cases it is economic to cease the fructose-1, 6-diphosphate production which irreversibly consumes ATP molecules. On the contrary, the deficit of energy means that most of the ATP is converted to ADP and AMP. The accumulation of these nucleotides warns that ATP resources are exhausted and the key step of ATP production, phosphorylation of fructoso-6-phosphate, should be accelerated.

Still more complicated enzymes are those which catalyze the formation of new nucleic acid molecules with definite nucleotide sequences. The most common enzymes are DNA-dependent DNA polymerases (EC 2.7.7.7) and RNA polymerases (EC 2.7.7.6). The reaction at each step of growth (elongation) of the polynucleotide chain may be written as follows:

(1)

B (base) = Thy, Cyt, Ade, Gua; R = H for DNA polymerases; B = Ura, Cyt, Ade, Gua; and R = OH for RNA polymerases.

The main peculiarity of these processes is that at each elongation step the enzyme must decide which of four NTPs present in the reaction mixture should be incorporated into the nascent nucleic-acid chain. The decision is taken with the help of a special double-stranded DNA molecule moving along the enzyme. This DNA is usually referred to as a template. Nucleotide residues of one of the strands participate step-by-step in the selection of complementary NTPs which phosphorylate 3'-OH groups of the growing chain. After the incorporation of the chosen nucleotide residue in this chain, the displacement of both template and growing chain by one nucleotide residue (translocation) takes place. In this case the enzyme and template participate simultaneously in the recognition of substrate. Obviously, to do this, the template should be recognized by a definite area of polymerase. The synthesis of new DNA or RNA molecules directed by a DNA template and catalyzed by DNA or RNA polymerases are usually referred to as DNA replication and transcription, respectively.

As these processes are of crucial significance for the multiplication and expression of genetic information; nucleic acid polymerases are extensively studied by different methods, including affinity modification. Therefore, a special section (Chapter 5, Section II.C) will be devoted to this question and additional details concerning the mechanism of action of these enzymes will be given.

E. Recognition in the Regulatory Processes

The regulation of biosynthesis is an indispensable feature of biochemical processes in living cells and organisms. There are several levels of regulation. In the preceding section we have considered the regulation of enzyme activity by inhibitors and activators. By phosphorylation of OH groups of definite serine, threonine, or tyrosine residues of these proteins by protein kinases, the regulation of functional properties of proteins by enzymatic modification is widespread. One such example will be given in the next section.

On a higher level is the regulation of the production of definite enzymes which influences the amounts of these enzymes in the cell. The most studied is the regulation of the synthesis of mRNAs coding for proteins in prokaryotic cells (regulation at the level of transcription).

RNA synthesis starts at a definite point of template DNA. Therefore, RNA polymerase should recognize this point in a manner which permits the selection of the first monomer molecule by the first coding nucleotide residue of the template. After the second NTP is selected with the help of the second coding nucleotide residue, the first phosphorylation step (initiation) proceeds.

$$(2)$$

Although the beginning of RNA synthesis does not necessarily mean that it will be safely finished, this step is crucial for the production of new RNA molecules. The region of template DNA recognized by RNA polymerase prior to initiation is called the promoter. In prokaryotes one promoter often precedes several functionally related genes. For example, there exists a special lac promoter which governs the transcription of several genes coding for proteins necessary for the utilization of lactose. Among them, lactose permease participates in the transfer of lactose and other galactosides via the cell membrane and β-galactosidase catalyzes the hydrolysis of lactose and other β-galactosides. Another well-studied promoter is the trp promoter governing the transcription of five genes coding for enzymes participating in the conversion of anthranilic acid to tryptophane.

Both promoters are under the control of special regulatory proteins called repressors. If the growth medium does not contain lactose, there is no need in the expression of the respective genes. Lac repressor has a high affinity to lac promoter and the latter is almost completely blocked in the presence of the former. At the same time, lac repressor has a high affinity to β-galactosides. The binding of β-galactosides leads to a conformational change depriving the repressor of the affinity towards the lac promoter. Therefore, as soon as lactose appears in the growth medium, the lac repressor dissociates off the promoter thus permitting RNA polymerase to start transcription.

The opposite is the case with the protein regulating trp promoter. In the absence of tryptophane, microorganisms should produce tryptophane and, therefore, need the enzymes synthesizing this amino acid. In these conditions, a regulatory protein called the trp aporepressor does not interact with the promoter. However, as soon as tryptophane appears in the growth medium or in the cells it becomes economic to cease the production of the enzymes of tryptophane biosynthesis. This occurs due to a specific complex formation between tryptophane and the trp aporepressor accompanied by a conformational change leading to the appearance of a protein form with a high affinity to the trp promoter. Consequently, the latter is blocked and further initiation of the respective mRNA biosynthesis is prevented.

Once more we see that recognition, in this case recognition by repressors of definite regions of DNA as well as recognition by these proteins of definite organic molecules, is the original cause of a very important biological phenomenon, the regulation of metabolism.

F. Recognition Processes at the Cell Surface

Many recognition processes of great biological significance proceed at the cell surface with the participation of specific membrane proteins. Among functions of numerous proteins embedded in the cell plasma membrane, three types should be mentioned. One is the transport of different substances across the phospholipid membrane which is impermeable to ions and polar molecules. The second type is the reception of different organic molecules serving as external signals from other cells, organs, or even from other organisms. Third, enzymatic activities are sometimes involved in these two functions and, therefore, respective enzymes are also embedded in the membrane.

In the preceding section we have presented an example of a transport protein, lactose permease, coded by one of the genes of lactose operon. Although a detailed mechanism of action of transport proteins remains unknown, it is evident that the process starts from the recognition of protein by molecules which must be transferred across the membrane. This means that the recognition site resides at the outer surface of the membrane. It is believed that some conformational change follows which makes the inner side of the cell accessible to the bound molecule.

Transport of some substances may proceed against a concentration gradient. Certainly, this transport requires some energy supply and this is achieved by conjugated ATP hydrolysis. The most striking example is the transport of Na^+ ions outside the cell and simultaneous transport of K^+ ions inside the cell by a Na^+, K^+ pump. The process is conjugated with ATP hydrolysis. Therefore, the system is known as Na^+, K^+-dependent ATPase (EC 3.6.1.3). This system is found in the plasma membrane of all animal cells. The hydrolysis of one ATP molecule is accompanied by the transfer of three Na^+ ions outside the cell and the countertransfer of two K^+ ions inside the cell. This process creates the constant electro-chemical potential between the interior of the cell and exterior medium, the former being charged negatively respective to the latter (resting potential). In some cells there exist special Na^+ channels which, being opened, permit Na^+ to move across the membrane from the exterior medium to the interior part of the cell thus decreasing temporarily the potential difference and creating a so-called action potential. After closing the channel, the Na^+, K^+ pump restores the stable resting potential. The propagation of the nerve impulse along neurons is the propagation of the action potential.

A similar system of the outstanding biochemical significance exists in the inner mito-chondrion membrane and is known as F_1 ATPase. The enzyme catalyzes ATP hydrolysis conjugated with the proton translocation across the membrane. F_1 ATPase participates in the production of the main amounts of ATP and therefore is the essential component of ATP synthase. Most organic compounds (e.g., glyceraldehyde phosphate, pyruvate, the inter-mediates participating in the conversion of citric acid to oxaloacetic acid in the tricarbonic acid cycle) are oxidized with the same compound, nicotineamideadenine dinucleotide NAD^+. The latter is converted to reduced form NADH.

$$(3)$$

At the inner mitochondrion membrane NADH is (1) oxidized by O_2 in the electron transport chain. In this chain, NADH is (2) oxidized with ubiquinone with the regeneration of NAD^+ and the formation of respective hydroquinone; the latter is (3) oxidized with two molecules of ferricytochrome c and ferrocytochromes formed reduce O_2 molecules (four cytochromes per one O_2 molecule). Each of three oxidation processes in the inner mitochondrion membrane are conjugated with the condensation of one ADP molecule with inorganic phosphate (oxidative phosphorylation).

$$\text{O-P-O-P-O-Ado} + \text{O-P-OH} \rightleftharpoons \text{O-P-O-P-O-P-O-Ado} + \text{OH}^- \tag{4}$$

F_1 ATPase is one of the essential parts of the oxidative phosphorylation system. It is believed that the flow of electrons from NADH to O_2 across the inner mitochondrion membrane creates the proton gradient which is used by F_1 ATPase to perform ATP formation. It is easily seen that the process is simply a reversal of the ATP hydrolysis-driven proton translocation and is therefore inherent to F_1 ATPase.

It is obvious that both Na^+, K^+ and F_1 ATPase operate as all enzymes and the recognitions of ATP and ADP and of Na^+ and K^+ ions in the first case are prerequisites of the function of these membrane enzymes.

The most intensively studied receptor system is that for acetylcholine, which operates at synapses — the regions of contact either between nerve cells or between nerve and muscle cells.

The arrival of an action potential in the presynaptic terminal of a nerve cell results in the release of the neurotransmitter acetylcholine.

$$CH_3CO\text{-}OCH_2CH_2\overset{\oplus}{N}(CH_3)_3 \tag{V}$$

This compound binds to specific acetylcholine receptors of the postsynaptic nerve or muscle cell. The binding results in the generation of a voltage of this cell which is either propagated as an action potential or, in the case of a muscle postsynaptic cell, induces muscle contraction. To cease the action of the arrived signal, acetylcholine is hydrolyzed by a special enzyme present in the region of contact, acetylcholinesterase (EC 3.1.1.7). The enzyme catalyzes the conversion of acetylcholine to acetate and choline. Due to this process, the duration of one signal is of the order of 1 msec.

In higher organisms, there are many substances which are produced in one part of the organism and recognized by another part of the same organism by specific target cells. These substances are known as hormones and are produced by special endocrine glands. Many of them are hydrophilic compounds and do not penetrate the cell membrane. They are recognized by respective receptors of the target cells. Recognition is followed by activation of the membrane enzyme adenylate cyclase which catalyzes the formation of the intracellular mediator adenosine-3′,5′-cyclophosphate (cAMP).

$$\text{ATP} \longrightarrow \text{cAMP} \tag{5}$$

cAMP is an activator of protein kinases which catalyze the specific phosphorylation of definite proteins of the target cell. For example, the hormone epinephrine

$$\text{HO-}\underset{\text{HO}}{\text{C}_6H_3}\text{-CHOH-CH}_2\text{-}\overset{\oplus}{N}H_2CH_3 \tag{VI}$$

produced by adrenal medulla is recognized by receptors of muscle or liver cells and stimulates the activation of glycogen phosphorylase (EC 2.4.1.1.). This enzyme catalyzes the con-

19

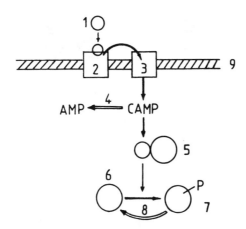

FIGURE 10. The scheme of regulation of glycogen phosphorylase: (1) epinephrine, (2) epinephrine receptor, (3) adenylate cyclase, (4) hydrolysis of cAMP by cAMP-phosphodiesterase, (5) phosphorylase kinase (catalytic and regulatory subunits), (6) dephosphorylated glycogen phosphorylase, (7) phosphorylated glycogen phosphorylase, and (8) hydrolysis of (7) by phosphorylase phosphatase.

version of polysaccharide glycogen to glucose-1-phosphate by transferring terminal glucose residues to inorganic phosphate. Subsequent isomerization leads to glucoso-6-phosphate, supplying the starting compound of glycolysis (Figure 9). Glycogen is the stored form of glucose and should be consumed in a specific physiological situation. The appearance of epinephrine is a signal indicating that this situation has appeared. An epinephrine receptor, being occupied with epinephrine, stimulates cAMP formation which activates the enzyme catalyzing phosphorylation of glycogen phosphorylase, called phosphorylase kinase (EC 2.7.1.37). The latter catalyzes the transfer of γ-phosphate from ATP to definite serine residues of glycogen phosphorylase, converting it to an enzymatically active form. To prevent prolonged action of the signal, two additonal enzymes are involved: phosphorylase phosphatase (EC 3.1.3.17) which catalyzes the hydrolysis of phosphate residues attached to glycogen phosphorylase and, therefore, inactivates the enzyme and cAMP-phosphodiesterase (EC 3.1.4.17) which removes cAMP by hydrolysis to AMP and, therefore, prevents further activation of phosphorylase kinase (Figure 10).

We see that extremely important physiological processes such as the transmission of nerve excitation, induction of muscle contraction, and hormone regulation have recognition processes which in these cases start at the cell surfaces with the participation of the membrane-embedded protein receptors.

G. Recognition in the Immune System

Recognition is used most perfectly in the immune system. This system is common in vertebrates and is especially highly developed in birds and mammals. The administration of some polymeric molecules, especially carbohydrates and proteins which are not inherent to the organism under consideration (these molecules are called antigens), results in the production by the immune system (bone marrow, spleen, thymus, etc.) of special proteins, antibodies with a specific affinity to the administered foreign molecules. Antibodies belong to a special class of proteins, immunoglobulins. The most widespread immunoglobulins G (IgG) consist of two identical pairs of polypeptide chains, heavy (H) and light (L), bound together by disulfide bridges. The structure of IgG is presented schematically in Figure 11.

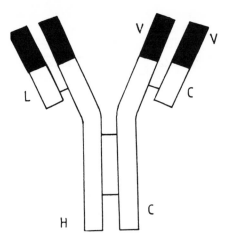

FIGURE 11.   Schematic drawing of immunoglobulin molecule. H, heavy chains; L, light chains; and
V, variable domains forming recognition site. Connecting lines represent disulfide bridges.

The area-recognizing antigen is formed by N-terminal polypeptide moieties of H and L chains and differs for different antigens. Antibodies are produced by special cells, B lymphocytes. It is commonly accepted that B lymphocytes capable of producing definite antigens already preexist in a very complicated population of lymphocytes and the appearance of antigens only stimulates the development and multiplication of these lymphocytes originating from one cell and, therefore, forming one clone. Besides B lymphocytes, the immune response to the administration of some antigens leads to the multiplication of definite T lymphocytes bearing at the outer membrane proteins specific to the same antigen.

Immunoglobulins do not interact with whole polymeric antigen molecules. Interaction proceeds with a definite area of antigen called antigenic determinants. Therefore, antibodies towards some foreign polymer do not represent the homogeneous protein but are the complicated mixture of immunoglobulins specific to different determinants. Moreover, a number of different immunoglobulins specific to the same determinant may be formed.

Low molecular weight molecules do not induce antibody formation. However in some cases, being conjugated to polymer (e.g., serum albumun), they stimulate the formation of antibodies which recognize these molecules. Among these molecules, dinitrophenyl residue is most thoroughly studied. Similarly, the conjugation of oligonucleotides and subsequent immunization of animals by these conjugates sometimes results in the production of immunoglobulins specific to definite nucleotide sequences. The small molecules which may induce the specific antibody formations being conjugated to polymers are referred to as haptens.

It is obvious that immunoglobulins isolated from a living organism represent a highly complicated mixture of different molecules with a specificity to numerous antigens, to sets of antigenic determinants of the individual antigens, and to the same determinants and haptens with different structures of the recognition area.

The preparation of homogeneous immunoglobulins and, consequently, the establishment of amino acid sequences and further investigations at a molecular level became possible due to the existence of a special disease called myeloma. In the course of this disease the malignant multiplication of a clone of B cells occurs, and an antibody typical of this clone strongly predominates among other immunoglobulins and may be purified to a homogeneous state.

It became possible to fuse myeloma cells with normal lymphocytes thus making the latter capable of malignant, unrestricted multiplication either in a cell culture or in the animal.

Taking normal lymphocytes from animals immunized by a definite antigen and selecting among the number of clones of fused cells obtained those which bear a specificity to this antigen, one can now produce and support (for a long time) monoclones which synthesize homogeneous antibodies (monoclonal antibodies).

The biological significance of the immune system is the elimination of viruses and foreign cells, including malignant cells, which appear due to occasional mutations and bacteria cells.

Due to high specificity and diversity, antibodies are widely applied for analytical purposes. Among numerous methods, modern radioimmunoanalysis (radioimmunoassay), enzyme immunoanalysis, and fluorescent immunoanalysis should be mentioned. In the most widely used version, antibodies to a substance to be analyzed are prepared. The labeled antigen is used. For low molecular weight compounds, radioactive antigens (haptens) are prepared, e.g., radioactive hormones. For polymeric compounds, the label may be introduced by chemical modification (e.g., iodination of protein with ^{125}I), conjugation with some enzyme (peroxidase is mainly used for this purpose), or introduction of chelated Eu^{3+} which may be determined with extremely high sensitivity by time-resolved fluorescence. A definite amount of labeled antigen is added to the probe to be analyzed and then both unlabeled natural and labeled artificial antigens are precipitated by the deficit of antibodies. Then either radioactivity, catalytic activity (e.g., by the reaction with 4-chloro-1-naphtol, in the case of peroxidase-labeled antigens, resulting in the appearance of a blue color), or fluorescence level are measured and compared with those added to the sample. Thus, the dilution of an introduced antigen and, consequently, the amount of antigen in the initial sample are easily calculated.

The described procedure of immunoassay represents only one of the various experimental procedures developed which are based on the fundamental property of antibodies to recognize certain molecules. Exhaustive information about immunoanalysis and applications can be found in the literature.[17]

In this section we have presented only the most essential phenomena based upon molecular recognition. To go into details of molecular organization, functions of macromolecules, and biological phenomena readers are advised to use general courses of biochemistry,[18] biophysical chemistry,[19] and molecular biology of the cell.[20]

II. CHEMICAL MODIFICATION OF BIOPOLYMERS

The chemical reactions of biopolymers which do not change the main chain of a polymer (polypeptide in proteins and polynucleotide in nucleic acids) are usually referred to as chemical modification.

Each biopolymer contains a number of identical monomer residues. If a biopolymer is denatured, i.e., lacking a specific three-dimensional structure, all identical residues should exhibit approximately equal reactivities. Thus, at pH 7 all lysine residues of a protein can be modified with 2,4-pentandione. At pH 8 the same reagent modifies both lysine and arginine residues.[21] Since lysine residues in contrast to arginines can be demodified by hydroxylamine treatment, this modification can be used for the preparation of both lysine- or arginine-modified proteins.

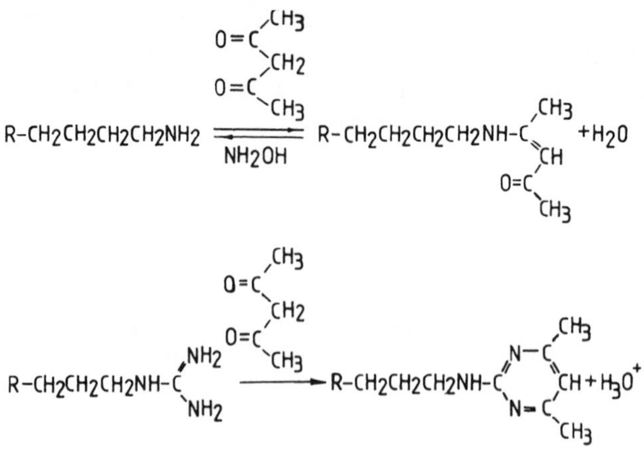

$$R-CH_2CH_2CH_2CH_2NH_2 \;\rightleftharpoons\limits_{NH_2OH}\; R-CH_2CH_2CH_2CH_2NH-C \begin{smallmatrix} CH_3 \\ CH \\ O=C-CH_3 \end{smallmatrix} + H_2O$$

$$O=C\diagdown CH_3,\; O=C\diagdown CH_2,\; O=C\diagdown CH_3$$

(6)

$$R-CH_2CH_2CH_2NH-C \begin{smallmatrix} NH_2 \\ NH_2 \end{smallmatrix} \longrightarrow R-CH_2CH_2CH_2NH-C \begin{smallmatrix} N=C-CH_3 \\ N=C-CH_3 \end{smallmatrix} CH + H_3O^+$$

Cytosine residues react with *O*-alkyl hydroxylamine derivatives giving rise to the formation of two types of products.[22,23]

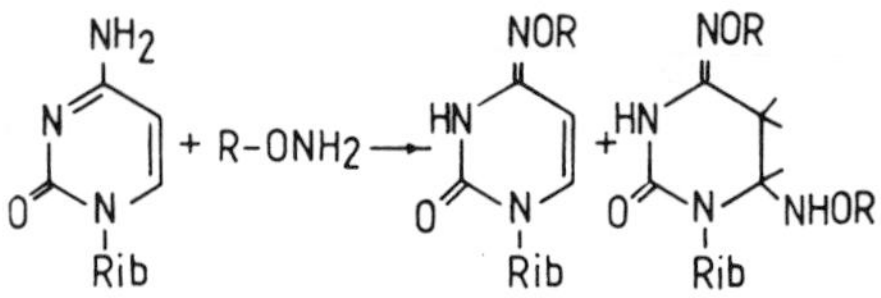

$$\text{(cytosine, NH}_2\text{)} + R-ONH_2 \longrightarrow \text{(NOR product)} + \text{(NOR, NHOR product)}$$

(7)

In native structure some residues may escape modification, being shielded from the attack of a reagent due to the specific macromolecular structure of a biopolymer. Thus, at pH 6.4, only three of the six tyrosine residues of pancreatic RNAase are iodinated by I_2.[24] This reaction will be additionally considered in Section C.

In special cases, the reactivity of a residue in a biopolymer may be strongly changed due to the specific influence of the adjacent groups. One of the most striking examples is the enhancement of the nucleophilicity of a hydroxy group of serine residue in the active center of serine proteases — trypsin, chymotrypsin, elastase, and others. This is due to the specific arrangement of three definite residues of aspartic acid, histidine, and serine forming a so-called charge relay system.

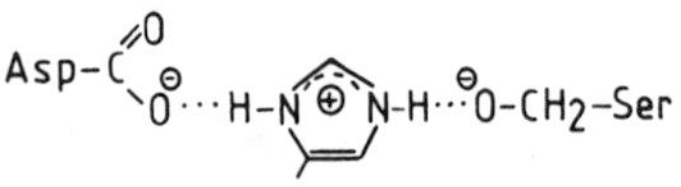

$$Asp-C \begin{smallmatrix} O \\ O^{\ominus} \end{smallmatrix} \cdots H-N^{\oplus} \diagup N-H \cdots ^{\ominus}O-CH_2-Ser$$

The functional significance of this system is to perform a nucleophilic attack on the peptide bond to be hydrolyzed. For example, trypsin recognizes peptide bonds formed by the positively charged amino acid residues, namely lysine or arginine, as C components. Within specific complex nucleophilic serine residue attacks this bond and splits the substrate into two fragments, one being attached by recognized lysine or arginine residue to the reactive serine of the enzyme via an ester bond. The C-terminal moiety of the polypeptide subjected to hydrolysis is liberated in this step. For example:

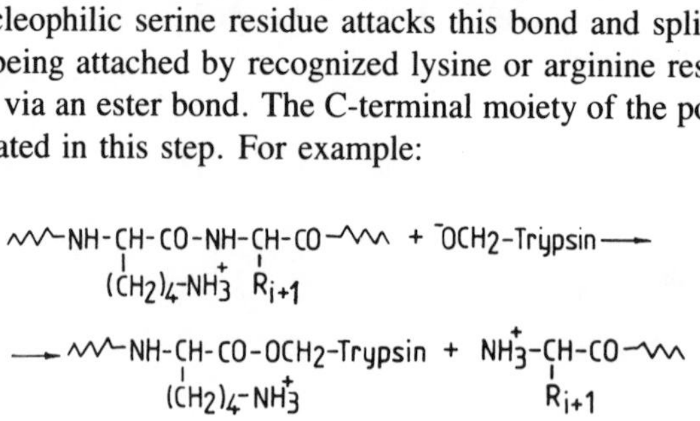

$$\text{⋀⋀⋀-NH-CH-CO-NH-CH-CO-⋀⋀⋀} + {}^-OCH_2\text{-Trypsin} \longrightarrow$$
$$\underset{(CH_2)_4-NH_3^+ \quad R_{i+1}}{}$$

$$\longrightarrow \text{⋀⋀⋀-NH-CH-CO-OCH}_2\text{-Trypsin} + NH_3^+\text{-CH-CO-⋀⋀⋀}$$
$$\underset{(CH_2)_4-NH_3^+ \qquad\qquad\qquad R_{i+1}}{}$$

(8)

This step is followed by the hydrolysis of the ester formed with the regeneration of the enzyme and the liberation of the N-terminal fragment.

Due to enhanced nucleophilicity, the same serine may be selectively phosphorylated with di(iso-propyl)fluorophosphate and similar reagents.[25]

$$(i-C_3H_7)_2-P(O)F + {}^-OCH_2-Enzyme \longrightarrow$$

$$\longrightarrow (i-C_3H_7)_2-P(O)-OCH_2-Enzyme + F^-$$

(9)

Chemical modifications of biopolymers are widely used in almost all areas of investigation and application, ranging from the isolation of biopolymers and the investigation of the structures and functions to industrial application. A book on general principles and techniques of chemical modification of proteins is available.[26] Further, we shall briefly consider the main fields using chemical modification.

A. Derivatives of Biopolymers

The first group of applications of biopolymer chemical modification is formed from those that made biopolymer properties more suitable for subsequent manipulations. Many problems are solved by this method.

In some cases, the conversion of a biopolymer to a derivative facilitates the isolation of the biopolymer. For example, tRNAs are obtained by traditional separation procedures as a mixture of tRNAs specific to different amino acids. These specificities are realized in aminoacylation catalyzed by aminoacyl(AA)-tRNA-synthetases (EC 6.1.1). Each enzyme of this group catalyzes the attachment of a definite amino acid to a definite tRNA (or to several tRNAs specific to the same amino acid) which proceeds in two steps: activation of the amino acid and transfer of the activated aminoacyl residue to tRNA.

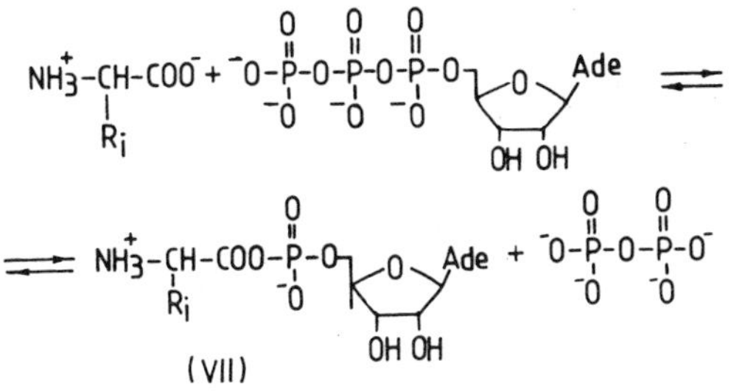

(10)

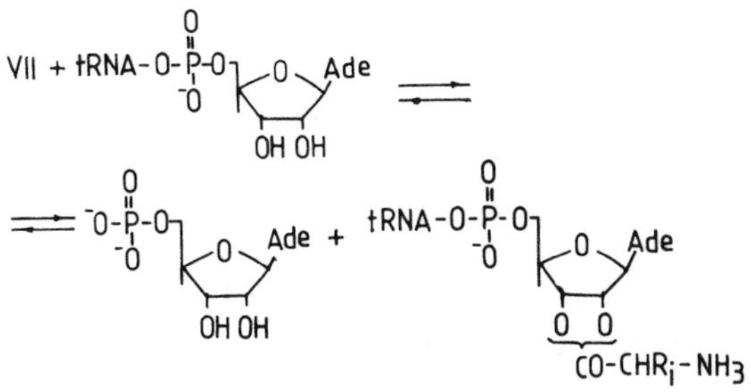

(11)

Due to the first step, the addition of ^{32}P-labeled inorganic pyrophosphate in the reaction mixture containing the enzyme, the respective amino acid, and ATP is followed by a ^{32}P exchange between pyrophosphate and ATP. Aminoacyl residue binds to 3'-terminal nucleotide residue of tRNA which is always represented by AMP. Acylation proceeds either at a 3′- or 2′-OH group. As it is easily seen from the general structure (II) of RNAs, the latter has a unique *cis*-diol group at the 3′-terminal nucleotide. This group may be selectively oxidized by periodate. AA-tRNA does not contain the *cis*-diol group. Therefore, if the mixture of tRNAs obtained from a biological source is incubated with one amino acid and respective AA-tRNA-synthetase in the presence of ATP and then is subjected to periodate treatment, the mixture is formed which contains one definite AA-tRNA and other tRNAs with oxidized 3′ ends. The latter may be easily separated by a sorbent bearing hydrazide groups according to reaction:

$$
\text{tRNA-O-}\overset{\displaystyle O}{\overset{\|}{P}}\text{-O} \underset{\overset{|}{\text{O}}}{} \quad \text{...} \quad \text{Ade} \quad + \text{ NaIO}_4 \longrightarrow
$$

OH OH

(12)

$$
\longrightarrow \text{tRNA-O-}\overset{\displaystyle O}{\overset{\|}{P}}\text{-O} \quad \text{...} \quad \text{Ade} \quad \xrightarrow{\text{NH}_2\text{-NH-polymer}} \quad \text{tRNA-O-}\overset{\displaystyle O}{\overset{\|}{P}}\text{-O} \quad \text{...} \quad \text{Ade}
$$

NH-polymer

Using this procedure yeast valyl tRNA was separated on the polyacrylhydrazide gel.[27] Similarly, AA-tRNA may be selectively acylated with *N*-hydroxysuccinimide ester of some hydrophobic acid, e.g., benzoic acids:

$$
\text{NH}_3^+\text{CHR}_i\text{CO-tRNA}^{(i)} + \begin{array}{c} \text{CH}_2\text{-CO} \\ | \\ \text{CH}_2\text{-CO} \end{array}\!\!\!\!\diagup \!\!\!\! \diagdown \text{NO-COC}_6\text{H}_5 \longrightarrow
$$

(13)

$$
\longrightarrow \text{C}_6\text{H}_5\text{CO-NHCHR}_i\text{CO-tRNA} + \begin{array}{c} \text{CH}_2\text{-CO} \\ | \\ \text{CH}_2\text{-CO} \end{array}\!\!\!\!\diagup \!\!\!\! \diagdown \text{NH}
$$

Thereafter, this tRNA derivative is easily separated from the bulk nonacylated tRNA using chromatography on hydrophobized cellulose (benzoylated or naphthoylated diethylamino-ethyl cellulose).[28]

The application of biopolymer chemical modification widely used is the immobilization of biopolymers on insoluble supports. The two most important fields are immobilized enzymes and affinity sorbents.[29-31]

The covalent attachment of enzymes to a polymeric support permits the use of an enzyme in the most technological version, as heterogeneous catalysts in flow systems. With sufficiently stable preparations of immobilized enzymes prepared without a significant loss of catalytic activity, the utilization of enzymes enhances manyfold thus making the application of enzymes economic in spite of the high price compared with simple inorganic and organic catalysts.

The immobilization of biopolymers with a specific affinity towards some partner permits easy separation of this partner from a complicated mixture by affinity chromatography. It is known that most eukaryotic mRNAs contain polyadenylic tails at the 3′ part of the polynucleotide chain. These RNAs may be easily separated from the bulk RNA by chromatography on immobilized poly(U) or oligo(dT), complementary to this tail, e.g., on poly(U)-Sepharose® or oligo(dT)-cellulose. Immobilization is easily carried out by the treatment of the mixture of poly- or oligonucleotide and carbohydrate carrier with some condensing reagent, e.g., water-soluble carbodiimide.

Among numerous possibilities presented by specific proteins, immunochromatography[31] should be mentioned. The immobilization of antibodies specific to a definite antigen, especially of monoclonal antibodies, provides a unique possibility to isolate this antigen from extremely complicated mixtures obtained from biological sources.

Chemical modification is also widely used to introduce different labels of biopolymers — fluorescent, paramagnetic, radioactive, etc. Fluorescence as well as the signals of electron paramagnetic resonance are sensitive to the macromolecular structure of biopolymers, and respective labels may be introduced into a biopolymer to follow conformational changes accompanying various transformations of this biopolymer. Radioactive labels are introduced into biopolymers to make the subsequent registration more sensitive. In particular, it is essential for radioimmunoanalysis of proteins (see Section I. G). Thus, ^{125}I may be incorporated into nucleic acids by treatment with ^{125}I-labeled KI[32] under oxidative conditions or into proteins by treatment with ^{125}I-labeled Bolton-Hanter reagent[33] (VIII).

$$\text{HO}-\!\!\!\bigcirc\!\!\!-CH_2-CH_2-\overset{O}{\overset{\|}{C}}-O-N\overset{\overset{O}{\diagup}C-CH_2}{\underset{\underset{O}{\diagdown}C-CH_2}{\big|}} \qquad \qquad \text{(VIII)}$$

The first reagent iodinates pyrimidine residues, the second one acylates nucleophilic groups of proteins. ^{14}C and ^{3}H labels may be easily incorporated into the proteins by treatment with the mixture of CH_2O and $NaBH_4$ using either $[^{14}C]CH_2O$ or $[^{3}H]NaBH_4$.[34,35] Lysine residues are methylated by this mixture without a loss of basic properties and positive charge in the neutral and weak alkaline medium.

B. Chemical Modification and Sequencing of Biopolymers

The monomeric composition of the biopolymers is well known. Problems arise only with some minor components which are sometimes scarcely identified and some new modified monomers never before encountered by investigators are present. Therefore, the main problem of biopolymer chemical structure is the elucidation of the sequences of monomeric residues in the polymer chains. All currently used methods of sequencing of biopolymers are based upon the selective splitting of biopolymers at a definite point or number of points. In many procedures used for this splitting, chemical modification is involved.

The key procedures in protein sequencing are the specific fragmentation of a protein to a mixture of peptides, separation of these peptides, and stepwise degradation of each of them followed by the identification of eliminated monomers. Splitting is usually performed by enzymes, e.g., trypsin and chymotrypsin, which cleave peptide bonds formed by carboxyl groups of aromatic amino acids: phenylalanine, tyrosine, and tryptophane. In several cases, the chemical modification of proteins permits the protection of some points against enzymatic cleavage or vice versa to form new cleavable points. Thus, proteins may be treated with either trifluoroacetic or citraconic anhydride[36,37] which modify ε-amino groups of lysine. For example:

$$\underset{CH_3-\overset{\|}{\underset{\|}{C}}-CO}{\overset{CH-CO}{\diagdown}}O + NH_2\text{-Protein} \xrightarrow{pH9} \underset{CH_3C-COO^-}{\overset{CH-CO-NH-Protein}{\|}} \qquad \qquad (14)$$

Lacking a positive charge, modified lysine residues lose the affinity to a trypsin active center and the enzyme attacks only those peptide bonds formed by arginines. After the separation of oligopeptides formed by the selective splitting at arginine residues, ε-NH_2 groups may be restored by mild acid treatment (pH 3).

The reaction with ethylenimine results in the formation of aminoethylated cysteine residues.

$$\text{Protein}-CH_2SH + \underset{CH_2}{\overset{CH_2}{|}}NH \longrightarrow \text{Protein}-CH_2-S-CH_2-CH_2-NH_3^+ \tag{15}$$

Due to the hydrophobicity of the sulfur atom, this residue strongly resembles the lysine side chain, $-(CH_2)_4NH_3^+$, and is recognized by trypsin. Thus, splitting at cysteine residues by trypsin occurs after such modification.[38]

The modification of the N-terminal NH_2 group with phenylisothiocyanate results in a strong labilization of the amide bond:

$$\text{(O)}-N=C=S + NH_2-CHR_1-CO-NH-CHR_2-CO-\text{\textbackslash\textbackslash\textbackslash} \longrightarrow$$

$$\text{(O)}-NH-\underset{S}{\overset{\|}{C}}-NH-CHR_1-CO-NH-CHR_2-CO-\text{\textbackslash\textbackslash\textbackslash} \tag{16}$$

A mild acid treatment results in the selective disruption of the peptide bond formed by first N-terminal residue with the liberation of phenylthiohydantoine bearing the R_1 radical:

$$\text{(O)}-NH-\underset{S}{\overset{\|}{C}}-NH-CHR_1-CO-NH-CHR_2-CO-\text{\textbackslash\textbackslash\textbackslash} \xrightarrow{CF_3COOH}$$

$$\text{(O)}-N \overset{\overset{S}{\overset{\|}{C}}}{\underset{\underset{O}{\overset{\|}{C}}-\underset{R_1}{\overset{|}{CH}}}{\qquad}} NH + NH_2-CHR_2-CO-\text{\textbackslash\textbackslash\textbackslash} \tag{17}$$

The identification of this thiohydantoine automatically results in the identification of the N-terminal amino acid, and the polypeptide shortened by one residue may be subjected to the next step of degradation. This procedure known as Edman's degradation is the main procedure for the sequencing of relatively short peptides.[39]

A pure chemical method of specific degradation of a polypeptide chain at methionine residues is based upon the labilization of a peptide bond following the methionine residue after the reaction with BrCN:[40]

$$\ldots-NH-\underset{\underset{CH_2-SCH_3}{\overset{|}{CH_2}}}{\overset{|}{CH}}-CO-NH-\underset{R_{i+1}}{\overset{|}{CH}}-CO-\ldots \xrightarrow{BrCN} \ldots-NH-\underset{\underset{CH_2\underset{C\equiv N}{\overset{\diagup CH_3}{\diagdown S}}}{\overset{|}{CH_2}}}{\overset{|}{CH}}-CO-NH-\underset{R_{i+1}}{\overset{|}{CH}}-CO-\ldots \longrightarrow$$

$$\longrightarrow \ldots-NH-\underset{\underset{CH_2}{\overset{\diagdown}{}}}{\overset{\diagup}{}}\underset{CH_2}{\overset{|}{C}}=NH-\underset{R_{i+1}}{\overset{|}{CH}}-CO-\ldots + CH_3SC\equiv N \tag{18}$$

The imidate formed instead of peptide bond is alkali labile and selective splitting may be carried out at these points.

The selective labilization of definite internucleotide bonds lies in the foundation of the Maxam-Gilbert method of DNA sequencing.[41] As a similar approach is widely used in the investigations of the modification points, we shall describe the procedure in some details.

The internucleotide bond in DNA is labilized when the glycosidic bond of a deoxyribose moiety with a heterocyclic compound is disrupted. In this case, the linear form of deoxyribose with the aldehyde group is easily formed. Hydrazones or Shiff bases may be easily obtained at these points. The presence of a double bond facilitates proton abstraction off the CH_2 group with the simultaneous elimination of a 3'-phosphate residue. The resulting derivative is able to eliminate 5'-phosphate by a similar mechanism.

$$(19)$$

The selective or at least strongly preferential disruption of glycosidic bonds with definite bases may be achieved by special chemical procedures. Thus, acid treatment results in the liberation of purines from DNA under conditions where pyrimidylic nucleosides are stable. The specific elimination of guanine bases from DNA can be achieved by alkylation with dimethylsulfate followed by piperidine treatment.

$$(20)$$

Pyrimidines are easily split off the nucleoside residue by treatment with hydrazine. For example:

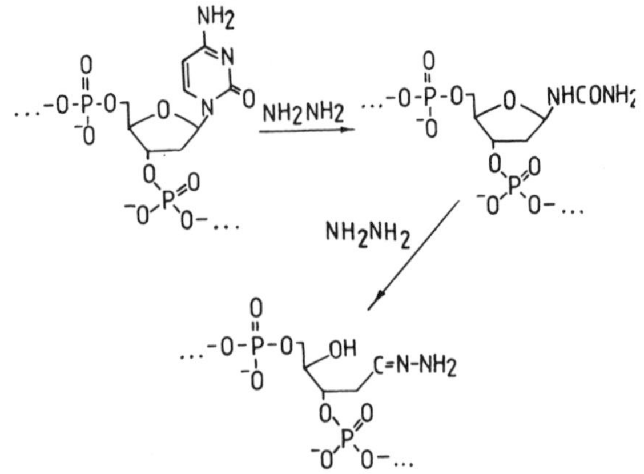

In low salt conditions both thymines and cytosines are liberated. In 1 *M* salt the reaction proceeds preferentially at cytosine residues.

DNA pieces subjected to sequencing are selectively labelled with ^{32}P at one of the ends, usually at the 5' end by treatment with [γ-^{32}P]ATP and the enzyme polynucleotide kinase (EC 2.7.1.78) introducing 5'-[^{32}P]-phosphate residue. Then, hydrolytically unstable apurinic or apyrimidinic sites are introduced in place of definite bases. The extent of base elimination should be low in order to result in the formation on the average of one break per polynucleotide chain. In this case, the set of 5'-end-labeled fragments is formed after hydrolysis at the sensitive sites. The lengths of these fragments correspond to the positions of the base subjected to elimination. These fragments are easily separated by gel electrophoresis, and the pattern of fragments gives the pattern of localization of certain type of base along the DNA sequence. Complete elucidation of the sequence is achieved by parallel analysis of four samples of nucleic acid under investigation subjected to four different fragmentation procedures specific to certain nucleotides. Specific reactions currently used in the chemical sequencing of DNA are given in Table 4.

C. Mapping of Secondary and Tertiary Structure of Biopolymers by Chemical Modification

As mentioned, reactivities of functional groups in macromolecules are affected by the microenvironment. Therefore, the determination of reactivities of individual functional groups can be an approach for the investigation of biopolymer structures providing that it is possible to translate information available as reactivities pattern into structural terms. Indeed, this translation is difficult and quantitative information available from chemical experiments is poor compared to that available from X-ray analysis. However, data on reactivities of functional groups are related to the most intimate properties of a biopolymer structure and are in ready-to-use form for projects related to derivatizations and various applications of biopolymers. Obvious advantages of chemical methods are that they can be applied to macromolecules in solution, under conditions where they are in the native conformation, and that the experiments can be performed in microscale, when appropriate labeling is used. Since chemical methods usually need minor modification when transformed for application to more complicated systems, they are extensively used to investigate macromolecular complexes and processes which cannot be studied at present by physical methods. The objective of this section is to describe the main chemical aporoaches used for the investigation of biopolymer structures and to give the reader ideas about the present state of development of chemical techniques.

Table 4
REACTIONS FOR SPECIFIC CLEAVAGE OF DNA

Cleavage specificity	Reagent	Conditions of elimination of modified bases	Cleavage conditions
G + A	Acid treatment	Acid treatment	Piperidine
G	Dimethyl sulfate	Piperidine treatment	Piperidine treatment
C + T	Hydrazine	Piperidine treatment	Piperidine treatment
C	Hydrazine + NaCl	Piperidine treatment	Piperidine treatment

From Maxam, A. M. and Gilbert, W., *Methods in Enzymology*, Vol. 65, Langone, J. J. and Vunakis, H. V., Eds., Academic Press, New York, 1980, 499.

In the primitive version, the chemical modification study of a biopolymer structure can answer the question of which residues are located at the biopolymer surface ("exposed" residues) and which residues are hidden within the molecular core ("buried" residues). The concept of exposed and buried residues was popular in early studies of biopolymers. In these first experiments, molecules were subjected to extensive modification under conditions stabilizing native biopolymer conformation, and it was assumed that all potentially reactive exposed groups are modified completely and the buried groups are not modified by the time the reaction kinetics reaches the plato. To a certain extent, it was found to be true for many proteins. Thus, in the iodination of ribonuclease at pH 6.4, only residues 73, 76, and 115 are iodinated while tyrosines 25, 92, and 97 are unreactive.[24] A similar pattern of tyrosine reactivity was observed when the enzyme was subjected to modification with tetranitromethane and other reagents.[42] The identity of the unreactive tyrosines correlates well with the X-ray crystallographic data which show that, in ribonuclease A, tyrosines 73, 76, and 115 have hydroxyl ends exposed, two tyrosine residues (25 and 97) are truly buried within the enzyme structure, and tyrosine 92 (although situated on the surface of the molecule) should be poorly reactive since the hydroxyl is involved in a hydrogen-bond formation.[43]

In this selected example and in some other cases the "all or none" concept of exposed and buried residues is true, taking into account the ambiguity of term "buried"; the well-reacting groups of biopolymers can be safely considered as the exposed groups. The groups which do not react can be either truly buried or be at the biopolymer surface but unreactive due to the interactions with surrounding groups. In most cases, however, between the free-reacting and inaccessible groups, intermediate groups apparently exist that show quite varied chemical properties according to local interactions. Therefore, it was recognized that a heavy substitution of biopolymers is disadvantageous since information about differences in reactivities of the exposed groups is lost when they are totally modified. Moreover, it was found that the modification of certain residues in biopolymers may well affect the reactivities of neighboring groups. As a result, in some cases, a heavy substitution leads to the formation of molecules modified at different positions and to different extents. The most clear-cut illustration of this effect is the modification of ribonuclease A with iodoacetic acid at pH 5.5. The reaction is accomplished when one histidine residue is modified per protein molecule, but, in fact, two products are formed in unequal amounts, which are ribonucleases modified either at His 12 or at His 119.[44,45] Alkylation in any ribonuclease molecule of one of these residues which are close together in the active center precludes the action at the other histidine. The most obvious effects of chemical modification resulting in conformational changes and opening of the initially "buried" regions in the biopolymer structure are observed in reactions of nucleic acids. Thus, the formaldehyde modification of single-stranded regions in DNA is followed by unzipping of the flanking helical regions and modification of residues which were initially unreactive.[46]

For the described reasons, contemporary chemical modification techniques are based on partial modification of biopolymers, when on the average less than one functional group is reacted per biopolymer molecule. A low substitution extent allows measuring initial rates of functional group modification which provides information about the reactivities in the native biopolymer structure which is not disturbed by the derivatization. This approach is kinetic in nature and avoids modification of truly buried residues which can be sometimes modified when prolonged modification times are used, due to fluctuations of biopolymer structure. Thus, a constant (C_2) fragment of the immunoglobulin light chain contains a buried intrachain disulfide bond. This bond, however, can be reduced when sufficient incubation times with a reducing agent are used, since a C_2 fragment molecule undergoes a global unfolding transition even in water.[47]

The analysis of partially modified nucleic acids and proteins is quite laborious since usual analytical approaches based on enzymatic digestions of biopolymers give rise to complex mixtures of unmodified and modified oligomers, the latter being in minor proportion. However, using efficient analytical techniques accomplishes analysis of such mixtures. The study of yeast tRNA[Phe] alkylation with a reagent MepUCHRCl[48] (2′,3′-*O*-[4-(*N*-2-chloroethyl-*N*-methylamino)benzylidene]-uridine-5′-methylphosphate) represents an example of structural investigation where directly approaching the problem of analysis of a trace-labeled biopolymer by usual oligonucleotide analysis was used. A nucleotide derivative of aromatic 2-chloroethylamine MepUCHRCl mainly modifies guanosine residues in tRNA. Cleavage of the acetal bond of the reagent by mild acidic treatment results in the formation of the dialkylaminobenzaldehyde moiety absorbing at 350 nm (E = 28.8 × 10^3 M^{-1} cm^{-1}).

(22)

Therefore, the alkylated products can be easily registered and quantitized even in tRNA samples containing on the average one alkylated residue per tRNA molecule.[49] tRNA[Phe] was alkylated with MepUCHRCl under conditions stabilizing the tRNA structure and where the molecule was unfolded. To identify and quantitize the alkylated residues, tRNA[Phe] was digested with ribonuclease A, and the produced oligonucleotides were resolved by diethylaminoethyl cellulose microcolumn chromatography and quantitized in the course of chromatographic runs using a multiwavelength UV detector. The data on reactivities of guanosine residues in tRNA[Phe] are summarized in Table 5. Alkylation with aromatic 2-chloroethylamines proceeds via an intermediate formation of an ethyleneimmonium cation as a rate-limiting step of the overall process; therefore, absolute rate constants for the alkylation by an ethyleneimmonium cation cannot be determined from the data available from the alkylation experiments.[50] Therefore, comparing guanosine reactivities, relative rate constants r_i are

Table 5
RELATIVE REACTIVITIES OF
GUANOSINES r_i IN YEAST tRNA[Phe]
TOWARD ALKYLATING REAGENT
MepUCHRCl IN THE NATIVE AND IN THE
UNFOLDED tRNA[Phe]

Guanosine	r_i (M^{-1})		Interactions of N7 atoms of guanosines
	A[a]	B[b]	
G1	8	24	
G3, G4[c]	2	24	
m^2G10	17	35	
G15	3	22	Mg^{2+} binding site ?
G18, G19[c]	21	37	
G20	1	29	Mg^{2+} binding site
G22	3	16	Hydrogen bond with N1 of m^7G46
G24	18	15	
m_2^2G26, G71[c]	8	35	
G30	21	26	
G34	39	80	
G42	2	14	
G43, G45[c]	2	15	Cation binding ?
G51, G53, G57[c]	3	22	N7 of G57: Hydrogen-bond with 02′ ribose of 55; N7 of G53: Mg^{2+} binding site
G65	1	23	

[a] Reactivities in the native tRNA[Phe] (37°C, 0.1 M Na$^+$, 5.8 mM Mg^{2+}, 0.04 M Tris-Cl$^-$, pH 7.5).

[b] Reactivities in the unfolded tRNA[Phe] (37°C, 0.04 M Tris-Cl$^-$, pH 7.5).

[c] For guanosines occurring in redundant fragments or in those fragments that have more than one copy of a guanosine residue, average values are given.

used, which are ratios of bimolecular rate constants for the reaction of guanosines G_i with an ethyleneimmonium cation to the sum of rates of by-reactions of the cation with nucleophiles present in solution.[49]

From the data presented in Table 5, it is seen that guanosines in the native tRNA[Phe] are essentially different with respect to reactivity and the pattern is too complicated to be interpreted in terms of "exposed" and "buried" residues. Differences in guanosine reactivities are created by the tRNA tertiary structure since, in the unfolded molecule, they react with similar rates. Low reactivities are characteristic for guanosines in double-stranded regions of the tRNA, in particular in oligoguanylic sequences. However, some guanosines in single-stranded regions are also low-reactive due to the tertiary interactions. Taking into account interactions of guanosines observed in the crystal tRNA[Phe] [12] and data of a computer analysis of the crystal tRNA[Phe] structure[51] which evidences that N7 atoms of guanosines G22 and G57 should be poorly available for chemical probes, one can conclude that the data on guanosine reactivities are in accordance with the pattern expected from the crystal tRNA[Phe] structure, and that this structure is very similar or identical to the structure of tRNA[Phe] in solution.

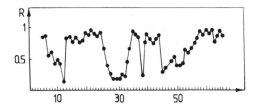

FIGURE 12.   Pattern of phosphate reactivities in tRNA^{Asp} complexed with aspartyl-tRNA-synthetase as compared to the free tRNA. R values are ratios between relative reactivities of phosphates in complexed and free tRNA^{Asp}. A ratio R < 1 means that the reaction of a given phosphate is inhibited by complex formation. (From Romby, P., Moras, D., Bergdoll, M., Dumas, Ph. Vlassov, V. V., Westhof, E., Ebel, J. P., and Giege, R., *J. Mol. Biol.*, 184, 455, 1985.)

D. The Study of the Structure of Multisubunit Proteins and Complexes of Biopolymers

Binding of a ligand to a biopolymer inevitably affects the reactivities of some functional groups in the biopolymer structure. The most obvious effect of complex formation is the shielding of the contact area of interacting molecules from chemical modification. A comparison of the modification pattern of a free biopolymer with that of the biopolymer complexed with a specific ligand leads to conclusions about the binding area and possible conformational changes induced by the interaction. Experimental techniques developed for the identification of biopolymer residues perturbed by complex formation will be described in Chapter 3, Section IV.B.

A typical example of a chemical modification study of a macromolecular complex is the study of yeast tRNA^{Asp} phosphates in contact with aspartyl-tRNA-synthetase.[52] The picture that emerges from studies on the interaction of tRNA and AA-tRNA-synthetases is a multipoint interaction scheme. Therefore, ethylnitrosourea modifying phosphates of tRNA provides an efficient probe for investigating the tRNA/AA-tRNA-synthetase interaction since it is small and reacts with residues scattered over the tRNA molecule. Since tRNA can be selectively cleaved at ethyl phosphotriester fragments, produced by the reaction with this reagent, a phosphate alkylation pattern can be easily observed by gel electrophoresis analysis of the cleavage products of the ^{32}P end labeled partially ethylated tRNA. Figure 12 shows the pattern of phosphate reactivities in tRNA^{Asp} complexed with aspartyl-tRNA-synthetase. It is seen that the complex formation perturbs reactivities of several phosphates, the most striking example concerns phosphates 27 to 33 where seven contiguous residues are strongly shielded by the enzyme. It is also seen that probably no conformational changes unfolding the tRNA structure occur, since no phosphates show enhanced reactivity in the enzyme-bound tRNA. Regions of tRNA^{Asp} protected by aspartyl-tRNA– synthetase are shown in Figure 13. It is seen that in the three-dimensional tRNA^{Asp} structure, protected phosphates are located in three contiguous stretches. It appears that one side of the L-shaped molecule is in contact with the enzyme.

Chemical modification with nonsequence-specific DNA-cleaving reagents is often used for determining locations, size, and relative affinities of binding sites for proteins and small molecules of DNA. DNA-binding molecules protect the binding sites from modification and hence from cleaving. The protected areas in end-labeled DNA fragments can be easily identified using Maxam-Gilbert methodology described in Section II.B. By analogy with a similar technique based on using DNA-cleaving enzymes,[53] these experiments are known as "footprinting". A typical example of a chemical footprinting experiment is the investi-

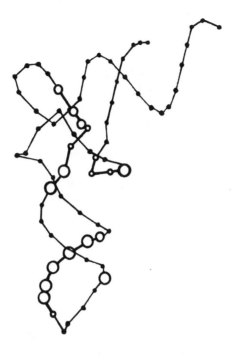

FIGURE 13. Three-dimensional structure of ribose-phosphate backbone of yeast tRNA[Asp]. Phosphates are shown by points, phosphates protected by aspartyl-tRNA-synthetase from alkylation are indicated by spheres, the diameter of each sphere is proportional to the extent of protection. (From Romby, P., Moras, D., Bergdoll, M., Dumas, Ph., Vlassov, V. V., Westhof, E., Ebel, J. P., and Giege, R., *J. Mol. Biol.*, 184, 455, 1985.)

gation of drug binding to DNA. To study sequence specificity of binding of tripeptide distamysin

$$(X)$$

to DNA, DNA restriction fragments were cleaved with (methidiumpropyl-EDTA)iron(II) [MPE.Fe(II)] in the presence of distamycine.[54] Strand cutting of DNA by MPE.Fe(II) is thought to result from hydroxyl radical-mediated oxidative degradation of the deoxyribose ring followed by base elimination and cleavage of the phosphodiester backbone, the mechanism of this reaction will be considered in Chapter 2, Section II.H. The cleavage proceeds with low sequence specificity and the reagent is an efficient probe for mapping the DNA sequences protected by a complex formation. From the obtained results, it was concluded that distamycin interacts with five-base-pair sites with a preference for poly(dA) · poly(dT) regions. The results are consistent with the model of distamycin binding to the minor groove of DNA with four amide NH furnishing hydrogen bonds to bridge DNA adenine N3 and thymine 02 atoms occurring on adjacent base pairs and opposite the helix strand.[55]

Indeed, the effect of a complex formation on the reactivity of functional groups in a biopolymer is not always only by direct steric hindrance. In some cases, a complex formation can induce conformational changes of biopolymers followed by a decrease or enhancement of reactivity of some residues located outside the interaction area. The former case is hardly distinguishable from the steric hindrance effect, while the latter case is especially informative suggesting appreciable changes in the biopolymer structure. Thus, at pH 7 in the absence of nucleotides, only one of the two essential sulfhydryl (SH) groups of myosin reacts with *N*-ethylmaleimide while another group is practically inaccessible to the reagent. The addition of MgADP to the protein drastically increases the reactivity of the second SH group most likely due to a conformational change of myosin.[56]

For some enzymes, hyperreactivity of side chains was demonstrated in the course of catalytic action of the proteins. Specific modifications of these residues can be achieved simultaneously with catalysis, when enzymes are in the "working state"; therefore, these modifications are known as syncatalytic modifications. A well-documented case of syncatalytic modification is the modification of SH groups in the enzyme aspartate aminotransferase (EC 2.6.1.1) which catalyzes the transfer of an amino group from glutamate to oxaloacetate-forming aspartate.

$$^-OOCCH_2CH_2\overset{\overset{\displaystyle NH_3^+}{|}}{C}HCOO^- + {}^-OOCCH_2\overset{\overset{\displaystyle O}{||}}{C}COO^- \rightleftharpoons$$

$$^-OOCCH_2CH_2\overset{\overset{\displaystyle O}{||}}{C}COO^- + {}^-OOCCH_2\overset{\overset{\displaystyle NH_3}{|}}{C}HCOO^- \tag{23}$$

In the presence of the substrate pair, glutamate + α-ketoglutarate, this enzyme forms covalent enzyme-substrate intermediates. In this working form of the enzyme, SH reagents modify the cysteine residue Cys 390 two orders of magnitude more readily compared to the free pyridoxal form of the protein.[57]

Being illustrated by simple examples, chemical approaches can be used with minor modifications for the investigation of much more complicated systems such as chromatin, ribosomes, and cell membranes. The latter system, being the most sophisticated, is intensively studied by various chemical techniques. Chemical modifications provide a direct approach in studying the spatial arrangement of components within the membrane, for determination of the transmembrane distribution or "sidedness"* of the membrane components. To identify components exposed on the outer surface of the cell membrane, membrane-impermeable chemical probes are used which cannot penetrate the cells. Membrane-permeable reagents will label both groups of molecules exposed on the outer and inner surfaces. Special approaches have also been developed to deliver reagents within cells for specific labeling of the components exposed on the inner surface of the cell membrane. Labeling of membrane-embedded proteins is achieved using lipophilic reagents which accumulate within membranes. The latter reagents will be considered in Chapter 2, Section III.A. Modification of membrane components is performed usually with reagents of low specificity such as photochemically generated nitrenes capable of modifying various biopolymers. However, when the role of some specific group is investigated, group-specific reagents are used. A typical example of chemical studies designed to investigate the sidedness of a membrane component are the experiments on labeling of a β chain of platelet glycoprotein Ib[58] which plays an important role in platelet physiology being involved in the process of adhesion of platelets to the surface of blood vessels and in the interactions with a number of plasma proteins. As chemical probes, in this study, highly fluorescent SH-specific probes were used, since there were evidences that the protein SH groups may contribute to processes involving interactions between the membrane glycoproteins and cytoplasmic constituents modifying the membrane events. The membrane-impermeable probe used in this study was a positively charged

* This means at what side of membrane modified residue of membrane protein is located.

reagent, monobromo(trimethylammonio)bimane (XI). Two membrane-permeable probes were monobromobimane (XII) and *N*-[5-(dimethylamino)-1-naphthalenyl]sulfonyl aziridine (*N*-dansyl-aziridine) (XIII).

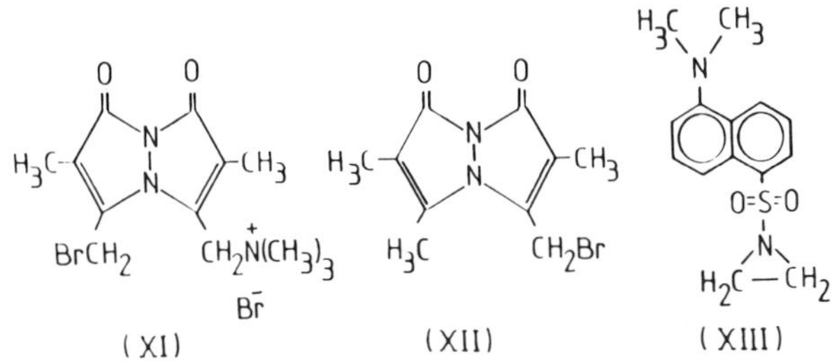

(XI)   (XII)   (XIII)

CH$_2$Br and aziridine groupings of the labels alkylate SH groups under mild conditions. When intact platelets were reacted with these probes, monobromobimane and *N*-dansyl-aziridine readily labeled glycoprotein Ibβ. The membrane-impermeable reagent did not react with glycoprotein Ibβ in intact platelets, but it did modify the glycoprotein in sonicated platelets, demonstrating that reactive SH groups are located on the cytoplasmic side of the membrane surface. These results together with data on the exposure of carbohydrate moiety of the glycoprotein on the outside membrane surface provide the evidence that glycoprotein Ibβ has a transmembrane orientation.

E. Bifunctional Reagents in the Study of the Structure and Complexes of Biopolymers

Chemical cross-linking is one of the most versatile approaches in investigating the structure of biopolymers and the arrangement of macromolecular complexes. Neighboring groups in biopolymers can be covalently connected via molecular bridges by bifunctional reagents which possess two reactive groups in one covalent construction. After the identification of the groups to which the two ends of the bridge are attached, conclusions can be drawn about distance between these groups in the biopolymer structure and the relative orientation. In this way, cross-linking reagents are used as molecular rulers for the determination of tertiary and quaternary structures of biopolymers in solution and in membranes. The accuracy of the determination is limited by the rigidity of the reagent structure and by the mobility of the biopolymer side chains subjected to cross-linking. The more rigid and smaller is the reagent, the more definite conclusions can be drawn about the mutual positioning of the groups which were cross-linked. However, in some cases, when difficulties in the cross-linking experiment are met due to a low content of the reactive biopolymer groups or improper orientation, longer reagents possessing flexible spacers between the reactive groups are helpful. Using these reagents, high yields of cross-linking can be achieved even with a sterically unfavorable positioning of the groups to be interjoined.

In the cross-linking experiments, both homobifunctional reagents bearing identical reactive groups and heterobifunctional reagents with chemically different reactive groups are used. An example of a homobifunctional reagent with a rigid backbone capable of modifying properly oriented neighboring nucleophilic residues such as Lys, Cys, or Tyr is 1,5-difluoro-2,4-dinitrobenzene is represented below.

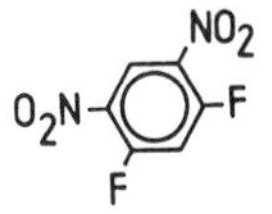

(XIV)

In ribonuclease, this reagent cross-links two lysines, Lys 7 and 41, which are closely positioned near the active site of the enzyme.[59] Probably the most popular in studying the quaternary structure of proteins are bifunctional imidoesters of the general formula:

$$\underset{\displaystyle CH_3\text{-}O\text{-}\overset{\displaystyle \overset{+}{N}H_2}{\overset{\|}{C}}\text{-}R\text{-}\overset{\displaystyle \overset{+}{N}H_2}{\overset{\|}{C}}\text{-}O\text{-}CH_3}{} \qquad\qquad (XV)$$

capable of selective cross-linking of amino groups. Spacer region R of these reagents can be of various length and rigidity.

p-Azidophenacyl bromide

$$N_3\text{-}\langle\bigcirc\rangle\text{-}\overset{O}{\overset{\|}{C}}\text{-}CH_2Br \qquad\qquad (XVI)$$

is a typical heterobifunctional reagent possessing an alkylating α-bromoketone group and photoreactive arylazido group. By reacting with the first group, a reagent can be coupled in the dark to nucleophiles such as the cysteine side chain. A reaction of the other group can be switched on when necessary to form the cross-link by UV irradiation. Under photolysis of the group, a very reactive nitrene species is formed capable of instant labeling of any juxtaposed biopolymer side chain.

The choice of cross-linking reagents is determined by the system under investigation. Reagents of high and moderate specificity are used for mapping protein and nucleic acid complexes since these biopolymers contain many reactive nucleophilic groups. Reagents with photoreactive groups are required when low reactive molecules are studied, in particular in membrane studies, for the determination of contacts between proteins and lipid fatty acid chains. A great number of bifunctional reagents is available commercially. For detailed information on synthesis, properties, and applications of bifunctional reagents the reader is referred to comprehensive reviews.[60-67]

A special group of cross-linking reagents is represented by the so-called ''contact site'' or ''zero length'' cross-linking reagents[68] which share the common property of being able to cross-link only very closely juxtaposed residues, giving rise to cross-links of a length so short as to imply that the cross-linked atoms must have been capable of coming into virtual contact with the native biopolymer structure. A typical representative of a contact site cross-linking reagent is formaldehyde which introduces one carbon atom bridge between the cross-linked amino groups and is referred to as a one-atom-bridge cross-linker. In this case, the distance between the covalently bridged atoms is similar to the minimal Van der Waals separation of the atoms before the modification. Formaldehyde is characterized by a very broad specificity allowing this reagent to be used for protein-protein and protein-nucleic acid cross-linking. It reacts readily with amino groups of protein side chains and heterocyclic bases of nucleic acids and cross-links them to amino groups or side chains of Glu, Asp, Cys, Tyr, Trp, His, or Arg. The main pathway of cross-linking includes hydroxymethylation of lysine residues followed by reactions with nucleophilic amino acid side chains.

$$R_1NH_2 + CH_2O \longrightarrow R_1NHCH_2OH \xrightarrow{\overset{+}{H},\,-H_2O}$$

$$R\text{-}\overset{+}{N}H=CH_2 \xrightarrow{R_2H} R_1\text{-}\overset{+}{N}H_2\text{-}CH_2R_2 \qquad\qquad (24)$$

Another example of one-carbon-bridge cross-linkers are monofunctional alkyl imidates.[68] These reagents exclusively modify primary amino groups and are used for protein-protein cross-linking. Modifying amino groups, they give rise to both N-alkyl amidines and imidates which react further with amino groups forming cross-links.

$$R_1-NH_2 + NH_2^+ = \overset{OR}{\underset{}{C}} - R_2 \longrightarrow R_1 - NH - \overset{OR}{\underset{NH_2}{C}} - R_2 + H^+ \tag{25}$$

$$\xrightarrow{-NH_3} R_1 - \overset{+}{NH} = \overset{OR}{\underset{}{C}} - R_2 \longrightarrow \quad N\ \text{alkyl imidate}$$

$$\xrightarrow{-ROH} R_1 - NH - \overset{}{\underset{NH_2^+}{C}} - R_2 \longrightarrow$$

$$N\text{-alkyl amidine}$$

$$\xrightarrow{NH_2-R_3} R_1 - NH - \overset{}{\underset{NH^+-R_3}{C}} - R_2 \quad + \quad ROH$$

$$\xrightarrow[\text{Slow}]{NH_2-R_3} R_1 - NH - \overset{}{\underset{NH^+-R_3}{C}} - R_2 \quad + \quad NH_3$$

Besides the obvious applications of cross-linking as an approach for intramolecular distance determination, this method can be used for the registration of conformational changes of biopolymers. Thus, reactions with cross-linking reagents have allowed registering and studying the conformational change of myosin resulted from the interaction with ADP.[69] The interaction of myosin with ADP was investigated using three cross-linking reagents: p-N,N'-phenylenedimaleimide (alkylates SH groups), 2,4-dinitro-1,5-difluorobenzene, and 4,4'-difluoro-3,3'-dinitrodiphenyl sulfone with cross-linking spans 12 to 14, 3 to 5, and 7 to 10 Å, respectively. In the absence of nucleotides, only the longest reagent could bridge the two essential SH groups of the protein. In the presence of MgADP, all three cross-linkers could bridge the SH groups indicating the close proximity in the protein structure when it binds the nucleotide ligand, suggesting a conformational change of myosin upon ADP binding.

Cross-linking is extensively used for the determination of the number and arrangement of subunits in oligomeric proteins and for the identification of amino acid residues within the subunit contact areas. Reagents used in these studies should be long enough to span the distance between reactive groups in the contacting subunits. Cross-linked oligomers are analyzed by SDS gel electrophoresis. The number of cross-linked products, the molecular weight, and relative contents give sufficient information to draw conclusions about number of protein subunits in the oligomer and the relative orientation.[70,71] Thus, SDS electrophoresis analysis of dimethyl suberimidate-cross-linked bovine lens leucine aminopeptidase reveals a species corresponding to a dimer, tetramer, and hexamer in much higher amounts as compared to trimers and pentamers. Therefore, it can be concluded that the enzyme is a hexamer arranged as a trimer of dimers.[72] Only this arrangement is consistent with the cross-linking pattern.

Applications of cross-linking techniques for the investigation of nucleoprotein complexes can be illustrated by the study of ribosome structure. For the identification of neighboring proteins within the ribosome a variety of bifunctional reagents was used, the most popular of which were bifunctional imidoesters (XV) and 2-iminothiolane (XVII). The latter reagent modifies primary amino groups of proteins forming amidine derivatives and introducing SH groups.

$$\text{R-NH}_2 + \underset{\substack{\diagdown \\ S}}{\overset{\overset{\displaystyle CH_2-CH_2}{\diagup \qquad \diagdown}}{CH_2}} \diagup C=\overset{+}{N}H_2 \overset{-}{Cl} \longrightarrow \text{R-NH-CH-CH}_2\text{-CH}_2\text{-CH}_2\text{-SH} \qquad (26)$$
$$\underset{\overset{|}{\overset{+}{N}H_2 \overset{-}{Cl}}}{}$$

(XVII)

The oxidation of ribosome modified with iminothiolane promotes disulfide bond formation between the modified residues and, consequently, cross-linking of the derivatized proteins. Since the formed cross-links can be cleaved by reducing reagents, the cross-linked proteins can be easily identified by electrophoretic methods. Extensive studies of both subunits as well as the whole 70S ribosome with various cross-linking reagents provided sufficient information for drawing a spatial map of protein positions within the ribosome[73] (Figure 14). Cross-linking studies have contributed much in understanding the tertiary folding of the rRNAs within the ribosomal body.[73,74] Computer analysis of RNA primary structure permits the revealing of complementary sequences and therefore probable hairpins which are the main elements of the secondary structure of RNAs. Cross-linking experiments may identify spatially juxtaposed elements of the RNA secondary structure and permit the study of tertiary folding of RNAs which is hardly available for investigation with other techniques. The mutual positioning of protein and RNA components within ribosomes has been also extensively studied using cross-linking techniques. Modification with bifunctional reagents has revealed many RNA-protein contacts. Thus, proteins S4, S7, S8, and S17 which first bind to 16S RNA in the course of self-assembly of the 30S ribosomal subunit (Section I.C) were found to be readily cross-linkable to the 16S RNA with various reagents[75,76] suggesting the direct contact with the RNA within the ribosomal structure.

Modification with bifunctional reagents proceeds in most cases as a two-step process. The first step is usually a direct chemical modification. In the second step, the reactive species is the derivatized biopolymer and the reaction proceeds within the specific complex. The direction of modification at this step is determined not only by the nature of the second reactive group of the bifunctional reagent. It is guided by interactions of the derivatized biopolymer. This step should be considered as affinity modification with an *in situ* generation of affinity reagents. This question will be considered in more detail in the sections where modification with macromolecular affinity reagents will be discussed.

F. Chemical Modification in the Study of the Structure-Function Relations of Biopolymers

Chemical modification experiments yield considerable information concerning the structural basis of biopolymer function and activity. In particular, this kind of experimentation has facilitated the understanding of the relation between enzyme structure and activity.[77-79] A classic example of a structure-function investigation of a biopolymer by chemical means is Chapeville's[80] experiment in which cysteinyl-tRNA^{Cys} was chemically converted to alanyl-tRNA^{Cys} by reduction of the cysteinyl residue with Raney nickel. It was demonstrated that the prepared alanyl-tRNA^{Cys} is recognized by the ribosomal machinery as cysteinyl-tRNA^{Cys}. This proves that it is tRNA moiety and not the aminoacyl residue which is recognized by the ribosome-bound codon in the course of translation. In this experiment chemical modification was used for nonenzymatic conversion of unique amino acid residue into another one.

The usual approach to study structure-function relations of biopolymers consists in the chemical damage of a certain type of groups in biopolymer structures. When this damage is accompanied by a loss of activity of the biopolymer, conclusions can be drawn regarding the details of the biopolymer function simply through knowledge of the types and number of the residues involved. The latter information can be obtained by a quantitative study of

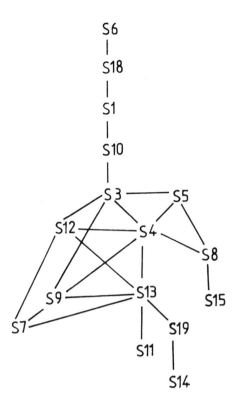

FIGURE 14. The survey of the cross-linking experiments performed with 30S subunits of *Escherichia coli* ribosomes. Proteins connected in the scheme by solid lines were found to be cross-linked.

modification and inactivation kinetics. The identification of groups, modification of which inactivates the biopolymer and which are usually regarded as ''essential groups'' required for a particular property or function of biopolymer, may suggest useful ideas concerning the organization of the enzyme active sites, areas in the biopolymer structure which are involved in an interaction with other molecules, and structure elements which are crucial for maintenance of the molecule in active conformation. Studying the protection of biopolymers from modification by ligands and the identification of residues within the ligand binding sites serves as a useful supplement to the inactivation experiments.

Easy identification of essential residues is usually achieved when only a limited number (one to two) of specific residues is attacked by the reagent and modification results in complete inactivation of the biopolymer. Thus, very specific modification of a single SH group in the proteinases papain and ficin inactivates the enzymes demonstrating the essential role of this group for catalysis.[81] Usually the results of chemical modification-inactivation studies are not clear-cut. Inactivation is achieved at rather heavy biopolymer derivatization; further analysis of modified residues is difficult and interpretation is uncertain. A typical example of studying the functional role of enzyme side chains is the investigation of the role of positively charged residues of *Escherichia coli* phenylalanyl-tRNA-synthetase in the aminoacylation of tRNA[Phe 82,83] (see Section II.A).

Substrates tRNA and ATP are anionic molecules with a high charge density and it is known that electrostatic interactions play an important role in binding tRNAs to AA-tRNA-synthetases. Therefore, one can expect that positively charged arginine and lysine residues of the enzymes are involved in the interaction. For investigation of the role of these residues in the enzyme function, modification at lysine or arginine residues with 2,4-pentandione (6) was used. It was found that both the aminoacylating ability of the enzyme and the ability to form a complex with tRNA^{Phe} decrease when increasing the extent of lysine modification. Modification of ~25 lysines per enzyme molecule resulted in a 90% loss of aminoacylating ability of the enzyme. On the other hand, interaction of the enzyme with phenylalanine and ATP was obviously unaffected since the modification did not influence the enzymatic activity with respect to ATP-[^{32}P]-pyrophosphate exchange (see Section II.A, Equation 10). In the presence of tRNA^{Phe} the enzyme was strongly protected against inactivation while the amino acid, ATP, and analogues of aminoacyl adenylate showed no protective effects. It was concluded, therefore, that some lysine residues of phenylalanyl-tRNA-synthetase are essential for the enzyme-tRNA interaction and that these residues are not important for functions of the sites for other enzyme substrates. Modification of arginines does not influence the complex formation of phenylalanyl-tRNA-synthetase with tRNA, but inactivates both enzyme-catalyzed reactions, ATP-[^{32}P]-pyrophosphate exchange and aminoacylation. ATP protected from the modification $8 \div 10$ arginine residues in the enzyme molecule, while phenylalanine was ineffective in the protection experiments. It was concluded that arginine residues of the enzyme are essential for the ATP binding.

In theory, the best way to study the role of residues of biopolymers in activity would be to investigate the properties of biopolymer molecules modified at certain single residues. However, in most cases modified biopolymer preparations represent complex mixtures of molecules modified to different extents, at different sites or at a combination of sites. Isolation of individual components from such mixtures is usually related to great experimental difficulties.

To avoid this laborious experimentation, an elegant damage-selection approach was proposed.[84] According to this approach, the biopolymer under study is modified to the extent of providing partial inactivation. Then a selection procedure is used which can resolve inactive biopolymer molecules from the molecules that retain activity. A comparison of modification patterns in the two fractions obtained, one of which contains inactivated molecules and the other active ones, yields information about essential residues in the biopolymer. The key step in this approach is the selection step which is based on physical or chemical differences between active and inactivated biopolymers. Thus, when aminoacylation of tRNA is studied,[84] the modified tRNA can be enzymatically charged with a corresponding amino acid and then AA-tRNA can be separated from the nonaminoacylated molecules by one of the procedures described in Section II.A. When the biopolymer property under study is a specific interaction, the modified biopolymer is allowed to interact with the molecular partner, and the complex is separated from the unbound molecules. An example of such experiments is the investigation of tRNA$_f^{Met}$ binding to the bacterial elongation factor EF-Tu.[85] This protein binds AA-tRNAs in the presence of GTP forming a ternary complex EF-Tu·GTP·AA-tRNA, which is a transport form of AA-tRNA for the delivery to ribosomes for polypeptide synthesis. EF-Tu binds all AA-tRNAs except for the initiator Met-tRNA$_f^{Met}$, which is used to start the protein chain building and requires another protein factor to be transferred to a ribosome. When tRNA$_f^{Met}$ was modified with sodium bisulfite, which converts cytidine residues to uridine residues at pH 8, it was found that the modified tRNA acquires the property to bind to the Tu factor. To identify the residues responsible for this effect, the tRNA was modified to a limited extent, aminoacylated, and mixed with excess of GTP and EF-Tu (Figure 15). The complex EF-Tu·Met-tRNA$_f^{Met}$·GTP was separated from unbound Met-tRNA$_f^{Met}$ by gel chromatography. Met-tRNAs from both fractions were subjected to analysis and it was found that protein binds only those tRNAs whose 5'-terminal cytidine residue was converted to uridine. Due to this modification, the first base pair of the acceptor stem of this tRNA, noncomplementary pair CA, was converted to a complementary one UA. The experiments demonstrated that the presence of the unpaired base at the 5'-terminus

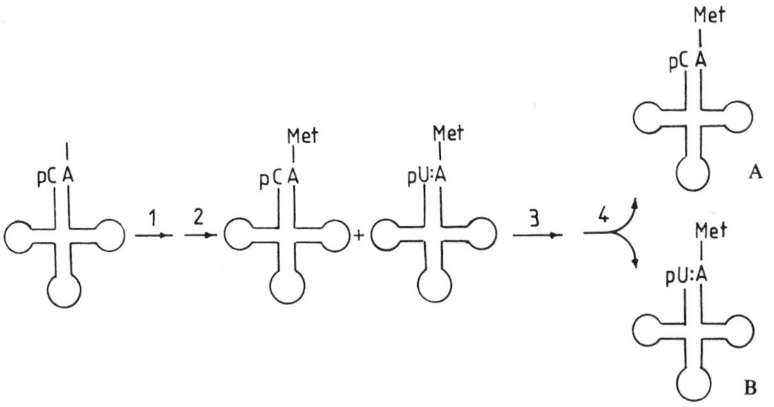

FIGURE 15. The scheme of the damage-selection experiment with *Escherichia coli* Met-tRNA$_f^{Met}$. (1) Enzymatic aminoacylation, (2) treatment with NaHSO$_3$, (3) formation of the ternary complex with EF-Tu·GTP, and (4) separtion of Met-tRNA$_f^{Met}$ nonbound with EF-Tu·GTP (A) from the ternary complex (B). Only 5′-terminal nucleotide and the corresponding adenosine of 3′-terminal part of tRNA are shown. The dots designate hydrogen bonds.

of *E. coli* tRNA$_f^{Met}$ is a feature allowing the elongation factor EF-Tu to discriminate Met-tRNA$_f^{Met}$ from the other AA-tRNAs.

Another typical example of damage-selection approach application is the investigation of contacts between *E. coli* RNA polymerase and the early A3 promoter of phage T7.[86] RNA polymerase was complexed to the promoter fragment modified at phosphodiester bonds with ethylnitrosourea. The fragments bearing phosphotriesters at positions that interfere with the complex formation could be separated from the others complexed to the enzyme by filtration through nitrocellulose which traps the protein-DNA complex. The filtrate contained only those promoter molecules modified at positions which prevent polymerase binding. A determination of positions of alkylted phosphates in the DNA fragment from the filtrate has revealed the phosphates whose modification interfered with the enzyme binding. The same experimental procedure was used for identifying purines essential for RNA polymerase binding. Alkylation with dimethyl sulfate was used in this case. These modification-interference experiments pinpointed phosphates of the DNA backbone and purines directly interacting or closely juxtaposed with the polymerase. Together with experiments on the protection by the enzyme of DNA purines from alkylation with dimethyl sulfate, these experiments identified regions crucial for both polymerase recognition and binding: the Pribnow Box and -35 region (Figure 16).

Indeed, the damage-selection approach works well when modification of a single residue totally inactivates the biopolymer under study. In this case the fraction of molecules which retains functional activity does not contain at all the molecules modified at this residue and the latter is easy to identify. However, when inactivation occurs only due to simultaneous modification of certain sets of residues and single modifications are insufficient for inactivation, the approach yields uncertain results. An example is given by the study of the effect of guanosine alkylation with reagent MepUCHRCl (IX) on the interaction of *E. coli* phenylalanyl-tRNA^{Phe} with ribosomes and poly(U)[87] (see Chapter 5, Section II.D). It was found that average incorporation of five reagent residues per tRNA molecule results in 50% inactivation of the latter in binding with ribosomes. Modified Phe-tRNA^{Phe} molecules lacking the ability to form a ternary complex with ribosomes and poly(U) are easily separated from the active molecules involved in the complex formation by gel chromatography. An analysis of modification patterns in the fraction of active Phe-tRNA^{Phe} molecules and in the inactive fraction has revealed that in both preparations all guanosine residues were modified to some

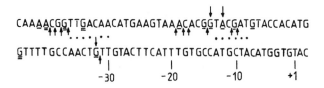

FIGURE 16.   The region of contacts of A3 promoter of phage T7 DNA with *Escherichia coli* RNA polymerase. Arrows indicate phosphates (lower arrows) and purine bases (upper arrows) essential for binding as revealed by damage-selection experiments. The underlined nucleotides are protected by RNA polymerase from methylation of dimethylsulfate. Double underlined nucleotides enhance the reactivity towards dimethylsulfate being in contact with RNA polymerase. Dotted lines indicate −35 region TTGAC and Pribnow Box TATAAT. (From Siebenlist, U. and Gilbert, W., *Proc. Natl. Acad. Sci. U.S.A.*, 77, 122, 1980.)

extent. The only difference between the fractions was that the inactive one contained on the average more heavily modified tRNA molecules. It was concluded therefore that in the tRNA there are no single guanosine residues crucial for the interaction with ribosomes and that inactivation of the biopolymer resulted from simultaneous modification of some sets of guanosine residues.

Chapter 2

AFFINITY MODIFICATION — ORGANIC CHEMISTRY

I. REACTIVE GROUPS OF BIOPOLYMERS

Since affinity modification makes use of specific recognition and binding of molecules to biopolymers, this type of reaction should be performed under conditions where biopolymers can realize the potential of molecular recognition: in aqueous solutions at ambient temperature and pH values not far from neutral (so-called "physiological conditions"). The consequence is that only few of the reactions known to organic chemists can be used in affinity modification. These are predominantly heterolytic reactions. Homolytic reactions are met in affinity modification only as rare, although important, exceptions when reactive groups of reagents generate free radicals as does the EDTA·Fe(II) group or when biradicals carbenes and nitrenes are formed upon photolysis of reagents carrying diazo and azido groups. The great majority of affinity reagents are represented by electrophiles since reactive side-chain groups as well as some groups of the main chains of biopolymers are nucleophilic.

Reactive groups in proteins are represented by sulfhydryl (SH) groups of cysteine, the imidazole ring of histidine, amino groups (terminal amino group and that of lysine), carboxyl groups (terminal and those of aspartate and glutamate), phenol groups of tyrosine, and the SCH_3 group of methionine. In nucleic acids the nucleophilic centers are in the nitrogen atoms of heterocyclic bases and the terminal phosphomonoester group. Weak nucleophiles are the phosphodiester group, the hydroxyl group of ribose, and oxygen atoms of heterocyclic bases (Table 1). The mentioned residues are ones present in the major constituents of proteins and nucleic acids. It should be noted that some biopolymers contain minor components which are characterized by chemical properties unusual for given types of biopolymers. Thus, some transfer (t) RNAs contain rare bases such as 4-thiouridine which is a quite powerful nucleophile.

Since the basic form of functional groups is nucleophilic, the efficient modification can be achieved only at pH values close to or above the pK_as. Main nucleophilic centers of proteins and nucleic acids and the ionization constants are given in Table 1. The presented values for definite compounds bearing respective groups should be regarded as approximate since ionization constants of residues in biopolymers are influenced by many factors such as proximity of charged groups, participation in hydrogen bonding, and hydrophobicity of the environment. A detailed discussion of these effects can be found in books on organic chemistry of nucleic acids[88] and bioorganic chemistry of proteins.[89]

Differences in nucleophilicities of reactive groups of biopolymers and the ranges of predominant existence of the basic form are great enough to allow selective derivatization of certain groups. Thus, specific modification of cysteine residues in proteins is achieved with various reagents simply due to the outstanding chemical reactivity of the SH groups. The order of reactivity of biopolymer functional groups depends strongly on the nature of attacking reagents; therefore, this question will be discussed in Section II. A detailed discussion of all the factors influencing the reactivity of biopolymer residues toward usual chemical reagents can be found in special reviews and books on chemical modification[26,79,88,89] and is not advantageous to discuss since in many cases the predominant factor influencing the specificity of a reaction in the case of affinity modification is the selective adsorption of the reagent on the surface of the target biopolymer molecule. The reactive group of the bound reagent is in forced contact with the biopolymer residues neighboring the binding site and modifies these residues even if they are low-reactive while much more reactive groups

Table 1
NUCLEOPHILIC CENTERS OF BIOPOLYMERS

Nucleophilic center	pK$_a$ proteins	Compound bearing the center
–SH	8.33	Cysteine
–NH$_2$ (N terminal)	8.13	Glycylglycine
ε–NH$_2$	10.53	Lysine
(ART)	6.0	Histidine
–COOH (C terminal)	3.06	Glycylglycine
β–COOH	3.86	Aspartate
γ–COOH	4.25	Glutamate
NH ‖ –NHCNH$_2$	12.48	Arginine
(ART)	10.07	Tyrosine

Nucleic Acids

Cytosine (N3)	4.25	Deoxycytosine
Adenine (N1)	3.8	Deoxyadenosine
Uracil (N3)	9.2	Uridine
Thymine (N3)	9.8	Thymidine
Guanine (N7)	2.4	Deoxyguanosine
Guanine (N1)	9.3	Deoxyguanosine
Phosphomonoester group	6.4	Nucleoside-5'-phosphate (average for major nucleotides)

present in the biopolymer will escape modification being improperly positioned with respect to the bound reagent. This effect can be illustrated by results of experiments on affinity modification of ribosomal (r) RNA with alkylating derivatives of oligoadenylates carrying aromatic 2-chloroethylamino groups: 2′,3′-*O*[4-(*N*-2-chloroethyl-*N*-methylamino)-benzylidene]oligoadenylates

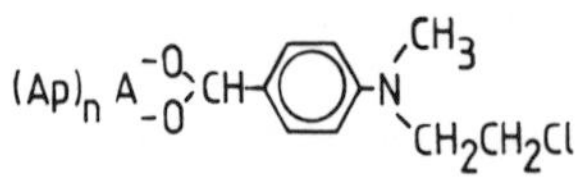

(XVIII)

$(Ap)_n A$ ... CH—⟨⟩—N(CH$_3$)(CH$_2$CH$_2$Cl)

It is known that aromatic 2-chloroethylamines predominantly alkylate guanosine residues in nucleic acids. Thus, alkylated guanosine was the main product in tRNA modified with the reagent IX shown in Chapter 1, Structure 31. However, when the analysis was performed for rRNA modified with XVIII, alkylated guanine, adenine, and cytosine represented 32, 25, and 21% of the alkylation products (for comparison, rRNA contains 32% of Gua, 25% of Ade, and 21% of Cyt). Some indications of the alkylation of phosphodiester groups were obtained.[90,91] This change in specificity means that the reactive group of the bound affinity reagent cannot move freely along the target sequence but rather is limited in mobility and thus is forced to react with the nearest nucleophile irrespective of chemical structure.

Since the aim of affinity modification is often the labeling and identification of biopolymer residues in the binding area, it is preferred to use affinity reagents carrying low-specific groups capable of modifying almost any residue they can meet on the biopolymer surface. Great hopes are associated with the development of photoactivatable affinity reagents diazo compounds and azides which form (under irradiation) the highly reactive species carbenes and nitrenes. In model experiments in organic solvents, nitrenes and carbenes were shown to be able to react even with hydrocarbons being inserted into C-H bonds. These reagents are discussed in Section II.E.

II. REACTIVE GROUPS OF AFFINITY REAGENTS

In the sections that follow the reactive groups used in the design of affinity reagents are described. Our intention was to characterize the groups which are well studied and have proved to be useful as reactive moieties of affinity reagents. Today it would be impractical to describe all the groups which have been used occasionally with limited success or which were proposed without any application in affinity reagents. Extensive information on this subject as well as information about synthetic procedures for the preparation of affinity reagents can be found in special reviews[4] and books.[26,92]

A. Alkylating and Arylating Groups

Alkylating groups are those capable of attaching an aklyl radical to a nucleophilic molecule through a saturated carbon atom. Since many of biopolymer residues possess nucleophilic centers, alkylating reagents show broad specificity in reactions with biopolymers. Thus, various affinity reagents carrying typical alkylating groups such as the haloketone group or amides and esters of haloacids were found to react with serine, tyrosine, methionine, histidine, cysteine, lysine, and glutamate residues in the target proteins.[93]

Alkylation in aqueous solutions can proceed according to two mechanisms,[50,94] S_N1 and S_N2: first-order nucleophilic substitution and biomolecular nucleophilic substitution, respectively. The S_N2 mechanism involves a transition-state complex formation containing both of the reagents.

$$R-S^- + ICH_2C\overset{O}{\underset{NH_2}{}} \longrightarrow \left[R-S\cdots\overset{H\ H}{\underset{\underset{NH_2}{C}}{C}}\cdots I \right]^- \longrightarrow$$

$$R-S-CH_2C\overset{O}{\underset{NH_2}{}} + I^-$$

(1)

In this case the reaction obeys second-order kinetics and the rate is proportional to a concentration of nucleophiles in solution. According to the S_N2 mechanism, haloketones, amides, and esters of haloacids react. The other mechanism involves the formation of a reactive intermediate which reacts further with a nucleophilic substrate to give the product of alkylation. In this case the reaction rate of the reagent consumption obeys first-order kinetics and does not depend on the concentration of nucleophiles.

$$(2)$$

This mechanism is typical of aromatic 2-chloroethyl-amines.[50,94] The typical reactivity orders for biopolymer residues toward alkylating reagents are the following: Cys,Met > His > Lys > Tyr > Asp,Glu for proteins and guanosine (N7) > adenosine(N1) > cytidine(N3), adenosine(N3) for nucleic acids at neutral pH. Thymidine does not react in these conditions. However, at pH above 8 ionized thymidine residue is readily alkylated at the nitrogen atom N3. The ionization of guanine residue in a weak alkaline medium results in the reaction at the nitrogen atom N1.

Products of alkylation are different in stability. While alkylated lysine, histidine, and cysteine residues are quite stable under conditions of analysis of modified proteins, the esters formed by the alkylation of carboxyl groups are labile and easily hydrolyze in alkaline and acidic media. As it was mentioned in Section II.B, alkylated heterocyclic bases are easily eliminated from polynucleotides. Phosphotriesters formed by alkylation at internucleotide phosphates of RNA decompose readily under physiological conditions. In DNA the phosphotriesters are more stable but can be degraded by alkaline treatment.

Amides and esters of haloacids were the first reactive groups used in affinity reagents. In their pioneer experiment, Baker et al.[95] prepared 4-iodoacetamido salicylic acid, the first reactive analogue of natural substrate lactate, and successfully applied it for modification of lactate dehydrogenase (EC 1.4.1.2).

$$(XIX)$$

In an attempt to achieve the same goal and to develop active site-directed reagents, another research group synthesized iodoacetylated derivatives of L-phenylalanine methyl ester and D,L-tryptophane methyl ester. These derivatives of the substrate and inhibitor of chymotrypsin were found to be capable of specific modification of the enzyme.[96]

Amides and esters of haloacids are the most popular reactive groups used in affinity reagents since they can be easily introduced into the structure of natural ligands by acylation of the amino or hydroxyl groups. The list of affinity reagents bearing these groups includes analogues of steroids and sugars, nucleic acids and components, peptides, and amino acids. An important advantage of these groups is related to the analysis of the modified proteins which yield upon the acid hydrolysis well-characterized carboxymethylated derivatives of amino acids which are easy to identify and quantitize.

The use of haloketone groups in affinity reagents was stimulated by successful experiments on affinity modification of chymotrypsin and trypsin with haloketone derivatives of N-tosyl-phenylalanine and α-N-tosyl-lysine, respectively.[97] Since haloketones very closely imitate the carboxyl end of amino acids

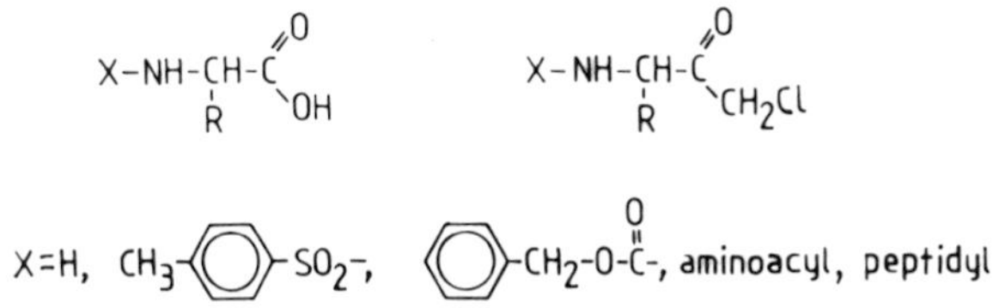

$$X-NH-CH-C\overset{\displaystyle O}{\underset{\displaystyle R}{\diagdown}}OH \qquad X-NH-CH-C\overset{\displaystyle O}{\underset{\displaystyle R}{\diagdown}}CH_2Cl$$

(XX)

$$X=H, \quad CH_3-\!\!\bigcirc\!\!-SO_2-, \quad \bigcirc\!\!-CH_2-O-\overset{\displaystyle O}{\overset{\|}{C}}-, \text{ aminoacyl, peptidyl}$$

they are widely used as peptide and amino acid analogues designed for modification of enzymes dealing with various peptides and amino acids as substrates. Thus, these derivatives were used for modification of corresponding aminoacyl(AA)-tRNA– synthetases.[98]

The chemistry of the 2-chloroethylamino group is studied in detail.[50,94] Various synthetic approaches for the introduction of these groups into biopolymers have been developed. Reactivity of the groups can be modulated easily by the choice of appropriate substituents. As it was shown in Equation 2, 2-chloroethylamines yield aziridinium ions by a unimolecular reaction. In aromatic 2-chloroethylamines, a nitrogen atom is not basic enough and the aziridinium cation is unstable. Therefore, this step is rate limiting.

Both monofunctional $>NCH_2CH_2Cl$ and bifunctional $-N(CH_2CH_2Cl)_2$ 2-chloroethylamino groups are used in affinity reagents. The latter can form cross-links between nucleophilic groups of biopolymers by consecutive alkylation reactions.

Aziridine groups can be considered formally as deprotonated cyclic intermediates formed by 2-chloroethylamino groups. Upon protonation these groups are converted to aziridinium cations which react further with nucleophiles Nu.

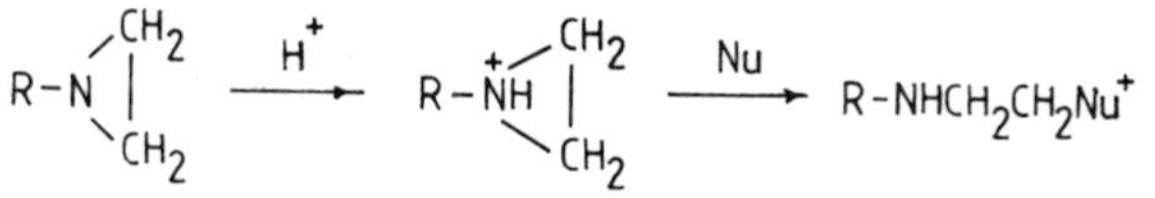

$$R-N\overset{\diagup CH_2}{\underset{\diagdown CH_2}{|}} \xrightarrow{H^+} R-\overset{+}{N}H\overset{\diagup CH_2}{\underset{\diagdown CH_2}{|}} \xrightarrow{Nu} R-NHCH_2CH_2Nu^+$$

(3)

In more alkaline media, aziridine groups can undergo a direct ring-opening reaction with nucleophiles. These groups can be built in the structure of synthetic analogues of natural ligands, as it was done in the case of oligothymidylate carrying one modified deoxycytidine residue which possesses an aziridine group instead of an amino group.[99]

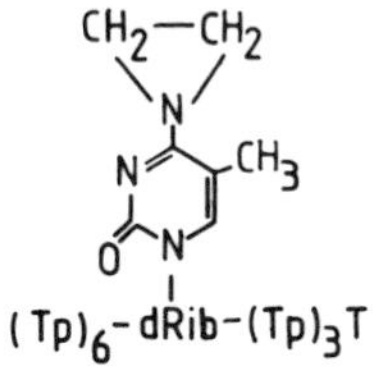

$$\begin{array}{c} CH_2-CH_2 \\ \diagdown N \diagup \\ \text{...} \end{array}$$

(Tp)$_6$-dRib-(Tp)$_3$T

(XXI)

Epoxide groups contain a three-membered oxygen ring of general structure $R_1-CH-CH-R_2$. As well as aziridines, epoxides can react with nucleophilic groups directly

$$R_1-CH-CH-R_2 \quad \underset{O}{\diagdown \diagup}$$

or via protonated intermediate formation. They can react at either of the carbon atoms of the ring. An example of an affinity reagent carrying an epoxide group is 2′,3′-epoxypropyl α-D-glucopyranoside[100]

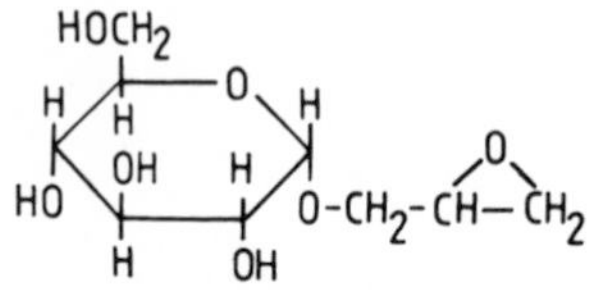

(XXII)

which readily inactivates soybean β-amylase (EC 3.2.1.2).

Diazoketones and diazoacetates which are used mainly as UV-light activatable precursors of carbenes, can serve as alkylating groups in the dark. The reactions proceed most probably according to the acid-catalyzed mechanism preferentially with weakly acidic nucleophiles via the formation of the intermediate carbonium ion:

$$R-CH=\overset{+}{N}=\overset{-}{N} + R_1-COOH \rightleftharpoons [R-CH_2-\overset{+}{N}\equiv N + R_1-COO^-] \longrightarrow$$

(4)

$$R-CH_2-O-\overset{O}{\overset{\|}{C}}-R_1 + N_2$$

The well-known affinity reagent with the reactive diazo group is a structural analogue of glutamine azaserine, an efficient inhibitor of

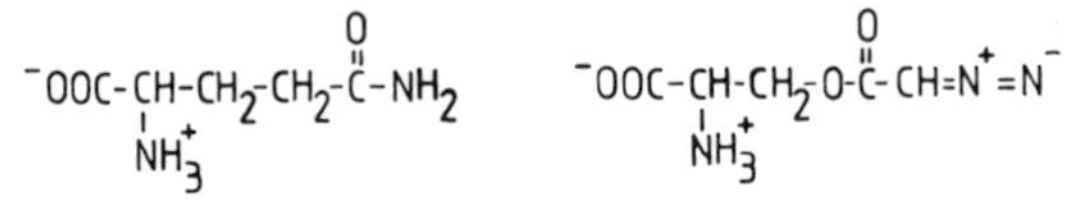

$$^-OOC-CH-CH_2-CH_2-\overset{O}{\overset{\|}{C}}-NH_2$$
$$\underset{\overset{|}{N}H_3^+}{}$$

glutamine

$$^-OOC-CH-CH_2-O-\overset{O}{\overset{\|}{C}}-CH=\overset{+}{N}=\overset{-}{N}$$
$$\underset{\overset{|}{N}H_3^+}{}$$

azaserine (XXIII)

enzymes dealing with glutamine utilization.[101]

Arylating groups, although being less reactive than alkylating groups, show similar broad specificity toward nucleophiles. These groups are represented in affinity reagents mainly by derivatives of fluorobenzene. A typical example is α-*N*-fluorodinitrophenyl-β-*N*-phospho-pyridoxyl-diaminopropionate which was used for the modification of aspartate aminotransferase (EC 2.6.1.1).[102]

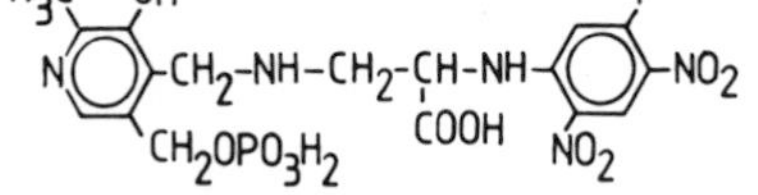

(XXIV)

This reagent is an analogue of the important coenzyme pyridoxal phosphate essential for the function of a great number of enzymes dealing with amino acids including numerous aminotransferases catalyzing the exchange of amino groups between α-amino and α-keto acids (e.g., see Chapter 1, Section II.D, Equation 23).

As arylating groups, the reactive groups of triazine dyes may be considered:

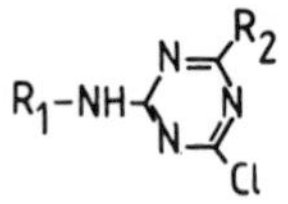

(XXV)

The dyes demonstrate affinity to nucleotide binding centers of various enzymes and modify them in a specific way.[103]

To the same group of reagents, L-(αS,5S)-α-amino-3-chloro-4,5-dihydro-5-isoxazole acetic acid (AT-125)

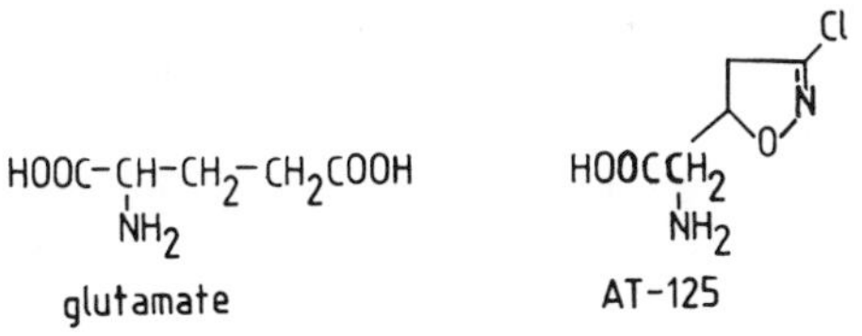

$$\text{HOOC–CH–CH}_2\text{–CH}_2\text{COOH}$$
$$|$$
$$\text{NH}_2$$

glutamate

$$\text{HOOC CH}_2$$
AT-125

(XXVI)

can be referred. AT-125 irreversibly inactivates γ-glutamyl transpeptidase (EC 2.3.2.2), the enzyme-catalyzing transfer of the γ-glutamyl moiety of glutathione to amino acids and dipeptides.[104]

B. Acylating, Sulfonylating, and Phosphorylating Groups

Acylating groups readily react with amino groups of proteins. The reaction proceeds more slowly at the imidazole of histidine, SH, carboxyl, and hydroxy groups. The products formed in the latter reactions are unstable in aqueous solutions.

The acylating function can be introduced in natural ligands by activation of carboxyl groups present in the structure. In the case of ligands possessing dicarbonic acid residues, an internal anhydride can be formed by carbodiimide treatment. An example is the reactive derivative of folic acid[105]

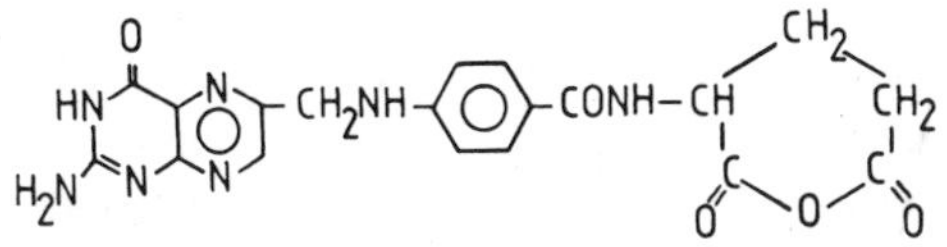

(XXVII)

which is an analogue of an important coenzyme tetrahydrofolic acid responsible for the transfer of monocarbon fragments in biosynthetic processes.

Carbodiimide modification can be also used for the activation of single carboxyl groups present in biopolymers. Usually this activation is performed simply by treatment of the interacting molecules with water-soluble carbodiimides: N-cyclohexyl-N′-β-(4-methylmorpholine)ethylcarbodiimide XXVIII or N-(3-dimethylaminopropyl)-N′-ethylcarbodiimide XXIX.

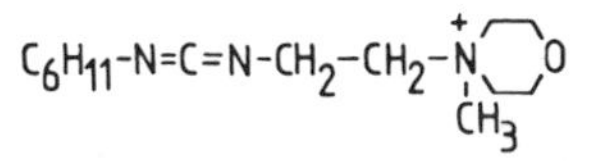

$$\text{C}_6\text{H}_{11}\text{–N=C=N–CH}_2\text{–CH}_2\text{–}\overset{+}{\text{N}}\text{O}$$
(XXVIII)

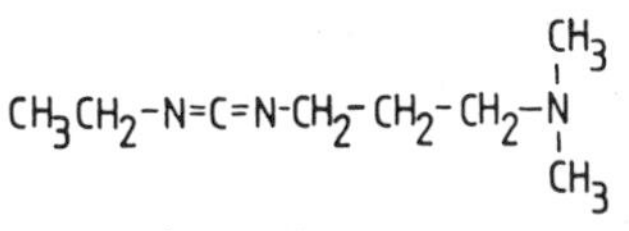

$$\text{CH}_3\text{CH}_2\text{–N=C=N–CH}_2\text{–CH}_2\text{–CH}_2\text{–}\overset{+}{\text{N}}$$
(XXIX)

This approach is often considered as cross-linking with a zero-length reagent.[68] The carboxylate anion attacks the protonated carbodiimide with the formation of O-acylisourea which either reacts with nucleophilic component HX giving rise to the formation of the condensation product with the liberation of the respective substituted urea or isomerizes to N-acyl urea.

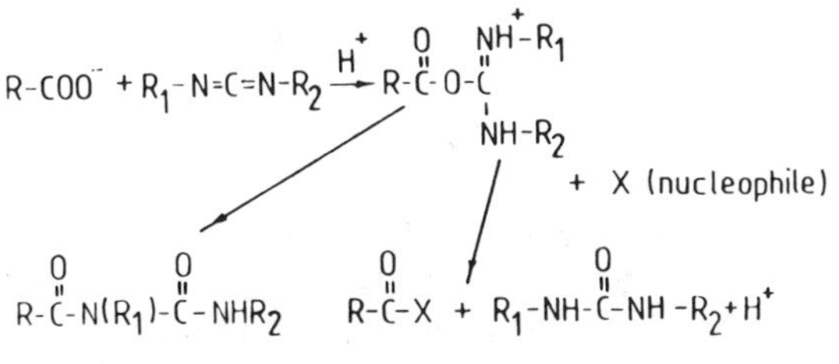

(5)

Nucleophile HX may be an amino, SH, or hydroxy group. The formation of *N*-acyl urea is interesting since it allows the reaction with carbodiimide for specific modification of carboxyl groups in proteins.

Sulfonylating groups which are usually represented by sulfonyl fluorides react as electrophilic reagents and are found to modify tyrosine, lysine, histidine, serine, and cysteine residues in proteins.

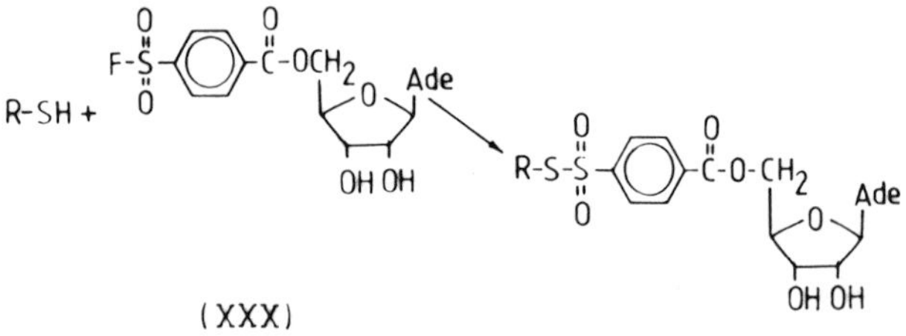

$$(6)$$

(XXX)

These groups were extensively used in various analogues of nucleotides. As shown in Equation 6, *p*-fluorosulfonylbenzoyl adenosine[106] XXX (FSO$_2$Bz-Ado) and the corresponding guanosine derivative (FSO$_2$Bz-Guo) are two of the most widely used affinity reagents and have played an outstanding role in affinity modification of purine nucleotide binding sites in proteins.[107] The former reagent possesses structural features that imitate adenosine-5'-diphosphate (ADP), adenosine-5'-triphosphate (ATP), and NAD in the interactions with enzymes and is an efficient affinity reagent for these enzymes. The carbonyl group of this reagent is structurally similar to the phosphate group of nucleotides and dimensions of the whole substituent place sulfonyl fluoride moiety in a position similar to that of the terminal phosphate of ATP. Under conditions of acid hydrolysis of proteins, the ester linkage between the benzoyl and nucleoside moiety is hydrolyzed.

In reactions of FSO$_2$Bz-Ado and FSO$_2$Bz-Guo with some enzymes, an interesting phenomenon was observed. Affinity modification resulted in the specific formation of disulfide bonds instead of covalent coupling of the reagent to the enzyme. It was proposed that the initial reaction with cysteine residue yields a thiolsulfonate and then neighboring the cysteine residue displaces the sulfinic acid moiety by attacking at the cysteinyl sulfur forming a disulfide.[108]

A number of residues in proteins can be phosphorylated: histidine, lysine, serine, threonine, and cysteine. Phosphorylating groups are widely used in the design of reactive derivatives of nucleic acid components.

Mixed anhydrides of nucleoside 5'-mono- and polyphosphates with sterically hindered carbonic acids have been shown to be moderately reactive phosphorylating reagents capable of reacting with proteins. Mixed anhydrides with mesitylene-carbonic acid were most thoroughly investigated.

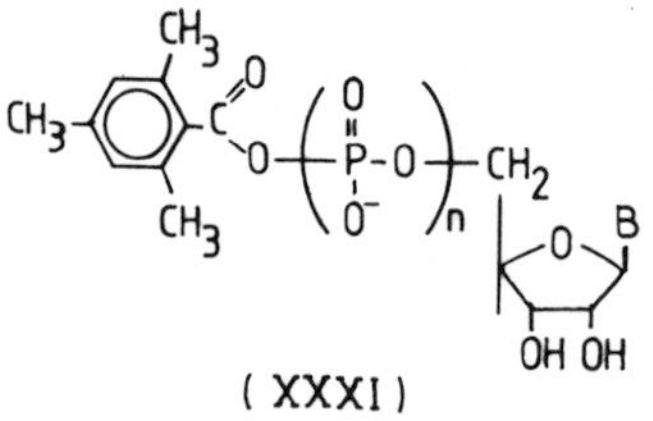

(XXXI)

These acylphosphates were used for the affinity modification of phosphatases from various sources and AA-tRNA-synthetases.[109,110] In the latter case they were found to be particularly efficient, probably imitating the structure of aminoacyl adenylate, an intermediate formed by the enzymes in the course of aminoacylation of tRNAs (see Chapter 1, Section II.A.).

An interesting phosphorylating group is the cyclic trimethaphosphate group which can be easily introduced in nucleoside triphosphates by treatment with carbodiimides.[111,112]

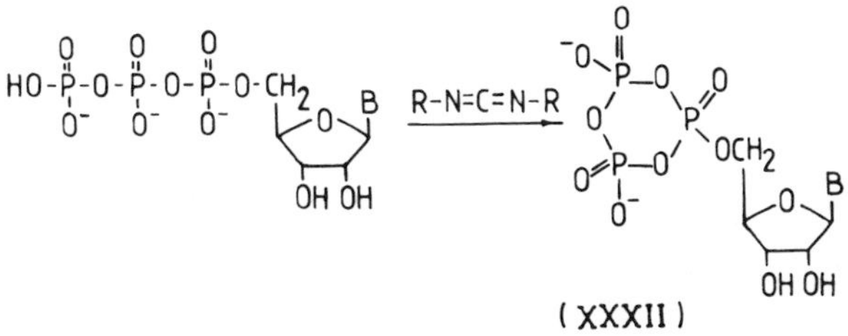

(XXXII)

(7)

These derivatives of nucleoside triphosphates are very similar to the parent molecules and readily react with the enzymes dealing with these compounds.

A perspective type of phosphorylating derivative of nucleic acid components is the derivative where the phosphate group is activated being converted to zwitterionic methylimidazolide phosphate XXXIII or dimethylaminopyridine phosphate XXXIV.[113]

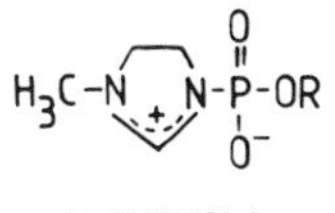

(XXXIII 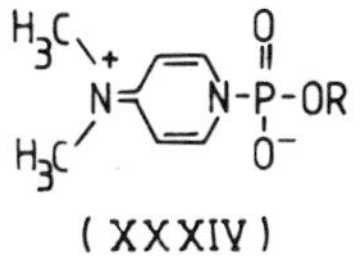)   (XXXIV)

The latter are the analogues of phosphopyridine derivatives formed in the course of oligonucleotide synthesis when the nucleotide component is activated in the pyridine medium.[114] However, in contrast to the phosphopyridine derivatives, the derivatives XXXIII and XXXIV are stable enough in aqueous solutions. An example of the application of such a reagent is the affinity modification of pancreatic ribonuclease with N-methylimidazolide of dinucleotide d(pTpA).[115]

Phosphorylating reagents which cause specific modification of enzymes are organophosphate and phosphonate esters that contain good leaving groups. These compounds readily react with a functionally important serine residue at the catalytic site of acetylcholinesterase (EC 3.1.1.7) and, therefore, are extremely toxic. The most potent inhibitors of this enzyme are known as nerve agents such as Soman XXXV.[116]

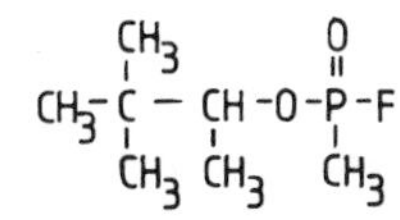

(XXXV)

It is not clear whether the specific reaction of these compounds with the enzyme should be considered as affinity modification or as a particularly efficient reaction with the hyperreactive enzyme residue.

C. Carbonylic Compounds

Aldehydes and ketones reversibly react with amino groups of lysine residues in proteins yielding Schiff bases.[26]

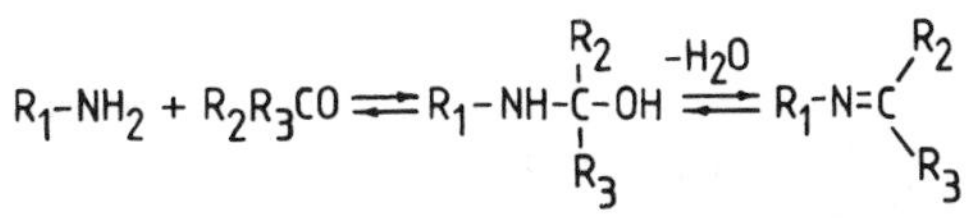

(8)

In the case of affinity modification the formed adducts are rather stable. To make the process irreversible, Schiff bases can be reduced to secondary amines with $NaBH_4$ or with sodium cyanoborohydride which is more stable under acidic conditions.

$$R_1-N=C\begin{smallmatrix}R_2\\ \\R_3\end{smallmatrix} \xrightarrow{NaBH_4} R_1-NH-CH\begin{smallmatrix}R_2\\ \\R_3\end{smallmatrix} \tag{9}$$

At this step radioactive label can be introduced by using the ^3H-labeled reductant.

α-Dicarbonyl derivatives R_1-CO-CO-R_2 react selectively with guanine residues in nucleic acids and arginine residues in proteins.

$$(10)$$

$$(11)$$

A similar reaction with pentandione yielding a more stable six-member ring was described in Chapter 1, Section II. Modified residues are stable under acidic conditions while at alkaline pH they decompose with the regeneration of the original compounds. Demodification of proteins can be achieved by treatment with 1,2-phenylenediamine. Borate reacts with the modified residues yielding more stable adducts.

α-Dicarbonyl compounds are extensively used in specific chemical modification of proteins at arginine residues for the investigation of the functional role of arginines. Although less frequently, α-dicarbonylic groups are used as reactive groups of affinity reagents. Thus, the purine derivative 9-(3,4-dioxopentyl)hypoxantine

was shown to be an arginine-directed affinity reagent for purine nucleoside phosphorylase (EC 2.4.2.1),[117] catalyzing the transfer of the ribosyl moiety from the nucleoside base to phosphate.

An important family of affinity reagents carrying dialdehyde reactive grouping is represented by the periodate-oxidized ribonucleosides, ribonucleotides, and oligo- and polynucleotides containing a ribose residue at the 3'-terminus. The dialdehyde group of the oxidized ribose readily reacts with amino groups of proteins.

$$(12)$$

Both morpholine derivatives and products of separate reactions of both aldehyde groups of the oxidized ribose can be formed.[118] The adducts with the dialdehyde derivatives are sufficiently stable to be isolated and characterized. Reduction of these adducts with NaBH$_4$ yields stable covalently linked compounds which can be subjected to structural analysis.

D. Sulfhydryl and Disulfide Groups
SH groups can react only with cysteine residues yielding mixed disulfides under mild oxidative conditions.

$$R_1-SH + HS-R_2 \xrightarrow{\text{oxidant}} R_1-S-S-R_2 \qquad (13)$$

Disulfide groups react with SH groups in an exchange reaction.

$$R_1-SH + R_2^- S-S-R_3 \longrightarrow R_1-S-S-R_2 + R_3^- SH \qquad (14)$$

The obvious limitation in the use of affinity reagents carrying SH and disulfide groups is related to the high specificity. They can be used only for modification of proteins containing cysteines in the target area. The disulfide bonds formed upon modification are unstable in the presence of thiols due to the disulfide exchange reaction (Equation 14) and are easily disrupted by oxidizing and reducing agents. An example of affinity reagents carrying SH groups is given by 5-hydroxy-3-mercapto-4-methoxy benzoic acid[119] which is a SH-containing analogue of dopamine.

This reagent is a potent inhibitor of the enzyme catechol O-methyltransferase (EC 2.1.1.6) which transfers the methyl group from S-adenosylmethionine to catechol substrates and is responsible for the detoxification of various xenobiotic catechols and inactivates catecholamines.

To make SH groups more reactive, they can be converted to mixed disulfides with good leaving groups. Thus, disulfide was formed between the single SH group of yeast iso-1-cytochrome c and 5,5'-dithiobis(2-nitrobenzoate).[120] The prepared derivative was used for affinity modification of yeast cytochrome c oxidase (EC 1.9.3.1) which is the last enzyme in the electron transport chain and transfers electrons from cytochrome c to cytochrome oxidase.

$$R_1-S-S-\bigcirc-NO_2 + HS-R_2 \longrightarrow R_1-S-S-R_2 + HS-\bigcirc-NO_2 \qquad (15)$$

R_1-SH: cytochrome c and HS-R_2: cytochrome c oxidase.

Another approach to the activation of natural SH groups was used to prepare a reactive derivative of coenzyme A (Chapter 1, III). The derivative was prepared by the oxidation of the coenzyme and the structure is presumably CoA-S-SO-CoA. In this compound, sulfinic acid represents a good leaving group. The derivative was used for the affinity modification of succinyl-CoA synthetase (EC 6.2.1.4),[121] an enzyme catalyzing an important reaction of the tricarbonic acid cycle, which yields the GTP molecules

$$\text{succinyl-CoA} + P_i + GDP \rightleftharpoons GTP + \text{succinate} + CoA \qquad (16)$$

Succinyl-CoA appears in this cycle as a result of oxidation of α-ketoglutarate proceeding in the α-ketoglutarate dehydrogenase complex

$$^-OOC-CO-CH_2-CH_2-COO^- + NAD^+ + CoASH \longrightarrow$$

$$CO_2 + ^-OOC-CH_2-CH_2-CO-SCoA + NADH \qquad (17)$$

E. Photoreactive Groups

Photoreactive (photoactivatable) groups have several advantages which make them indispensible in certain experiments. Due to the stability in the absence of irradiation, affinity reagents with photoreactive groups can be used as inert ligands in all preliminary studies such as binding experiments or even immunization for rising antibodies. The possibility to activate the groups of reagents at any desired moment allows the control of the process of modification: to form the complexes, to remove excess reagent in order to reduce nonspecific labeling, and then irradiate the reaction mixture. The possibility to generate very reactive forms of reagents almost instantly by a single UV-laser pulse opens the opportunity to use a photoaffinity modification technique for investigating fast kinetic phenomena such as macromolecular complex formation. Recently, interest in photoaffinity modification has grown steadily. Literature on this technique and on the chemistry of photomodification is available.[92,122-126]

The precursors of the highly reactive biradicals carbenes and nitrenes are the main types of currently used photoreactive groups. As carbene precursors, derivatives bearing diazo groups are used, the most popular of which are diazoacetates $R-O-CO-CH=N^+=N^-$, diazoketones $R-CO-CH=N^+=N^-$, and diazomalonates $R-CO-C(=N^+=N^-)-COOR$

$$R_1R_2C=N^+=N^- \xrightarrow{h\nu} R_1R_2C: + N_2 \qquad (18)$$

The UV spectrum of diazoketone and diazoacetyl derivatives has two absorption bands: a major band at 240 to 280 nm (E $\sim 10^4 M^{-1}$ cm^{-1}) and a broad weak band centered at about 340 to 350 nm (E $\sim 10 \div 40\ M^{-1}$ cm^{-1}). Irradiation at shorter wavelengths will produce higher amounts of carbenes; however, nucleic acids and proteins absorb in this UV region and can be damaged by irradiation. Therefore photolysis is performed usually at wavelengths >300 nm.

Other important carbene precursors are diazirine derivatives, three-membered cyclic isomers of diazo compounds.

$$R_1R_2C\overset{N}{\underset{N}{\|}} \xrightarrow{h\nu} R_1R_2C: + N_2 \qquad (19)$$

Although the synthesis is quite laborious, reagents carrying diazirino groups are considered the most prospective as they can be activated by 350 to 400 nm UV light (E $\sim$ 300 M^{-1} cm^{-1}).

Arylazido groups are most widely used in photoaffinity modification. Upon UV irradiation they decompose to yield biradical nitrenes.

$$R - N_3 \xrightarrow{h\nu} R - \ddot{N} + N_2 \qquad (20)$$

Arylazides with electron-withdrawing substituents absorb at 300 to 450 nm (E $\sim$ 5 $\cdot$ 10^3 M^{-1} cm^{-1}). Use of other types of carbene and nitrene precursors[122] is limited due to disadvantages related to insufficient stability, inappropriate photochemical properties (require short wavelength UV light for activation), or the tendency to rearrange to inactive compounds or species which are consumed in the light-independent reactions.

The carbon atom of carbenes and the nitrogen atom of nitrenes possess two electrons on the nonbonding orbitals. Spins of these electrons may be either parallel (triplet state) or antiparallel (singlet state). Lifetimes of carbenes and nitrenes are greatly varied for different precursors. They are also functions of the temperature and environment. Singlet nitrenes produced from arylazides have a lifetime of about 100 nsec. Triplet nitrenes live 10 to 100 times longer. Lifetimes of carbenes lie within the nanosecond range and for some singlet carbenes may be as short as 0.02 nsec.

Reactions of carbenes and nitrenes were investigated mainly in model systems where the species were generated in organic media.[122,125] More reactive singlet biradical species react in a single-step process acquiring an electron pair from the substrate and inserting into NH, SH, and even into CH bonds. Triplet species react in a two-step process. The first step is usually the proton abstraction leading to the formation of free radicals. The covalent coupling proceeds in the secondary addition reactions of these radicals.

Carbenes were shown to react with any bipolymer residue. Thus, when membrane-embedded proteolipid of the F$_o$ part of ATP synthase in mitochondria and chloroplasts was labeled with carbenes generated from 3-[trifluoromethyl-3(m-[^{125}I]phenyl)]diazirine, modification of methionine, valine, leucine, serine, threonine, isoleucine, and phenylalanine residues was observed.[127] However, products of the modification were not chemically characterized as in almost all cases of photoaffinity modification.

It should be noted that even the most reactive carbenes which can modify various biopolymer residues are not completely indiscriminating in the reactions and react with nucleophiles more readily than with hydrocarbons. Less reactive nitrenes only insert into CH bonds when they have no other option available. The most widely used aryl nitrenes show considerable selectivity in the reactions and are inefficient in the reaction with alkanes.

The appearance of carbenes and nitrenes in the parent molecules upon irradiation can result in intramolecular reactions. Thus when arylazides were irradiated in an aqueous solution they were converted to azepines presumably via the intermediate azirine formation[128]

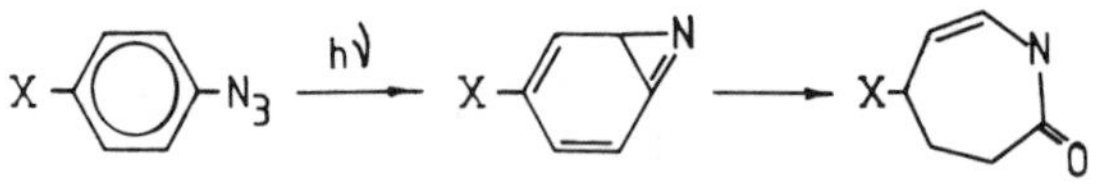

$$X-\bigcirc-N_3 \xrightarrow{h\nu} X-\underset{}{\bigcirc}^N \longrightarrow X-\bigcirc^N_O \qquad (21)$$

The investigation of the photolysis of ATP γ-(*p*-azidoanilide) has demonstrated that the nitrene formed upon irradiation of this compound in aqueous solution rearranges to a long-living electrophilic reagent, ATP γ-(*p*-benzoquinoneimineimide).[129]

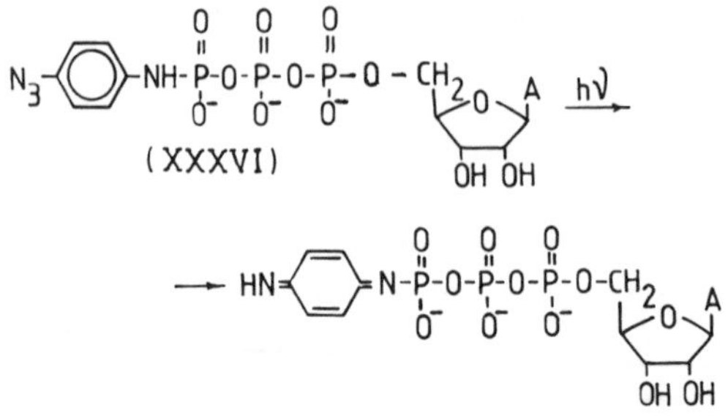

$$N_3-\bigcirc-NH-\overset{O}{\underset{O^-}{P}}-O-\overset{O}{\underset{O^-}{P}}-O-\overset{O}{\underset{O^-}{P}}-O-CH_2 \underset{OH\ OH}{\overset{}{\bigcirc}} A \xrightarrow{h\nu}$$

(XXXVI)

$$\longrightarrow HN=\bigcirc=N-\overset{O}{\underset{O^-}{P}}-O-\overset{O}{\underset{O^-}{P}}-O-\overset{O}{\underset{O^-}{P}}-O-CH_2 \underset{OH\ OH}{\overset{}{\bigcirc}} A$$

(22)

The investigation of the photomodification of pancreatic ribonuclease with ATP γ-(*p*-azidoanilide) has demonstrated that it is the *p*-benzoquinoneimineimide intermediate and not the nitrene species which reacts with the enzyme. It is clear that the formation of electrophilic reagents upon activation of photoreactive groups partially eliminates the expected advantages of the photoaffinity modification.

The drawback of diazo derivatives as carbene precursors is that they form carbenes which tend to rearrange to ketenes (Wolff rearrangement).

$$R_1-\overset{O}{\underset{}{C}}-\ddot{C}-R_2 \longrightarrow R_1-\overset{R_2}{\underset{}{C}}=C=O \qquad (23)$$

The electrophilic reagents, ketenes, acylate nucleophilic groups present in biopolymers, e.g.:

$$R_1R_2C=C=O + NH_2-R_3 \longrightarrow R_1R_2CHCONHR_3 \qquad (24)$$

Some other rearrangements are known for particular diazo reagents. For N^6-diazomalonyl cAMP, the reversible formation of a triazole form predominating at pH > 7 was observed[130] due to the Dimroth rearrangement which is characteristic of α-diazoamides.

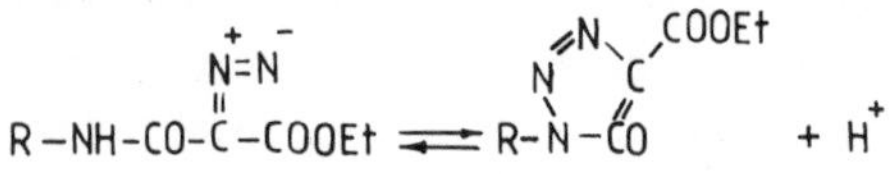

$$R-NH-CO-\overset{\overset{+}{N}=\overset{-}{N}}{\underset{}{C}}-COOEt \rightleftharpoons R-\underset{}{N}-\overset{\overset{N}{\underset{}{N}}\underset{COOEt}{C}}{CO} + H^+ \qquad (25)$$

This form is poorly susceptible to photolysis. In the case of diazirines, the by-reaction leading to a decreased yield of carbenes is the conversion to diazo compounds.

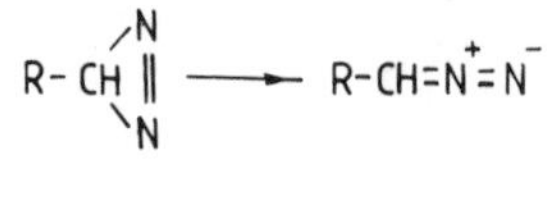

$$R-CH\overset{N}{\underset{N}{\|}} \longrightarrow R-CH=\overset{+}{N}=\overset{-}{N} \qquad (26)$$

The latter are also photolabile but require a short wavelength light to be activated and can react with nucleophiles in the dark.

The conversions in Equations 21 to 26 are some of the reasons for low efficiency of photomodifications with reagents producing carbenes and nitrenes. It is clear that reaction pathways of carbenes and nitrenes are greatly influenced by the substituents present in the parent compounds. The choice of appropriate substituents can overcome some of the disadvantages inherent to these reagents. Thus, electron-withdrawing substituents such as trifluoromethyl or carbethoxy groups decrease the tendency of carbenes toward the Wolff rearrangement and stabilize the reagents in the dark. In ethyl trifluorodiazopropionate

$$CF_3 \overset{\overset{\displaystyle N_2}{\|}}{\underset{\displaystyle C}{}} COOC_2H_5$$ the diazo group is quite stable even in an acidic medium and only 15%

of the produced carbene is subjected to the Wolff rearrangement.[131] 3-Trifluoromethyl-3-phenyldiazirine and 3-aryl-3-halodiazirines were prepared in order to develop groups with improved carbene-generating properties. Upon photolysis of these groups a 65% yield of carbenes was observed.[132]

Besides the widely applied carbene- and nitrene-generating groups, some other photo-reactive groups have proved to be useful. Aryl halides and bromo- and iodo-substituted pyrimidines are used as photoactivatable precursors of free radicals, which are formed upon excitation of these compounds with UV light >300 nm as a result of homolytic fission of the carbon-halogen bond.

$$\text{(27)}$$

The produced free radicals can form covalent bonds with biopolymer residues. If no appropriate groups are present in the vicinity of the free radical, it usually reacts with the neighboring ribose extracting a hydrogen atom from the sugar. This by-reaction is followed by subsequent degradation of the ribose and cleavage of the polynucleotide chain. The presence of oxygen in the reaction medium promotes photooxidative degradation of the halogenated nucleotide moieties. The details of the mechanisms of these degradations are to be elucidated.

The groups described above do not react directly with biopolymers upon excitation. Irradiation of the groups leads to the destructive transformation to reactive species which react with the targets and can be considered as the true affinity reagents. These species either form covalent bonds with biopolymers or are deactivated irreversibly in by-reactions. The latter process usually predominates in reactions of carbenes and nitrenes. In theory, more efficient groups should be those which directly react with biopolymers in an excited state.[133] In this case excitation is reversible. The examples of such groups are acetophenone and benzophenone derivatives which can be activated by UV light above 300 nm, and steroid derivatives with α,β-unsaturated ketone structures. Upon irradiation, the triplet excited state of the ketones is formed which can abstract a hydrogen atom from an available donor and yield two free radicals which can subsequently recombine.

$$R_1-\overset{\overset{\displaystyle O}{\|}}{C}-R_2 \xrightarrow{h\nu} R_1-\overset{\overset{\displaystyle O^*}{\|}}{C}-R_2 \xrightarrow{XH}$$

$$\longrightarrow R_1-\overset{}{\underset{\displaystyle OH}{C}}-R_2 + X\cdot \longrightarrow R_1-\overset{\overset{\displaystyle X}{|}}{\underset{\displaystyle OH}{C}}-R_2 \tag{28}$$

It was found that relatively weak α-CH bonds of amino acids and some CH bonds in nucleotides can be donors in these reactions. It is important that these excited triplet states do not react with water and, if they are not consumed by modification of a biopolymer, they relax to the ground state and can be repeatedly excited. This mechanism of reversible photoactivation can achieve a high modification yield to be nearly quantitative.[134]

It should be noted that photoaffinity modification with ketone derivatives sometimes results in site-specific damage of the target biopolymer without covalent coupling of the reagent. As an example, photoaffinity modification of 3-oxosteroid Δ^5-Δ^4-isomerase from the microorganism *Pseudomonas testosteroni* can be considered (EC 5.3.3.1). This enzyme catalyzes the isomerization of 5,6 double bonds in ketosteroids to 4,5 ones. UV irradiation of the isomerase in the presence of 3-oxo-4-estren-17-β-yl acetate (XXXVII) resulted in the inactivation of the enzyme, but the only damage observed was the conversion of aspartic acid residue 38 to alanine one.[135]

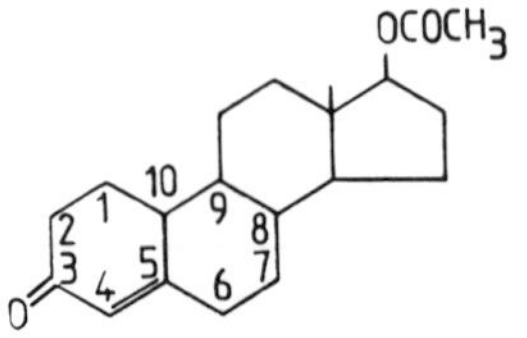

(XXXVII)

This result may be ascribed to the reaction of free radicals formed from the biopolymer residue (Equation 54) with the solvent.

Very efficient photoaffinity modification was observed in the cases of affinity reagents carrying nitroaryl groups. Although the chemistry involved in photoreactions of nitroaryl compounds is not completely understood and may be different for various nitroaryl derivatives, the high modification yields suggest that the groups are photoactivated reversibly. An example of such reagents is flunitrazepam (XXXVIII), a representative of the benzodiazepine

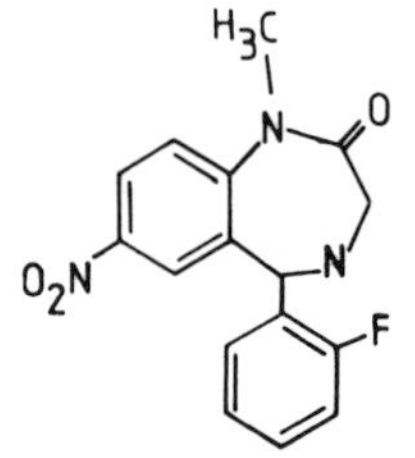

(XXXVIII)

group of drugs which elicit the diverse therapeutic effects (anxiolytic, hypnotic, anticonvulsant, and muscle relaxant) by interacting with a special target membrane proteins in the brain, the benzodiazepine receptors. These drugs are thought to facilitate the action of the neurotransmitter γ-aminobutyric acid, increasing the frequency of chloride channel opening in response to this transmitter. Upon UV irradiation flunitrazepam binds to the receptor irreversibly. The reaction proceeds efficiently and results in a substantial reduction of the number of the active receptors.[136]

Nitrophenyl ethers are another type of group which react directly with biopolymers upon excitation.[137] Upon UV-light (>300 nm) irradiation they undergo photochemical nucleophilic aromatic substitution.

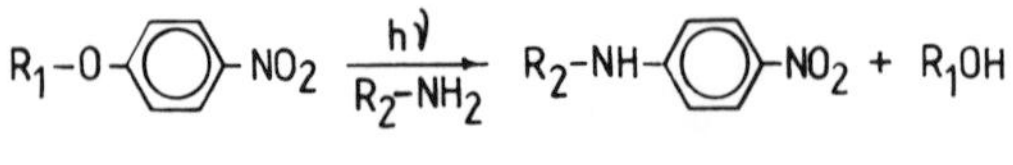

$$R_1-O-\!\!\left\langle\bigcirc\right\rangle\!\!-NO_2 \xrightarrow[R_2-NH_2]{h\nu} R_2-NH-\!\!\left\langle\bigcirc\right\rangle\!\!-NO_2 + R_1OH \qquad (29)$$

The reaction proceeds through the formation of a short-living triplet excited state and thus ensures that the activated species do not wander away from the juxtaposed residues before reacting. In the case of *p*-nitrophenyl ethers, modification results in the transfer of the nitrophenyl group of the reagent to the nucleophile and the liberation of the addressing structure (Equation 29).

The attachment of the entire reagent moiety R_1 to the target residue can be achieved when the second ether group R_2-O is present in the meta position of the nitrophenyl derivative, since hard nucleophiles such as aliphatic amines preferentially attack the meta position (Equation 30a).

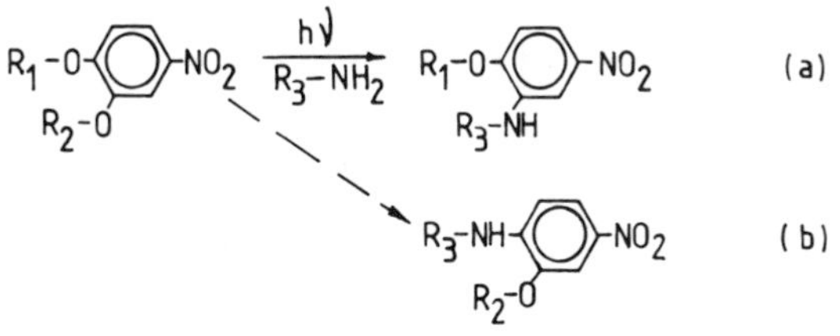

(30)

The possibility of such a reaction was demonstrated for the derivative of the antibiotic of the glutarimide family cycloheximide which interferes with protein biosynthesis on eukaryotic ribosomes. In model experiments when the derivative was irradiated in the presence of methylamine, it was found that substitution occurs predominantly at the meta position of the nitrophenyl grouping.[138]

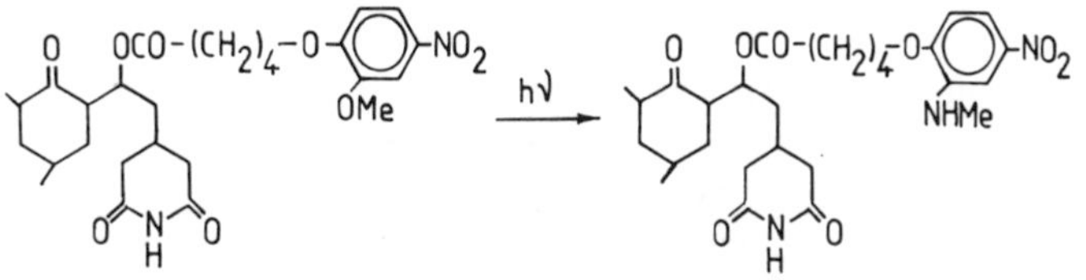

(31)

Cycloheximide 5-(2-methoxy-4-nitro)phenoxypentanoate

Interesting possibilities for the modification of nucleic acids are provided by the psoralen derivatives.[139] Psoralen is a linear furocoumarin which possesses two photoreactive sites located on the 3,4 and 4',5' double bonds. XL shows the general skeleton of the linear furocoumarin (psoralen). Only the position numbers essential for further consideration are given.

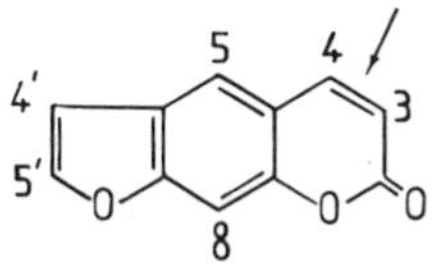

(XL)

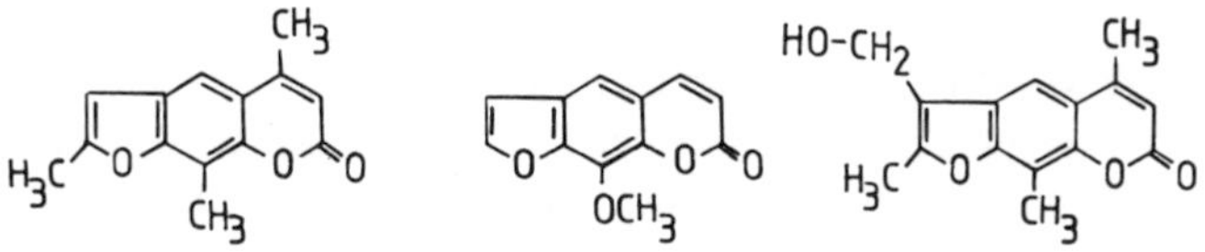

TMP (XLI)   8-MOP (XLII)   HMT (XLIII)

Currently in widespread use are 4,5′,8-trimethylpsoralen (trioxalen, TMP), 8-methoxypsoralen (methoxypsoralen, 8-MOP), and 4′-hydroxymethyl-4,5′,8-trimethylpsoralen (HMT). Since psoralens have two photoreactive sites they are widely used in cross-linking experiments for the investigating the three-dimensional structure of nucleic acids. An important advantage of psoralens is the ability to form the cross-linking products in a stepwise manner. Irradiation at 390 nm results in a monoadducts formation. These monoadducts can be further used as affinity reagents for respective targets activatable by irradiation at 300 nm.[140]

The reaction of psoralens with nucleic acids proceeds as cycloaddition between the 5,6 double bond of pyrimidines and the psoralen double bonds. Thymidine and uridine are much more reactive compared to cytidine. The covalent adducts formed are rather stable chemically. Monoadducts and diadducts to pyrimidines are photoreversed by UV irradiation at 250 nm thus facilitating the analysis of the modification products. Psoralen residues were used as reactive groups in protein derivatives designed for the modification of nucleic acids. Thus, phage λ protein was converted to reactive derivatives by acylation with a bifunctional reagent XLIV.

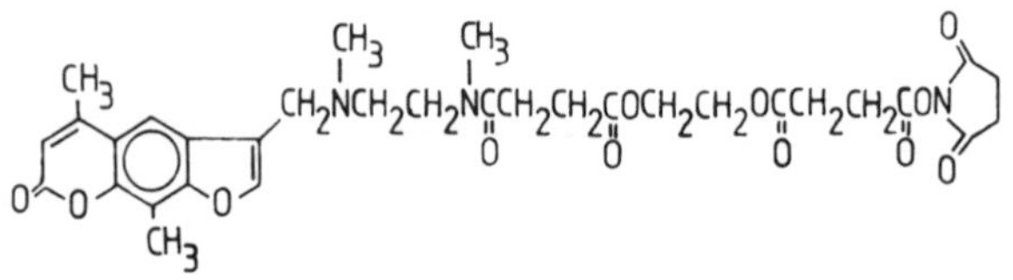

(XLIV)

Subsequent irradiation resulted in cross-linking of the protein and DNA components of the phage.[141]

Intercalating dye groups were proposed as DNA-cleaving photoactivatable groups.[142] The activation can be achieved by long-wavelength UV light ($\lambda > 300$ nm). At high irradiation intensity (> 100 mW/cm^2) the chromophores can be doubly exicted and the absorbed two-photon energy (5 to 7 eV) may be nonradiatively transferred to nucleic acids resulting in a number of photoconversions including chain scission. The possibility to use this approach for sequence-specific modification of nucleic acids was tested in experiments with a oligothymidylate derivative carrying the dansyl group XLIV.

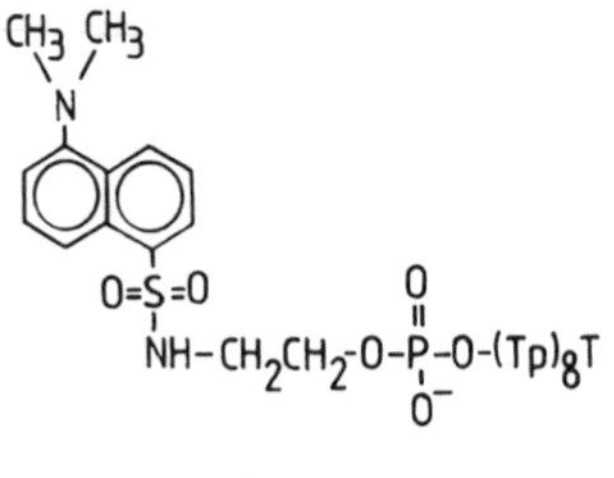

(XLV)

It was found that upon laser irradiation of the derivative with polyadenylic acid which binds oligothymidylate in a complementary complex, the degradation of the polynucleotide occurs. Noncomplementary polycytidylic and polyuridylic acid remained unchanged by the derivative under similar conditions.[143,144]

F. Direct Photocross-Linking of Biopolymers with Specific Ligands

UV-absorbing chromophores of nucleic acids, proteins, and a great number of natural low molecular weight ligands represent built-in photoreactive groups. Therefore, UV irradiation of any specific complexes containing UV-absorbing partners may result in the intermolecular covalent bond formation. This type of affinity modification may be considered as zero-length cross-linking.

The obvious advantage of this approach is simplicity since no derivatization of ligands is needed other than the introduction of radioactive label. The drawback of the method is related to the simultaneous generation of multiple reactive centers in biopolymers, most of which are outside the area involved in the complex formation. Various reactions of these centers may result in the damage of a biopolymer followed by severe changes of macromolecular structure. Therefore, special care should be taken to control the state of biopolymer structure upon irradiation. Otherwise, the reaction should be carried out as rapidly as possible to escape any conformational changes.

Special attention was paid to cross-linking of nucleic acids and components with proteins as a promising approach to elucidation of the mechanism of protein-nucleic acids recognition.

A few years ago, a systematic study of photocross-linking of homopolynucleotides with amino acids and peptides was undertaken.[145-147] The results obtained for polyribonucleotides and DNA with 19 amino acids (proline was not tested) are presented in Table 2. It is seen that most amino acids may participate in photocross-linking. The process is more efficient and less specific with pyrimidine polynucleotides, especially with poly(U) and poly(rT). At the same time poly(A) cross-links with significant efficiency only with glutamine, lysine, phenylalanine, and tyrosine.

In the case of tRNAs unique possibilities for UV-induced cross-linking are provided by the minor components 4-thiouridine, wybutosine, and 5-carboxymethoxyuridine. These nucleosides can be activated by UV light > 300 nm and form specific cross-links with components of tRNA binding sites.

Despite many investigations, knowledge of the actual chemical nature of the adducts formed in UV-irradiated biopolymers remains quite scant. The reason is that irradiation leads to diverse chemical reactions of biopolymers which include direct photoreactions, subsequent transformations of photoadducts, as well as numerous photooxidation processes. Therefore, an analytical problem arises since the products are usually presented by a very complex mixture of substances, many of which are chemically unstable. Recent knowledge in this field is presented in reviews.[148,149]

Among the few examples where the structure of photoproducts was elucidated, several are worth considering. The chemistry of thymidine-thymidine, thymidine-cytidine, and thiouridine-cytidine UV-induced cross-linking was investigated in detail. In these cases five to six double bonds of the pyrimidines were involved in the reaction which leads to the cyclobutane dimer formation, e.g., thymidine photodimer (XLVI):

(XLVI)

Knowledge of the chemistry involved in this process and properties of the dimers helped to elucidate the nature of cross-linking between N-acetylvalyl-tRNA$_1^{Val}$ and ribosomes which occurs upon irradiation with light at 300 to 400 nm. Structural analysis has revealed the residues involved in cross-linking: 5-carboxymethoxyuridine cmo⁵U-34 of the tRNA and C-1400 of the 16S rRNA.[150] Physicochemical properties of the adduct and the ability to be photoreversed at 254 nm could conclude that a cyclobutane dimer was formed between the minor tRNA base and cytidine residue at the tRNA binding site of the ribosome. Extensive studies were initiated on the photochemistry of nucleic acid bases in the presence of the basic amino acids lysine and arginine since they should be close to phosphates in protein-nucleic acid complexes due to the positive charge of the side chains. It was found that upon irradiation ($\lambda = 254$ nm) a photoaddition reaction can proceed readily between thymidine and lysine.[151]

Table 2

EFFICIENCY OF CROSS-LINKING OF AMINO ACIDS TO POLYNUCLEOTIDES WITH 254-nm UV LIGHT[145-147]

Amino acid	Poly(U)	Poly(rT)	Poly(C)	Poly(G)	Poly(A)	DNA
Alanine	+ +	+ +	+ −	+ −	−	−
Arginine	+ + +	+ + +	+	+	+ −	+ −
Asn	+ +	+ + +	+	+ −	+ −	+
Asp	+ −	−	−	−	−	−
Cysteine	+ + +	+ + +	+ +	+	+ −	+ + +
Gln	+ +	+ +	+ −	+ −	+	+ +
Glutathione	+	−	−	−	−	−
Glycine	+	+	+ −	+ −	−	+ −
Histidine	+ +	+ + +	+ −	+ −	+ −	+
Isoleucine	+	−	−	+ −	−	+ −
Leucine	+ +	+ +	+ −	+ −	+ −	+ −
Lysine	+ + +	+ + +	+ + +	+	+	+
Methionine	+ +	+ +	+	+ −	+ −	+ −
Phenylalanine	+ + +	+ + +	+ +	+ +	+ +	+ +
Serine	+ −	+	+	−	−	−
Threonine	+	+	+	−	−	−
Trp	+ + +	+ + +	−	+ −	−	+ + +
Tyrosine	+ + +	+ + +	+ +	+	+	+ +
Valine	+	−	−	+ −	+ −	+ −

Note: Reactions were performed at pH 7.0 and concentrations of amino acids were 50 mM. −: absence of crosslinking; + −: quantum yield $\phi < 1 \cdot 10^{-5}$; +: $\phi = (1 \div 4) \cdot 10^{-5}$; + +: $\phi = (4 \div 10) \cdot 10^{-5}$; and + + +: $\phi > 10 \cdot 10^{-5}$.

$$\text{(32)}$$

This reaction was shown to proceed in chromatin upon UV irradiation and explains the reversibility observed in experiments on histone-DNA cross-linking. Heating of UV-cross-linked chromatin results in the release of free histones that contain partially modified lysine residues. It should be noted that the formation of a ring-opened adduct according to Reaction 32 and the elimination of the base destabilize the corresponding DNA residue, and it is cleaved upon heating or under slightly alkaline conditions.

Although light absorption by pyrimidine bases is usually considered as the primary event in the photoreaction between DNA components and amino acids, there is a probability that the absorption of the aromatic amino acids tryptophane and tyrosine of proteins induces some of the cross-linking reactions. In model experiments, the formation of two anomeric photoproducts in a reaction between tryptophane and thymine was observed.[151] They are formed presumably via tryptophane excitation in a multistep process in which the tryptophane moiety is deaminated and decarboxylated.

$$\text{Trp} + \text{Thy} \xrightarrow{h\nu} \qquad \qquad \qquad \qquad (33)$$

One more example of the chemically characterized photoproduct is the photoadduct of NAD and diphtheria toxin fragment A. This toxin consists of two polypeptides, fragments A and B, which are linked by a disulfide bridge. The B chain is responsible for the binding of the toxin to the receptor-bearing cells and the transportation of subunit A into the cytosol. The A subunit of the toxin contains a catalytic center that transfers the adenosine diphosphate ribose moiety of NAD into a covalent linkage with the peptidyl elongation factor 2 (EF-2) of the eukaryotic ribosome.

$$\text{NAD} + \text{EF-2} \rightleftarrows \text{ADP-ribosyl-EF-2} + \text{nicotinamide} + H^+ \qquad (34)$$

This catalytic modification inactivating the elongation factor and suppressing protein biosynthesis is the reason for the extreme efficiency of the toxin in killing animal cells. Subunit A of the toxin has a single high affinity NAD binding site. Upon irradiation of the protein with NAD ($\lambda = 254$ nm), glutamic acid 148 of the polypeptide was found to be modified.[152] It was demonstrated that the photoproduct contains the entire nicotinamide moiety of NAD linked via the C-6 to the decarboxylated γ-methylene carbon of Glu-148.

$$(35)$$

Two important factors contributing to the diversity of the chemistry involved in the UV cross-linking should be mentioned. The first one is related to the obvious fact that the relative intensities of various UV-induced reactions depend on the wavelength of the exciting light. Thus, in chromatin irradiated with 254-nm light, DNA-histone cross-links dominate while irradiation with 280-nm light results mainly in cross-linking of histones 2A and 2B. Short-wavelength UV light, even commonly used 254-nm light, is absorbed, although poorly, by the sugar phosphate groups and this results in the cleavage of polynucleotides.[153]

Another factor is associated with the intensity of UV light used for excitation. At high intensities of irradiation, two-quantum activation converts any nucleic acid base to a highly reactive center. This leads to the appearance of reactive centers uniformly distributed along the nucleic acid sequence and provides the possibility to form additional cross-links which are unable to arise via single excited states. In accordance with the expected effect, a 20 to 100 times higher cross-linking efficiency was observed upon two-step excitation as compared to single-step excitation in experiments on protein-RNA cross-linking in *Escherichia coli* ribosomes.[154] The main advantage of high intensity UV irradiation is the possibility of obtaining reasonable levels of cross-linking at very short times.[154,155] The chemistry involved in reactions of high excited states of biopolymer residues remains to be elucidated.

The considered examples from the field of nucleic acids and protein chemistry do not exhaust the applications of UV-induced cross-linking. This approach is widely used for the investigation of various complexes of biopolymers with low molecular weight compounds, in particular for labeling of specific receptors. A great number of natural compounds absorb at longer wavelengths as compared to proteins and nucleic acids. They can be safely cross-linked to binding sites upon irradiation with light nondamaging to biopolymers. Since the usual aim of such experiments is to label the receptor or to localize the ligand binding site, no attempts were made to gain an insight into the chemistry involved in the photocross-linking process.

G. Suicide Reagents

In spite of the great variety of biochemical transformations accomplished by enzymes, the key catalytic steps performed by the protein components use very few fundamental chemical principles: general acid-base and nucleophilic catalysis. The variety of other types of chemical transformations represented in numerous biocatalytic processes are mainly carried out by metal ions or special organic molecules (coenzymes and prosthetic groups) entrapped by the protein moiety of the enzymes. Enzymes use acid-base and nucleophilic catalysis very efficiently, being masters of proton abstraction and addition and possessing specially organized nucleophilic centers reacting with substrates such as serine residues in the active site of serine proteinases (see Chapter 1, Section II).

In some natural compounds and deliberately designed substrate analogues, the usual conversions, such as proton abstraction in the enzyme active sites, trigger structural changes leading to the formation of highly reactive species which bind to the enzyme covalently and destroy the activity. According to various terminologies, these compounds are known as suicide reagents, suicide substrates, mechanism-based enzyme inhibitors, or K_{cat} inhibitors. The term "suicide" conveys the fact that enzyme molecules catalyze inactivation themselves. As suicide reagents also, substrate analogues are considered which are converted at the enzyme active sites to species capable to form very long-living covalent adducts or non-covalent complexes of unusual stability with the enzymes and, therefore, drastically reduce the catalytic activity. Since suicide reagents cannot damage other biopolymers except for the target ones, they are extremely specific and are of exceptional value for enzyme-pattern targeted chemotherapy. Several recent reviews on the chemistry of suicidal inactivation and the application of suicide reagents are available.[156-158]

A great number of suicide reagents are converted to reactive species due to the formation of carbanion-like structures in the course of enzymatic catalysis. These structures are formed when enzymes abstract protons from substrates or accomplish decarboxylation reactions. When an acetylenic group is present adjacent to the carbanion center, a propargylic rearrangement to allene occurs which is an active alkylator, in particular when conjugated to an appropriate neighboring group such as a carbonyl group.

$$-C\equiv C-CH_2^- \xrightarrow{-H^+} -C\equiv C-CH- \xrightarrow{+H} -C=C=C- \tag{36}$$

In substrate analogues containing halogen or other good leaving groups, the X formation of an adjacent carbanion-like center results in the facile elimination of this group to yield an olefinic system which is an active alkylating function set up for Michael addition when the formed double bond is neighboring the carbonyl or imino group.

$$-\overset{X}{\underset{|}{\overset{|}{C}}}-\overset{H}{\underset{|}{\overset{|}{C}}}- \quad\diagdown\quad\diagup \overset{+}{H},X^- \tag{37}$$
$$\diagup\quad\diagdown \overset{C=C}{\diagup}\diagdown$$

$$-\overset{X}{\underset{|}{\overset{|}{C}}}-\overset{|}{\underset{|}{C}}-COO^- \quad CO_2,X^- \tag{38}$$

The activation of an acetylenic group by neighboring proton abstraction can be illustrated by classical experiments of Bloch[159] on the inhibition of β-hydroxydecanoyl thiolester dehydrase (EC 4.2.1.60). This is the key enzyme in the bacterial biosynthesis of monounsaturated fatty acids. The enzyme dehydrates β-hydroxyacyl thiolesters to α,β-enolyl acyl-*S*-CoA and isomerizes it to β,γ-enoyl-CoA. To accomplish isomerization, the enzyme abstracts a proton from the C2 atom of the substrate to form a carbanion equivalent.

$$R-CH_2-\underset{\underset{OH}{|}}{CH}-CH_2-\overset{O}{\overset{\|}{C}}-SR' \xrightarrow{-H_2O} R-CH_2-CH=CH-\overset{O}{\overset{\|}{C}}-SR' \rightleftharpoons$$

$$R-CH=CH-CH_2-\overset{O}{\overset{\|}{C}}-SR' \tag{39}$$

The suicide substrates, β,γ-acetylenic thiolesters, according to Bloch, are recognized by the enzyme as substrates for isomerization but in this case the formation of C2 carbanion and isomerization lead to the formation of conjugated allene which readily alkylates histidine residue present in the enzyme active site.

$$R-C\equiv C-CH_2-\overset{O}{\overset{\|}{C}}-SR' \longrightarrow R-CH=C=CH-\overset{O}{\overset{\|}{C}}-SR' \longrightarrow$$

$$\longrightarrow R-CH=C-CH_2-\overset{O}{\overset{\|}{C}}-SR' \tag{40}$$

Enz

The same principle of suicidal inhibition was suggested for other enzymes involved in double-bond isomerization. A typical example is 3-oxosteroid Δ^5-Δ^4-isomerase from *Pseudomonas testosteroni*. The plausible mechanism of this enzyme action involves general acid-base catalysis. The enzyme abstracts the 4β proton from the substrate and donates a proton to the 6β position. A postulated intermediate in this process is the $\Delta^{3,5}$-dienol (XLVIII).

$$\tag{41}$$

(XLVII)   (XLVIII)   (XLIX)

The same enzymatic action, abstraction of the 4β proton in the suicide inhibitors acetylenic secosteroids,[160] pregnyne (L), and estryne (LI) results in the isomerization of the compounds to corresponding reactive conjugated allenic ketones which alkylate the enzyme.

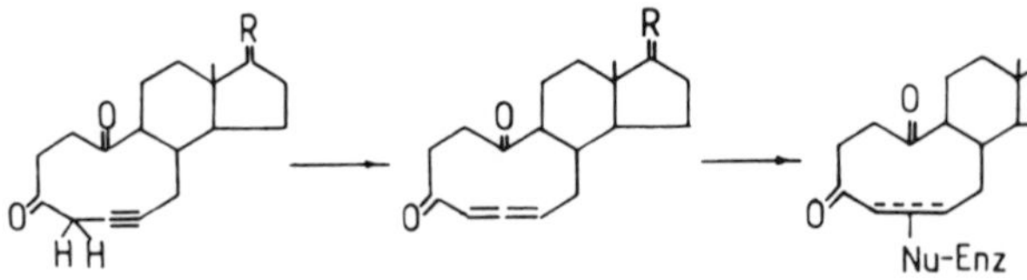

(42)

It is interesting that the latter reagent was found to modify asparagine residue 57 in the enzyme[161] demonstrating the capacity of amide groups to react with electrophilic reagents.

Proton abstraction-triggered suicidal reactions have been often considered as the reason for specific inactivation of enzymes using pyridoxal phosphate as the coenzyme. These enzymes catalyze a variety of transformations at the α, β, and γ carbon atoms of amino acids generating α-carbanion-like species during catalysis. A typical example is methionine γ-lyase, the enzyme involved in bacterial methionine metabolism which catalyzes the methionine breakdown to methane thiol and 2-ketobutyrate.

$$H_3C-S-CH_2-CH_2-\overset{\overset{+}{N}H_3}{\underset{|}{C}H}-COO^- \longrightarrow H_3C-SH \ + \ CH_3-CH_2-\overset{O}{\overset{\|}{C}}-COO^- +\overset{+}{N}H_4 \qquad (43)$$

The role of the pyridoxal phosphate cofactor is to form a Schiff base with amino groups of the substrates as the first step of the enzymatic reaction. In the course of the following steps, the enzyme facilitates the sequential abstraction of α and β protons resulting in the elimination of a γ substituent.

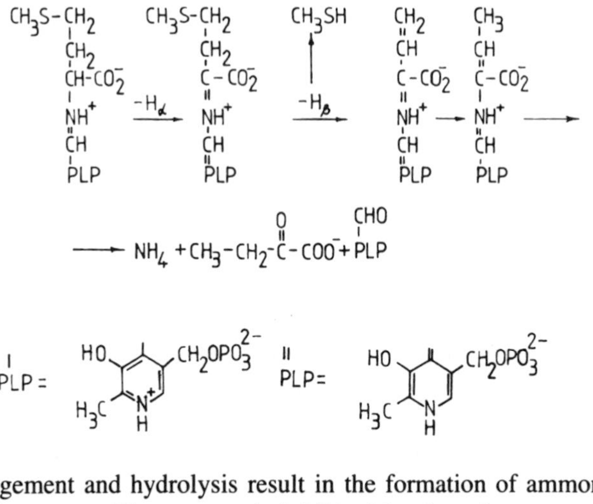

(44)

Further rearrangement and hydrolysis result in the formation of ammonia, butyric acid, and a regenerated enzyme. The enzyme is inhibited by the substrate analogue propargly-glycine which contains the acetylenic group and is believed to yield allene upon neighboring carbanion formation.[156]

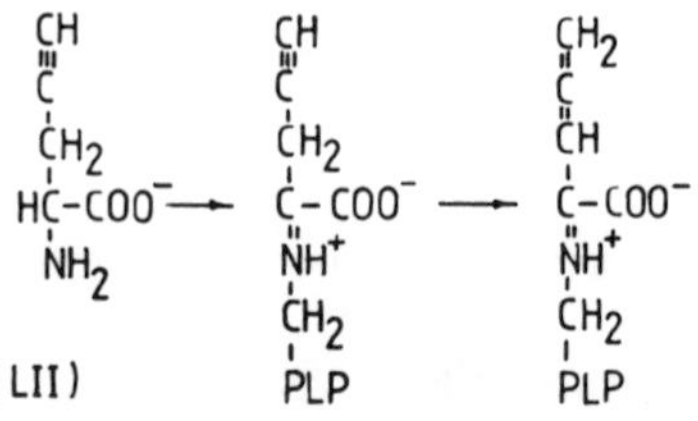

(45)

The same primary event, proton abstraction, can lead to the appearance of activated olefinic groups in the reagent molecules. Such groupings can be arranged by an oxidative mechanism. The typical example of this pathway to formation of reactive species is the conversion of unsaturated amino acids to alkylating compounds by pyridoxal phosphate-dependent enzymes such as aspartate aminotransferase.[158] The deamination reaction catalyzed by aspartate aminotransferase involves the formation of the aldimine intermediate of pyridoxal phosphate and the amino acid followed by α-hydrogen atom abstraction, isomerization of the double bond, and hydrolysis to a keto-derivative of the substrate and pyridoxamine form of the enzyme.

$$
\begin{array}{c}
R-CH-COO^- \rightleftarrows R-CH-COO^- \rightleftarrows R-C-COO^- \rightleftarrows R-C-COO^- \\
\quad\;\; NH_3^+ \qquad\qquad NH^+ \qquad\qquad NH^+ \qquad\qquad O \\
\qquad\qquad\qquad\quad CH \qquad\qquad CH_2 \\
\qquad\qquad\qquad\quad PLP \qquad\qquad PLP \qquad\qquad\quad +
\end{array} \tag{46}
$$

$$
\begin{array}{c}
CH_2-NH_2 \\
PLP
\end{array}
$$

When the suicide reagent vinylglycine (2-amino-3-butenoic acid) (LIII) is transaminated by aspartate aminotransferase, abstraction of α-hydrogen can result in the isomerization of the double bond present in the analogue molecule to yield the activated olefinic group, the Michael acceptor capable of reacting with nucleophilic residues Nu of the enzyme.

$$
\begin{array}{c}
CH_2=CH-CH-COO^- \longrightarrow CH_2=CH-CH-COO^- \longrightarrow CH_3-CH=C-COO^- \xrightarrow{\;Nu\;} \\
\qquad\quad NH_3 \qquad\qquad\qquad\qquad NH^+ \qquad\qquad\qquad\qquad NH^+ \\
\qquad\qquad\qquad\qquad\qquad\qquad CH \qquad\qquad\qquad\qquad\quad CH \\
(LIII) \qquad\qquad\qquad\qquad\qquad PLP \qquad\qquad\qquad\qquad\quad PLP
\end{array}
$$

$$
\begin{array}{c}
\qquad\qquad Nu \\
\longrightarrow CH_3-CH-CH-COO^- \\
\qquad\qquad\quad NH^+ \\
\qquad\qquad\qquad CH \\
\qquad\qquad\qquad PLP
\end{array} \tag{47}
$$

Another mechanism initiated by the proton abstraction activates the most well-studied types of suicide reagents for pyridoxal phosphate-dependent enzymes, β- or α-fluoromethyl amino acids. In this case the net elimination of H^+ and F^- generates alkylating species. Thus, alanine racemase is inhibited by synthetic antibiotic β-fluoroalanine.[156] Bacterial alanine racemase supplies the cells with the D-alanine isomer which is an important constituent of bacterial cell walls. Racemization proceeds most probably via α-carbanion formation. Similar proton abstraction in β-D-fluoroalanine (LIV) results in the loss of HF to yield an aminoacrylate compound which modifies nucleophiles in the enzyme active site.

$$
\begin{array}{ccc}
FCH_2-CH-COO^- & CH_2=C-COO^- & Enz-Nu-CH_2-CH-COO^- \\
\quad\;\; NH^+ & \quad NH^+ & \qquad\qquad\quad NH^+ \\
\qquad CH & \qquad CH & \qquad\qquad\qquad CH \\
\qquad PLP \;\longrightarrow & \qquad PLP \;\longrightarrow & \qquad\qquad\qquad PLP \\
(LIV)
\end{array} \tag{48}
$$

A similar mechanism of alkylating species formation operates when fluorinated amino acid analogues are decarboxylated by pyridoxal-dependent enzymes. In this case reactive compounds are formed due to net CO_2, Hal elimination (Equation 38). An example of such reagents is the well-known inhibitor of ornithine decarboxylase (EC 4.1.1.17), α,α-difluoromethyl ornithine (LV).[162] Ornithine decarboxylase catalyzes the initial and rate-limiting steps in the polyamine biosynthetic pathway. It decarboxylates ornithine to yield putrescine which is the precursor of spermidine and spermine, essential components of chromatin playing an important role in replication and cell proliferation. In the aldimine formed by the amino group of α,α-difluoromethyl ornithine and aldehyde group of pyridoxal phosphate, decarboxylation of the analogue leads to the elimination of the fluoride ion and the formation of the alkylating group.

$$
\begin{array}{c}
\underset{(LV)}{\overset{\displaystyle COO^-}{\underset{\displaystyle NH_2}{H_2N(CH_2)_3-\overset{|}{\underset{|}{C}}-CHF_2}}} \longrightarrow
\underset{}{\overset{\displaystyle COO^-}{\underset{\displaystyle \overset{NH^+}{\underset{\overset{CH}{PLP}}{\|}}}{H_2N(CH_2)_3-\overset{|}{\underset{|}{C}}-CHF_2}}} \xrightarrow{-CO_2,F^-}
\end{array}
$$

$$
H_2N(CH_2)_3-\underset{\underset{\underset{PLP}{CH}}{NH^+}}{C}=CHF \xrightarrow{Nu-Enz}
H_2N(CH_2)_3-\underset{\underset{\underset{PLP}{CH}}{NH^+}}{C}-CHF-Nu-Enz
$$

(49)

A different type of suicidal reagent activatable by proton abstraction is represented by compounds where a carbanion-mediated formation of a double bond in a position allylic to a aryl sulfoxide group leads to the structure which tends to rearrange according to the 2,3-sigmatropic rearrangement mechanism to allyl sulfonate ester which is a reactive electrophilic group.

$$
\underset{}{\overset{R\quad O}{S}} \rightleftharpoons \overset{R\quad O}{S} \xrightarrow{Nu} \overset{HO}{\diagup} + Nu-S-R
$$

(Nu-nucleophile)

(50)

This mechanism operates when the above-mentioned enzyme methionine γ-lyase (Equations 43 and 44) is inactivated by 2-amino-4-chloro-5-(p-nitrophenylsulfinyl)pentanoic acid (LVI). Enzymatic processing via carbanione intermediates and release of HCl lead to the compound possessing a β,γ double bond in a position allylic to the sulfoxide which can rearrange yielding reactive species. It was found that the reagent LVI inactivates the enzyme by the formation of mixed disulfide with the cysteine residue Enz-S-S-Ar.[163]

$$
\begin{array}{cccc}
\underset{(LVI)}{\overset{NO_2}{\underset{\overset{\displaystyle NH_3^+}{\overset{\displaystyle CHCOO^-}{\overset{\displaystyle CH_2}{\overset{\displaystyle CHCl}{\overset{\displaystyle CH_2}{\overset{\displaystyle S\to O}{\bigcirc}}}}}}}{}} &
\overset{NO_2}{\underset{\overset{\displaystyle CH}{\overset{\displaystyle PLP}{\overset{\displaystyle NH^+}{\overset{\displaystyle CHCOO^-}{\overset{\displaystyle CH_2}{\overset{\displaystyle CHCl}{\overset{\displaystyle CH_2}{\overset{\displaystyle S\to O}{\bigcirc}}}}}}}}}{} &
\overset{NO_2}{\underset{\overset{\displaystyle CH}{\overset{\displaystyle PLP}{\overset{\displaystyle NH^+}{\overset{\displaystyle CCOO^-}{\overset{\displaystyle CH}{\overset{\displaystyle CH}{\overset{\displaystyle CH_2}{\overset{\displaystyle S\to O}{\bigcirc}}}}}}}}}{} &
\overset{NO_2}{\underset{\overset{\displaystyle CH}{\overset{\displaystyle PLP}{\overset{\displaystyle NH^+}{\overset{\displaystyle CCOO^-}{\overset{\displaystyle CH-CH=CH_2}{\overset{\displaystyle O}{\overset{\displaystyle S}{\bigcirc}}}}}}}}{} \\
\longrightarrow & \longrightarrow & \longrightarrow & \longrightarrow
\end{array}
$$

(51)

$$
\begin{array}{c}
Enz-S-S-\bigcirc-NO_2 \\
+ \\
\underset{\overset{\displaystyle CH}{\overset{\displaystyle PLP}{\overset{\displaystyle NH^+}{\overset{\displaystyle CCOO^-}{\overset{\displaystyle CHOH}{\overset{\displaystyle CH}{CH_2}}}}}}{}
\end{array}
$$

There are several ways of damaging species formation when enzymes use O_2 for the oxidation of substrates. A well-known case is the inactivation of P-450 monooxygenases during the oxidation of olefinic and acetylenic compounds. Liver hemoprotein cytochrome P-450 monooxygenases catalyze the oxidation of various lipophilic compounds in the first phase of pathways of biotransformation and detoxification of drugs, pesticides, and other xenobiotics. They consist of a flavoprotein NADPH-cytochrome P-450 reductase and various isozymes of cytochrome P-450 that serve as general oxygen activators and oxene transferases. Cytochrome P-450 Fe_{450} contains a heme cofactor (LVII) which is coordinated to the apoenzyme and may bind an oxygen atom as the sixth ligand forming the heme-oxene complex $Fe_{450} = O$.

(LVII)

It is commonly accepted that the epoxidation of olefins and hydroxylation of C–H bonds carried out by the enzymes are achieved by using the process inserting an oxygen atom from the heme-oxene complex.

Suicide reagents, acetylenic and olefinic compounds, inactivate P-450 monooxygenases irreversibly primarily due to alkylation of the heme prosthetic groups. In some cases covalent binding to apoenzymes was also observed. A possible mechanism of the modifications is the reaction with initial oxidative products. Oxygen insertion into the π bond of acetylenes and olefins results in the formation of vinyl oxirans and epoxides, respectively. The former species are very unstable and could be considered as alkylating reagents capable of modifying the tetrapyrrole nitrogens. However, in the case of olefins it was proved that the reactive species are not the epoxides. Therefore, it was proposed that the inactivating reactions may proceed according to the mechanism including the initial formation of carbcation-like intermediates.[164]

$$\underset{\underset{Fe_{450}}{\overset{\parallel}{}}}{\overset{O}{}} + R_1-CH=CH-R_2 \longrightarrow \underset{\underset{Fe_{450}}{\overset{|}{O^-}}}{\overset{R_1^-CH-\overset{+}{C}H-R_2}{}} \qquad (52)$$

Free radical formation was proposed as the inactivating event in the reaction of mitochondrial monoamine oxidase (EC 1.4.3.4) with 1-phenylcyclopropylamine.[165] This FAD-containing enzyme catalyzes the oxidative deamination of biogenic amine neurotransmitters to the corresponding aldehydes. It is believed that the oxidation reaction proceeds via two one-electron transfers from the substrate to the flavin. In the course of the reaction, amine is oxidized to the iminium ion and the flavine is reduced. The iminium ion is hydrolyzed further to the aldehyde and reduced flavin is reoxidized enzymatically with dioxygen.

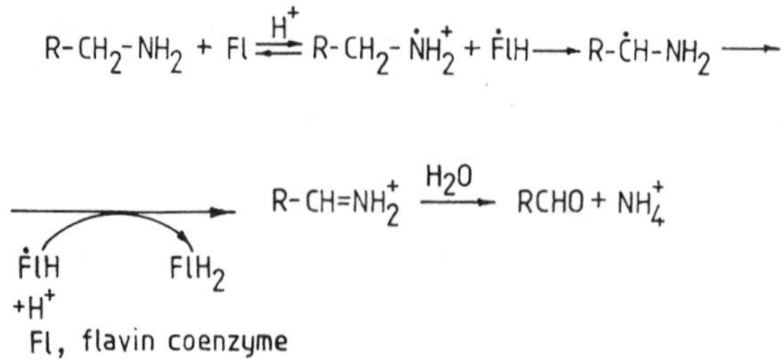

$$R-CH_2-NH_2 + Fl \overset{H^+}{\rightleftharpoons} R-CH_2-\overset{+}{NH_2} + \dot{F}lH \longrightarrow R-\dot{C}H-NH_2 \longrightarrow$$

$$\overset{\longrightarrow}{\underset{\underset{Fl, \text{ flavin coenzyme}}{\overset{\dot{F}lH \qquad FlH_2}{+H^+}}}{}} R-CH=\overset{+}{NH_2} \overset{H_2O}{\longrightarrow} RCHO + NH_4^+ \qquad (53)$$

In the case of the suicide reagent 1-phenylcyclopropylamine (LVIII) the enzyme is inactivated and covalent modification of the coenzyme and the apoenzyme is observed. The proposed mechanism of inactivation of the enzyme includes one-electron oxidation of the reagent by the flavin cofactor leading to the amine radical cation formation. The latter undergoes a homolytic cyclopropane ring opening yielding the reactive free radical which causes enzyme modification.

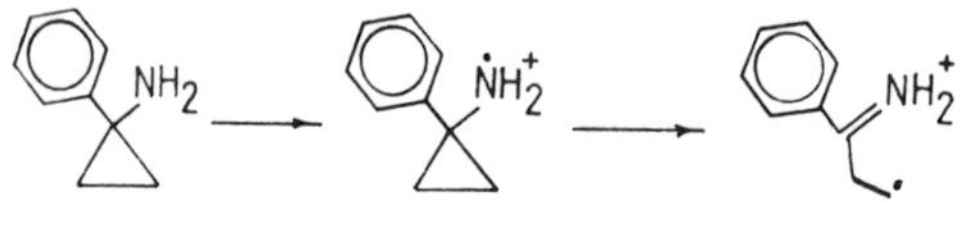

$$(54)$$

(LVIII)

Free radical reactive species were proposed to be involved in several other cases of suicide inactivation of enzymes. This may be the case for the enzymes involved in the biosynthesis of important mediators of the inflammatory cascade, leukotrienes and prostaglandins. Suicide reagents for these enzymes are acetylenic analogues of arachidonic acid.[166] Thus, 14,15-dehydroarachidonic acid (LIX) was found to be an irreversible inhibitor of soybean 15-lipoxygenase (EC 1.13.11.12), the enzyme converting arachidonic acid to 15-(hydroperoxy)-5,8,11(Z),13(E)-eicosatetraenoic acid. It is proposed that peroxidation of the analogue leads to a derivative in which the hydroperoxy group is unstable and undergoes O–O bond homolysis yielding radical species initiating the enzyme-inactivating process.

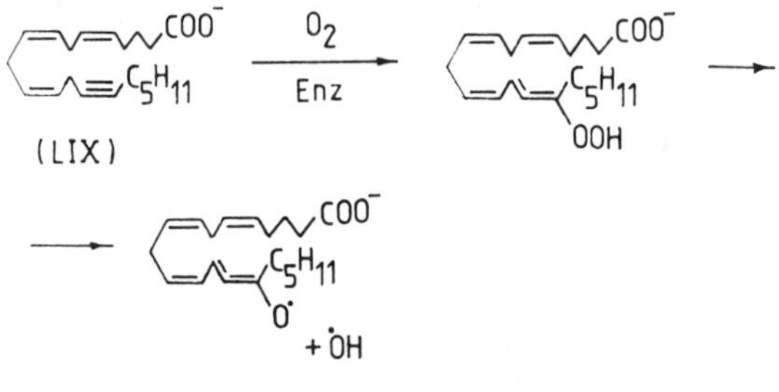

$$(55)$$

(LIX)

Enzymes which employ nucleophilic catalysis carry out the transfer of electrophilic groups of substrates first to nucleophilic centers present in the enzymes and then to the outer nucleophile, usually water. At the first stage of catalysis two chemical events, cleavage of a covalent bond present in the substrate and intermediate covalent adduct formation between the substrate and the enzyme, occur. Both these events can be inactivating for the enzyme molecules processing certain types of substrate analogues.

Examples of suicide reagents which are converted to reactive species upon enzymatic cleavage of ester and amide bonds are specific inhibitors of α-chymotrypsin (EC 3.4.21.1). Chymotrypsin hydrolyzes amide and ester bonds in carboxylic acid derivatives showing preference to derivatives of aromatic carboxylic acids according to the reaction mechanism similar to that presented by Equation 8 in Chapter 1. The hydrolysis proceeds via the formation of an intermediate acyl enzyme in which the carbonyl group of a substrate is linked to the hydroxyl group of a serine in the enzyme active site. This intermediate is normally hydrolyzed rapidly to regenerate chymotrypsin.

In the case of the suicide inhibitors haloenol lactones such as 6-iodomethylene-naphthyl-tetrahydropyran-2-one (LX),[167] the first step of catalysis should result in lactone ring opening, acyl enzyme formation, and the appearance of a reactive iodomethyl keto group in the enzyme-linked reagent moiety which can react with juxtaposed active site residues and inactivate the enzyme.

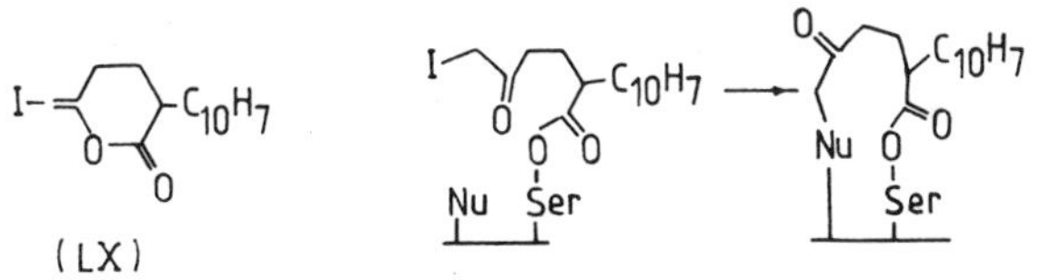

(56)

When suicide chymotrypsin inhibitor N-nitroso-1,4-dihydro-6,7-dimethoxy-3(2H)-iso-quinolinone (LXI) interacts with the enzyme, cleavage of the substrate amide bond at the stage of acyl enzyme formation results in the appearance of diazohydroxide grouping. Facile decomposition of the latter leads to the formation of an unstable diazonium ion and further to a benzylic carbcation. Since this very reactive alkylating group is covalently tethered to the acyl enzyme intermediate, it attacks the active side residues and destroys the enzyme activity.[168]

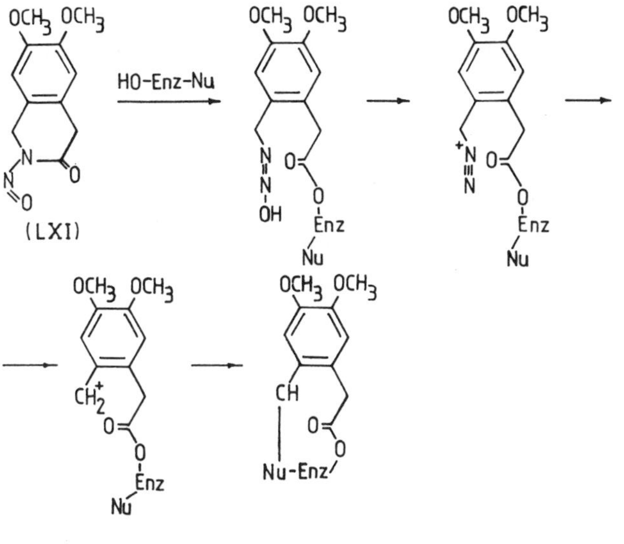

(57)

Nu: some additional (besides Ser-195) nucleophilic center of enzyme.

Cleavage of a covalent bond in a substrate analogue is the event triggering the suicidal reaction pathway in the case of important penicillin analogues inactivating the bacterial enzymes β-lactamases (EC 3.5.2.6). The existance of these enzymes is the most common basis for bacterial resistance to the β-lactam antibiotics penicillin and cephalosporin. The enzymes hydrolyze the β-lactam ring of these antibiotics as it is shown for benzylpenicillin (LXII) resulting in the formation of products which are harmless to bacteria.

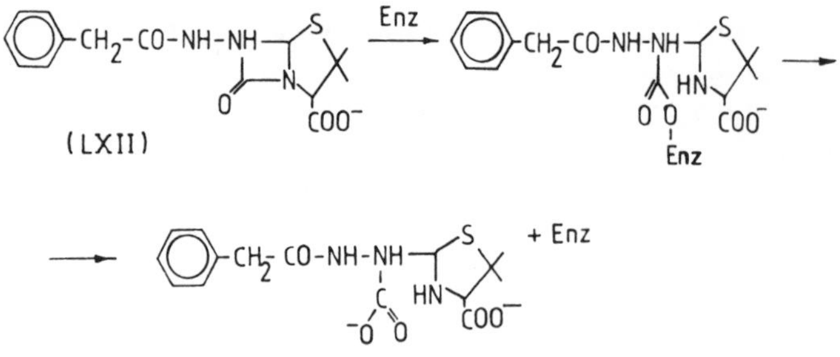

(58)

At the first step of the process the enzyme active site serine residue is acylated followed by hydrolytic breakdown of the acyl enzyme.

The suicide substrates oxapenicillin analogue clavulanic acid (LXIII) and some other derivatives such as penicillanic acid sulfone (LXIV) are capable of normal acyl enzyme formation during the β-lactam ring opening.

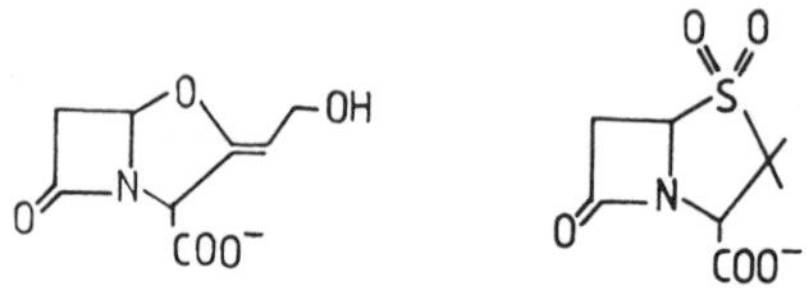

However, in contrast to normal substrates, the β-lactam ring-opened derivatives are unstable and are cleaved spontaneously at the four-five ring junction, e.g., for penicillanic acid sulfone:

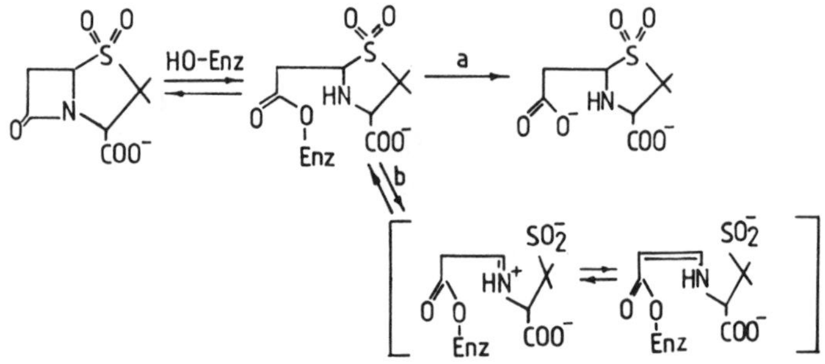

(59)

Therefore, in addition to the usual direction of the reaction, the hydrolytic breakdown of the acyl enzyme (path a), the formation of relatively stable intermediates, occurs (path b) which interferes with the enzyme action. Covalent modification of the enzyme was observed due to the reactions of the intermediate with the active site amino acid residues.[169]

The considered suicide β-lactamase inhibitors have shown therapeutic value as good synergists of standard penicillins against penicillin-resistant bacteria.

It should be noted that detailed mechanisms of many of the reactions which are considered as suicide inhibition due to the formation of reactive species by the enzymes remain to be investigated. The described mechanisms of formation of reactive species are in many cases hypothetical since experimental studies of the very short-living intermediates and structural analysis of the modified enzymes are related to considerable difficulties. First, detailed chemical and physicochemical studies of enzymes modified with substrate analogues have demonstrated that in some cases where the usual paradigm could suggest suicidal inhibition through the formation of highly reactive intermediate species; in fact, the usual affinity modification takes place. Thus, irreversible inhibitors of chymotrypsin, derivatives of 6-chloro-2-pyrone, can be thought to be converted to highly reactive acyl chlorides upon enzymatic lactone hydrolysis and acyl enzyme formation.

$$\tag{60}$$

This mechanism probably operates when the enzyme is inactivated by 3-benzyl-6-chloro-2-pyrone (LXV).

(LXV)   (LXVI)

However, when detailed chemical, nuclear magnetic resonance, and X-ray crystallographic studies were undertaken for characterization of the covalent adduct formed in the reaction of chymotrypsin with 5-benzyl-6-chloro-2-pyrone (LXVI), it was found that the pyrone ring of the derivative is intact and the atom C6 is bound covalently to the γ-oxygen of the enzyme active site serine. Therefore, it was concluded that no enzyme-catalyzed esterolysis occurs in the case of LXVI and modification of the enzyme proceeds according to usual nucleophilic substitution at the atom C6 of the reagent.[170]

$$\tag{61}$$

Suicidal inhibition at the stage of covalent intermediate adduct formation is achieved when the substrate analogue contrary to the normal substrate forms stable covalent bonds with the enzymes resulting in efficient trapping of the enzymes in midcatalytic cycle. An example of this mechanism is the inhibition of trypsin with (p-amidinophenyl)methanesulfonyl fluoride (LXVII).[171]

$$\tag{LXVII}$$

As it was mentioned in Chapter 1, Equation 8, in reactions catalyzed by serine proteases, acylated enzyme is formed which is readily hydrolyzed in the case of normal substrate processing. In the case of a suicide reagent, a sulfonylated enzyme is formed instead of the acylated derivative. Since the former derivative is rather stable and can hydrolyze only slowly the enzyme is inactivated.

$$\tag{62}$$

The same mechanism provides great inactivating efficiency of nerve agents (see Section II.B) phosphorylating the active site nucleophile in acetylcholine esterase.

Some suicide reagents halt the catalytic reactions completely at certain steps. This is the case for 5-fluorouracil, a currently used anticancer agent which inactivates thymidylate synthase (EC 2.1.1.45), the important enzyme of nucleotide biosynthesis converting uridylate to thymidylate. In the course of the normal reaction, the ternary complex of uridylate, enzyme, and N^5, N^{10}-methylene tetrahydrofolate cofactor is formed and the protein and cofactor consecutively form covalent bonds with C6 and C5 atoms of the nucleotide substrate, respectively.

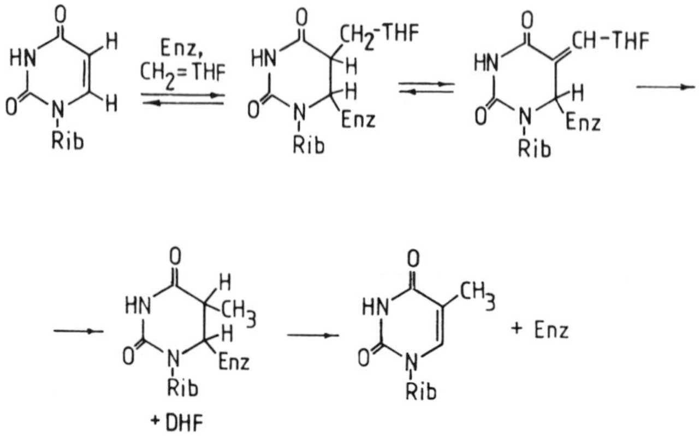

(63)

CH$_2$=THF, N^5,N^{10}-methylene tetrahydrofolate; THF, tetrahydrofolate; and DHF, dihydrofolate.

Further, in the covalent intermediate, an apparent 1,3-hydride shift occurs yielding the methyl group at C2 of the nucleotide substrate and releasing 7,8-dihydrofolate.

5-Fluorouracil, which is readily converted in vivo to 5-fluorodeoxyuridylate, interacts with the enzyme as well as the normal substrate up to the point when the C5–H bond is to be cleaved. Since the analogue contains fluorine instead of hydrogen at this position and the enzyme cannot break the very stable C–F bond, the process freezes at this step.[172]

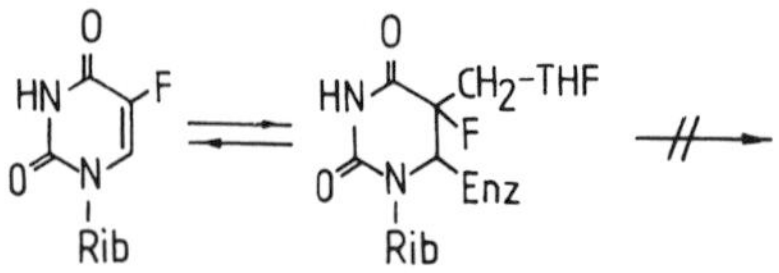

(64)

The enzyme is trapped in the midcatalytic cycle and the complex can break down only very slowly, half-time being of order of several hours.

There are several types of specific modifications of enzymes where they facilitate inactivation of themselves using mechanisms other than the direct conversion of substrates to reactive species or the formation of the dead-end complexes. However, these modifications also take the advantage of specific chemical properties of enzyme active sites and the chemistry involved in enzymatic catalysis. Therefore, these cases can be considered special types of suicide reactions.

A special type of modification of enzymes is paracatalytic modification. This type of modification is based on the ability of enzymes to convert substrates to intermediate species capable of reacting with extrinsic chemical reagents which normally do not touch the substrates. In some cases, these by-reactions of enzymatic processes lead to the formation of reactive compounds which modify the enzymes covalently and cause inactivation. The most thoroughly investigated paracatalytic reactions are those for the enzymes forming readily oxidizable carbanion intermediates.[173] The typical example is fructose 1,6-bis-phosphate aldolase, an intermediate enzyme of glycolysis (Chapter 1, Figure 9) catalyzing the reversible cleavage of fructose-1,6-bis-phosphate to glyceraldehyde-3-phosphate and dihydroxyacetone phosphate (DHP). In the course of catalysis the enzyme forms a carbanion, the resonanse structure of enamine intermediate Schiff base formed by DHP and the E-NH$_2$ group of lysine residue in the active center.

$$H_2\overset{\overset{O}{\|}}{C}-O-\overset{\overset{O}{\|}}{\underset{\underset{O^-}{|}}{P}}-O^-$$
$$\underset{\underset{\ominus}{|}}{\overset{|}{C}}=NH^+-Enz$$
$$H\overset{|}{C}-OH$$

This intermediate was found to be readily oxidized by suitable electron acceptors such as hexacyanoferrate $[Fe^{3+}(CN)_6]^{3-}$. In the course of the oxidation of the carbanion intermediate, the enzyme is inactivated. The exact reason for the inactivation is not yet elucidated.

Several cases are known when photolysis of photoreactive affinity reagents was facilitated upon a specific complex formation due to specific properties of the binding sites. These cases were referred to as photosuicide modification. An example of biopolymer-assisted photoaffinity modification is the photomodification of acetylcholinesterase with p-dimethylaminobenzene diazonium fluoroborate.[174] It was found that affinity modification of the enzyme with this reagent can be accomplished both by using direct photoactivation of the reagent with 410-nm light and by activation by energy transfer from the tryptophane present in the active site of the enzyme using 295-nm UV irradiation. In the latter case the quantum yield of the inactivation reaction was higher and nonspecific labeling by the nonbound reagent activated in solution was decreased.

Facilitation of the photoaffinity labeling by the target biopolymer according to different mechanism was observed when the same enzyme was subjected to affinity modification with N-nitroso analogues of acetylcholine, methyl(acetoxymethyl)nitrosamine (LXVIII) and methyl(butyroxymethyl)nitrosamine (LXIX).[175]

$$CH_3-\overset{\overset{O}{\|}}{C}-O-CH_2-\overset{\overset{CH_3}{|}}{N}-N=O \qquad CH_3-CH_2-CH_2-\overset{\overset{O}{\|}}{C}-O-CH_2-\overset{\overset{CH_3}{|}}{N}-N=O$$

(LXVIII) (LXIX)

Photolysis of N-nitrosamines yields highly reactive alkylating species. It is the protonated form of the derivatives which is susceptible to UV light, therefore photolysis proceeds poorly under physiological pH values. Since acetylcholinesterase facilitates the protonation of these reagents, UV irradiation in a neutral medium results in efficient modification of the enzyme.

In this section the main types of suicide reagents were considered. Many of other examples can be found in the Appendix.

H. Metal-Containing Groups

Thousands of papers have been published on the problem of interaction of metal ions with nucleic acids and proteins. Some cations play central roles in enzymatic catalysis being enzyme cofactors such as zinc in carboxypeptidase or part of enzyme prosthetic groups such as iron in hemoglobin. This branch of bioorganic chemistry is considered as a separate field of science, bioinorganic chemistry. However, the use of metal ions as reactive groups of affinity reagents is only beginning. Two essential properties of metal ions, namely the ability to coordinate functional groups of biopolymers as ligands and to act as oxidative-reducing catalysts, suggest some prospective applications of the metal ions in affinity modification. In this section we shall describe the recent experiments where complexes of some metals were used as affinity reagents, and attempts have been made to design catalytic affinity reagents.

Several cases are known where metal ions provide unusual stability to nucleotide-enzyme complexes. Thus, a number of enzymes involved in phosphotransferase or phosphohydrolase reactions recognize a vanadate ion as an analogue of inorganic phosphate. In contrast to the normal complexes, those formed by a vanadate, nucleotide, and enzyme are extremely tight. In the case of myosin ATPase (EC 3.6.1.3), which plays a central role in the mechanisms of muscle contraction, stoichiometric amounts of vanadate and ADP are tightly bound to the enzyme causing inactivation. The complex dissociates very slowly with half-lifetime of the order of days. Therefore, ADP plus vanadate were referred to as a "reversible affinity label" for this enzyme.[176,177] It was proposed that the enzyme-ADP-vanadate complex represents a stable analogue of some enzyme-ADP-phosphate intermediate which is formed in the course of enzymatic ATP hydrolysis.

Similarly, one can consider as affinity reagents complexes of Cr^{3+} and Co^{3+} with nucleoside di- and triphosphates which bind very tightly to some nucleoside triphosphate-interacting enzymes imitating initial ATP or final ADP + P_i as well as an intermediate or transition state formed in the course of catalysis. A typical example of such enzymes is glutamine synthetase (EC 6.3.1.2) which catalyzes the synthesis of glutamine from glutamate and ammonia in conjunction with the hydrolysis of ATP to ADP and inorganic pyrophosphate. The enzyme is irreversibly inactivated by β,γ-Cr(III) $(H_2O)_4$ ATP.[178] It was demonstrated that in complexes formed by enzymes and ATP + Cr(III) or Co(III) the metal bridges between the enzyme and the triphosphate chain of the nucleotide.[179]

Studying of platinum(II) complexes as potential reactive groups for affinity reagents has been started recently. The platinum compound *cis*-[Pt(NH$_3$)$_2$Cl$_2$] (cisplatin) is an efficient anticancer drug; therefore, the chemistry of platinum complexes and the reactions with biopolymers were intensively studied.[180]

Complexes of Pt(II) and amino ligands are kinetically inert under conditions of reactions with biopolymers. These ligands can serve to anchor the platinum groups to the addressing part of the affinity reagents. Good leaving groups such as NO_3^-, H_2O, SO_4^{2-}, Cl^-, and Br^- listed in increasing order of the affinity to platinum can be substituted in reactions with nucleophilic groups of biopolymers. In reactions with biopolymers, platinum reagents show a chemical specificity similar to that of alkylating reagents and attack the biopolymer residues in order of nucleophilicity. Although products of platination are stable under physiological conditions and under conditions used in the course of biochemical analysis, platinum groups may be removed from the modified residues easily by treatment with cyanide or thiourea.[181,182] It should be noted also that in contrast to alkylation, modification with platinum reagents does not destabilize the glycosyl bond in nucleotides.

Platinum(II) complexes can have several ligands which behave differently in the substitution reactions with biopolymer residues. Thus, in platinum reagent LXX

(65)

LXX

H_2O can be substituted very easily. The reaction of LXX with oligocytidylates yields adducts possessing the other ligand OH^- which can be substituted, although slowly. The oligocytidylates therefore are converted to affinity reagents bearing platinating groups. It was found that, upon binding of the oligocytidylate derivatives with complementary polyinosinic acid, affinity modification occurs due to the substitution of inosine N1 atoms for the OH^- ligands in the platinum groups.[183]

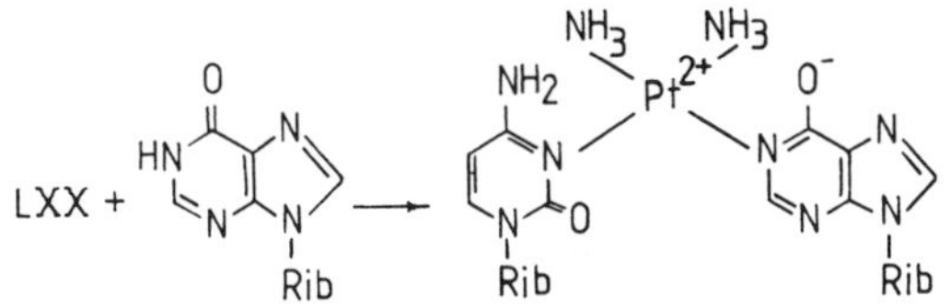

(66)

A binuclear platinum complex

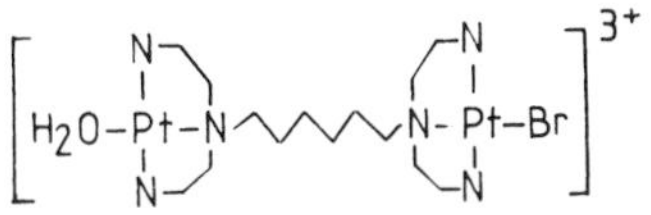

LXXI

was proposed for the derivatization of oligonucleotides in order to prepare the platinating derivatives. The more reactive $[H_2O\, Pt(dien)]^{2+}$ fragment of this compound may be used for coupling with the oligonucleotide carrier by substituting guanine for H_2O. The other ligand Br^- is much more difficult to substitute. The oligonucleotide derivatives bearing a low reactive $[(dien)PtBr]^+$ unit are capable of efficient substitution of Br^- by some nucleic acid base only being in the close contact inside the complementary complex. This was demonstrated in experiments with synthetic oligonucleotides. Treatment of the oligonucleotide LXXII with the reagent LXXI resulted in an anchoring of the compound to the N7 atom of the single guanosine present in this oligonucleotide. When the prepared derivative LXXIII formed a complementary complex with the oligonucleotide LXXIV the latter was efficiently platinated at guanosine residue G1.[184]

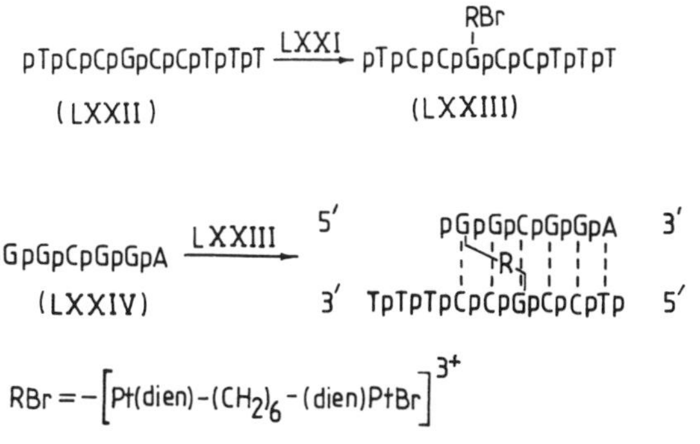

(67)

(68)

In the past few years attempts have been made to design affinity reagents with catalytic groups. Each reagent molecule with such a group could perform chemical modification of several target biopolymer molecules. This type of affinity reagent would be valuable in particular in the field of therapeutic applications of affinity modification since it may accomplish exhaustive modification of target molecules using low concentrations of affinity reagents. The starting point of these studies was the discovery that anticancer antibiotic bleomycin

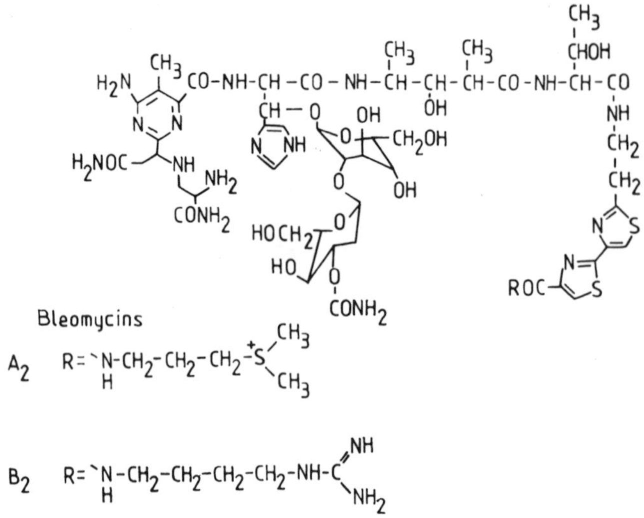

Bleomycins

A_2 $R = {}^{\backslash}\underset{H}{N}-CH_2-CH_2-CH_2-\overset{+}{S}\underset{CH_3}{\overset{CH_3}{<}}$

B_2 $R = {}^{\backslash}\underset{H}{N}-CH_2-CH_2-CH_2-CH_2-NH-C\underset{NH_2}{\overset{NH}{<}}$

degrades DNA using oxidative activity of ferrous ions via activated oxygen species.[185,186] Under certain conditions the number of DNA lesions produced in this reaction can substantially exceed the number of bleomycin molecules consumed.[187]

It is accepted that the degradation of DNA by the bleomycin-activated oxygen species occurs according to free radical mechanisms. One of the pathways of DNA degradation by bleomycin is suggested to be initiated by the formation of a C4′ deoxyribosyl radical. The latter is transformed to C4′-hydroperoxide and then the C3′-C4′ bond cleavage occurs resulting in the formation of base propenals and oligomers containing carboxymethyl groups attached to the 3′-terminal phosphate residues and free 5′-phosphates.[188] This type of damage is observed predominantly in the G pyrimidine sequences of DNA.

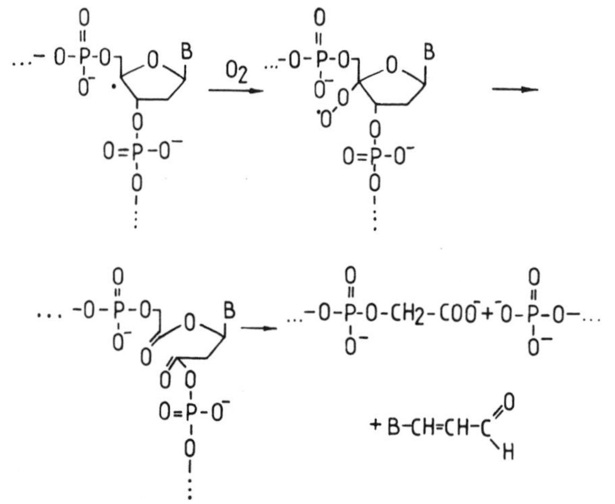

(69)

Another degradation pathway includes the direct release of free heterocyclic bases, preferentially pyrimidines, and the cleavage of polynucleotides at the sites of destroyed nucleotides.[189] Besides direct cleavage, bleomycin introduces into DNA a considerable amount of lesions which are developed by additional treatment such as "alkali-labile sites" of unclear chemical nature.

Attempts have been made to imitate the catalytic DNA-cleaving activity of bleomycin using simpler chemical compounds. The idea was to use chelated cuprous or ferrous ions (M) which can generate OH˙ radicals in a reaction occurring in the presence of oxygen and reducing agents.[190]

$$
\begin{aligned}
M^{(n-1)+} + O_2 &\longrightarrow M^{n+} + O_2^{\dot{-}} \\
2O_2^{\dot{-}} + 2H^+ &\longrightarrow H_2O_2 + O_2 \\
M^{(n-1)+} + H_2O_2 &\longrightarrow M^{n+} + OH^{\cdot} + OH^-
\end{aligned}
\tag{70}
$$

Reducing agents such as ascorbate recover the initial state of the metal oxidation. In principle, this reaction mechanism can generate several or many reactive species by the same metal group. Therefore some ferrous and cuprous complexes were attempted to be used as OH-generating groups. Two compounds,

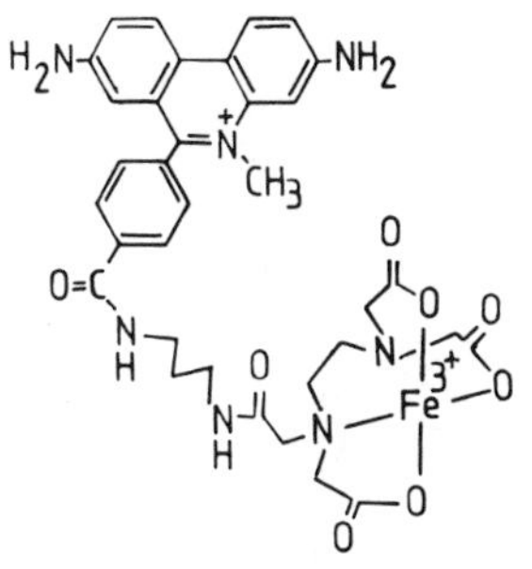

(LXXV) 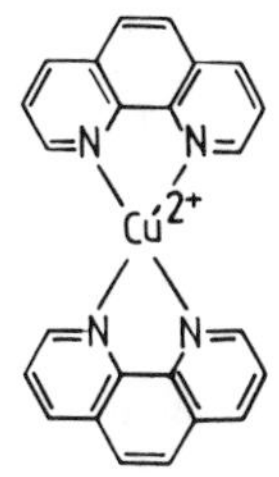   (LXXVI)

methidiumpropyl-EDTA·Fe(II) (LXXV)[191] and 1,10-phenanthroline cuprous complex (LXXVI),[192] were shown to degrade DNA in the presence of oxygen and reducing agents. Both reagents are currently used as nonspecific DNA-cleaving reagents.

A hemin group was also reported to cleave DNA in the oxygen-dependent reaction.[193] The reagent was a porphirine moiety attached to acridine (LXXVII) in order to stabilize a complex of the reagent with DNA due to acridine intercalation.

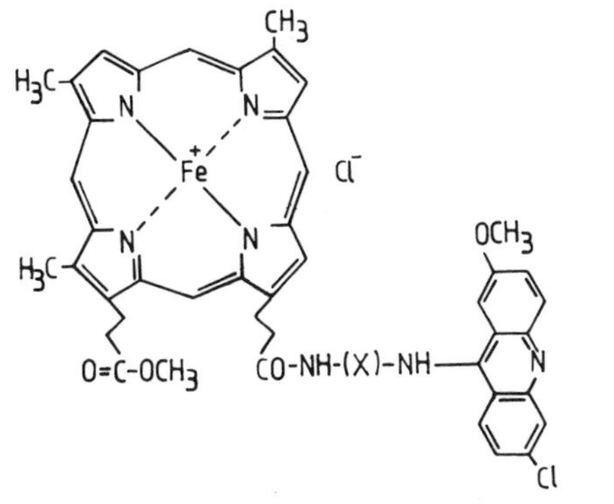

(LXXVII)

$X = (CH_2)_2, (CH_2)_3, (CH_2)_4, (CH_2)_2NH(CH_2)_2,$ or $(CH_2)_3NH(CH_2)_3.$

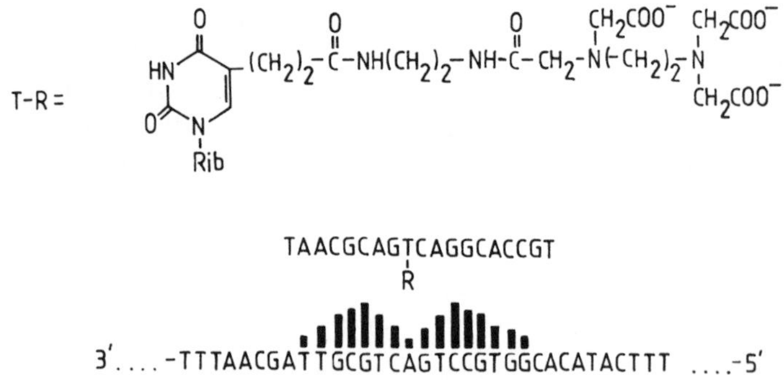

$$T-R =$$

TAACGCAGTCAGGCACCGT

3'.... -TTTAACGATTGCGTCAGTCCGTGGCACATACTTT ....-5'

FIGURE 1. Cleavage pattern of DNA fragment with complementary addressed reagent bearing EDTA·Fe(II) residue in the presence of dithiothreitol. The full bars indicate cleavage points and heights indicate the relative cleavage intensity. (From Dreyer, G. B. and Dervan, P. B., *Proc. Natl. Acad. Sci. U.S.A.*, 82, 968, 1985.)

The attractive prospect of developing methods of site-specific cleavage of DNA stimulated the synthesis of EDTA·Fe(II)-bearing analogues of distamycin X and similar oligopeptides capable of specific binding to the AT-rich sequences of DNA. It was found that these derivatives bring about the cleavage of DNA preferentially in the vicinity of the binding sites.[191] The cleavage did not occur as the pinpoint cuts at definite nucleotide positions but rather as several nucleotide units that were damaged within six to eight-nucleotides long sequences.

Sequence-specific degradation of polynucleotides was achieved also using oligonucleotide derivatives bearing EDTA·Fe(II) groups.[194-196] The typical example is the experiment on the cleavage of a single-stranded DNA fragment with a 19-nucleotides long oligonucleotide affinity reagent carrying the EDTA·Fe(II) group at C5 of thymidine in position 10. The reagent cleaved the target over a range of 16 nucleotides within the complementary site as it is shown in Figure 1.

Results of the described experiments demonstrated that the radicals produced by the EDTA·Fe(II) group do not migrate too far from the point of formation and result in local damage when delivered to a certain site of nucleic acid structure. The damage of extensive nucleotide sequences observed in experiments with oligopeptides and oligonucleotide derivatives can be attributed both to migration of the OH radicals and to the conformational mobility of the EDTA group of the reagents. In the experimental studies reported, the efficiency of DNA degradation was quite low, which is inconsistent with the catalytic character of the reactive group of the reagents. The obvious reason for this effect is the autocleavage of the affinity reagents by the free radicals which can diffuse from the reactive groups in many directions. Free radical-generating affinity reagents are under intensive studies and one may hope that in the near future a new generation of affinity reagents capable of catalytic modifications of biopolymers will appear.

III. DESIGN OF AFFINITY REAGENTS

A. Targeting (Addressing) Structures of Affinity Reagents

The design of an affinity reagent includes choosing a targeting (addressing) structure that provides the noncovalent specific binding of a reagent to a biopolymer, choice of the appropriate reactive group, and construction of the molecule possessing both structural units.

Up to now, no general methods for the rational design of targeting structures for proteins have been developed. Therefore, the main applications of affinity modification to these most intensively studied biopolymers are related today to the modification of functional proteins where natural structures such as substrates and various regulatory molecules which are recognized by specific enzymes are available. Affinity reagents targeted by these structures are excellent for exploring the active sites of the proteins. On the other hand, the rest of the biopolymer surfaces remain unavailable for investigation by affinity modification since there are no rational means to find the structure with a specific affinity to a definite area of protein molecules outside the active centers. Some nonfunctional regions of protein surface serve as antigenic determinants and one may consider respective antibodies or fragments as potential natural addresses. This promising possibility has not yet been realized in affinity modification although antibodies are widely used for targeting of specific molecules and liposomes to certain cells (see Chapter 5, Section III.E).

Contrary to proteins, the general mechanism of recognition operating in the specific binding of nucleic acids is understood. Specificity is provided here by interactions of complementary nucleotide sequences, and oligonucleotides can serve as addressing structures for corresponding regions in single-stranded polynucleotides.

After the targeting structure for a reagent is chosen, one must learn which of the functional groups present in this structure are not essential for recognition. In order to prepare the affinity reagent, these groups should be substituted by reactive groups or reactive groups should be coupled to them. The main approaches will be described in the next section. The inessential groups are usually identified in preliminary experiments with nonreactive analogues. In the case of enzymes this information can be obtained by comparing structures of substrates with those of competitive enzyme inhibitors. Thus, experiments with many analogues of amino acids have demonstrated that in accordance with the known function of AA-tRNA-synthetases the enzymes are highly sensitive to structural changes in side chains of the amino acids. A positively charged amino group is also important for recognition. At the same time, drastic changes introduced at the carboxyl group do not severely influence the affinity of the amino acids to corresponding enzymes.[197] These results suggested the synthesis of amino acid derivatives bearing reactive groups at the carboxyl end for the investigation of amino acid binding sites of AA-tRNA-synthetases. The most popular reagents of this type are aminoacyl methylene chlorides XX which were used for affinity labeling of a number of enzymes.[98] Even a more bulky substituted photoreactive phenylalanine derivative LXXVIII was successfully used for photoaffinity modification of phenylalanyl-tRNA synthetase.[198]

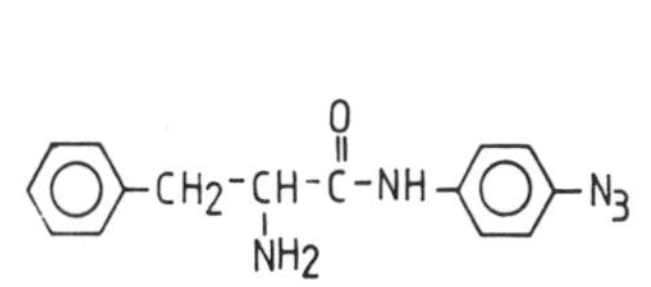

$$\langle O \rangle - CH_2-CH-\overset{O}{\overset{\|}{C}}-NH-\langle O \rangle - N_3 \qquad (LXXVIII)$$
$$\underset{NH_2}{}$$

A specific problem arises when the biopolymer to be modified deals with very small ligands. Sometimes it turns out possible to find a reactive molecule which imitates steric and physicochemical properties of the natural ligand. Thus, ferrate anion FeO_4^{2-} with a strong oxidative potential can be substituted for inorganic phosphate PO_4^{2-} in a considerable number of enzymes which interact with phosphoryl moieties. The bound ferrate destroys amino acid residues (Tyr, His, and Trp) in the enzyme active sites causing inactivation. Thus, (as was described in Chapter 1, Section I.F) glycogen phosphorylase catalyzing cleavage of glycogen with inorganic phosphate is modified at the active-site tyrosine residue by the ferrate.[199] Some other examples of the use of ferrate anion as a site-specific reagent for phosphoryl sites in proteins can be found in Table 2 in the Appendix.

Similarly, cyanogen C_2N_2 may be substituted for CO_2 molecule in the enzymes converting CO_2, e.g., carbonic anhydrase (EC 4.2.1.1), which catalyzes the following reaction.

$$CO_2 + H_2O \rightleftarrows H^+ + HCO_3^- \qquad (71)$$

The enzyme is irreversibly inhibited by cyanogen.[200] Chemical details of the process remain to be elucidated.

A hydrogen peroxide anion HO_2^- was proposed to be the species causing the inactivation of superoxide dismutase (EC 1.15.1.1) upon extended exposure to H_2O_2 at alkaline pH. This enzyme catalyzes the dismutation of $O_2^{\dot-}$ yielding O_2 and H_2O_2. It was proposed that the inactivation occurs due to the hydrogen peroxide anion HO_2^-. The latter imitates $O_2^{\dot-}$, binds to the copper ion in the catalytic center of the enzyme, and forms the reactive species which causes modification of the active-site histidine residue.[201]

An irreversible inhibition of nitrogenase (EC 1.18.2.1) observed in the presence of nitric oxide and nitrite can also be ascribed to the affinity modification process. This enzyme catalyzes the reduction of N_2 to NH_3 in the ATP-dependent reaction. It was proposed that the inactivating species is the NO molecule which imitates the enzyme substrate, binds to the protein, and disrupts in some way the essential for catalytic activity Fe_4S_4 cluster present in this enzyme.[202] Sometimes a small ligand can be imitated by part of the structure of a larger molecule. Thus, inorganic phosphate was imitated by the affinity reagents methylphosphate and even 4-azido-2-nitrophenyl phosphate. The first reagent efficiently alkylated active-site carboxyl residues in inorganic pyrophosphatase (EC 3.6.1.1).[203] The other reagent was used for photoaffinity modification of mitochondrial adenosine triphosphatase (EC 3.6.1.3) at the single inorganic phosphate binding site present in this enzyme.[204] In more complex molecules this situation is routine and very often only part of the reagent structure is involved in recognition. Thus, in the affinity reagent *p*-iodoacetamidosalicylic acid XIX synthesized by Backer in the very beginning of the affinity modification studies, only a small part of the structure imitates lactate, the substrate of lactate dehydrogenase (see Section II.A).

The addressing part of affinity reagents should not possess all the groups of natural ligands involved in recognition. Since the molecular recognition processes are based essentially on the multipoint interactions of molecules, even incomplete specific structures or parts can show considerable affinity to respective biopolymers. Thus, glucose phosphate isomerase (EC 5.3.1.9) catalyzing the isomerization of glucose-6-phosphate to fructose-6-phosphate in the course of glycolysis is readily inactivated by the affinity reagent *N*-bromoacetyle-thanolamine phosphate (LXXIX)

$$
\begin{array}{ll}
\begin{array}{l}
HC{=}O \\
HC{-}OH \\
HO{-}CH \\
HC{-}OH \\
HC{-}OH \\
CH_2O - PO_3^{2-}
\end{array}
&
\begin{array}{l}
\overset{\displaystyle O}{NH{-}\overset{\|}{C}{-}CH_2Br} \\
CH_2 \\
CH_2O{-}PO_3^{2-}
\end{array}
\\[1em]
\text{glucoso-6-phosphate} & \qquad\text{(LXXIX)}
\end{array}
$$

which imitates the phosphorylated fragment of the substrate, efficiently binds to the enzyme, and alkylates active-site histidine residue.[205]

Another example where the reactive compound imitates only part of the specific structure is given by thiourea dioxide (LXXX) which is the irreversible inhibitor of glutamine synthetase. This compound bearing a reactive sulfinic acid group was proposed to mimic a fragment of the γ-glutamyl phosphate (LXXXI) intermediate formed in the course of the enzymatic reaction.[206]

$$\underset{H}{\underset{|}{HN}}{=}C{-}\overset{O}{\underset{}{\overset{||}{S}}}{=}O$$
with NH$_2$ on C

(LXXX)

$$CH_2^- \overset{\overset{+}{NH_3}}{\underset{|}{CH}}{-}COO^-$$
$$\underset{|}{CH_2} \quad O$$
$$O{=}C{-}O{-}\overset{O}{\underset{OH}{\overset{||}{P}}}{-}O^-$$

(LXXXI)

Sometimes sufficient affinity can be provided even by noncovalent interactions of single groups of affinity reagents. Thus, it is established that the acetylcholinesterase active site possesses an important cation binding area. Due to this area, cationic molecules which have nothing to do with the acetylcholine structure in Chapter 1, (V) except for the positively charged groups that can interact with the enzyme. The typical example of such compounds is the inhibitor of the acetylcholinesterase-alkylating reagent N,N-dimethyl-2-phenylaziri-dinium cation.[207]

$$\text{(phenyl)}{-}CH{-}N^+ \underset{\diagdown CH_3}{\overset{\diagup CH_2 \diagup CH_3}{}}$$

(LXXXII)

Another well-known example is the general anion-binding site of alcohol dehydrogenase (EC 1.1.1.1). This enzyme catalyzes the reversible oxidation of various alcohols to the corresponding aldehydes using NAD as the coenzyme. The anion-binding site of the enzyme interacts with the pyrophosphate group of the coenzyme but it also appears to be able to bind other anionic structures.[208] Due to this interaction, the enzyme is inhibited by iodo- and bromoacetate according to the mechanism of affinity modification.[209]

The considered examples show that molecules nonrelated to real substrates can bind to functionally active sites of proteins due to a more or less pronounced partial complementarity to the structure of the enzyme functional sites. Therefore, one can expect the existence of such molecules for various enzymes, in particular in the cases when they possess binding sites for ionic compounds or hydrophobic or aromatic molecules. In these cases the recognition does not require very precise adjustment of interacting molecules as it happens when the binding is achieved by hydrogen bond formation or coordination to metal ions. Therefore, foreign molecules of appropriate shape and hydrophobic or ionic properties have good chances to imitate the natural ligands. Indeed, in the cases of the enzymes where many various molecules were tested such compounds were found.

Thus, a number of aromatic reactive compounds have been found to be affinity reagents modifying ATP-linked enzymes. One example is the deliberately synthesized reagent 2-[(4-azido-2-nitrophenyl)amino]ethyl triphosphate (LXXXIII) which was applied for the modification of the ATP binding site of myosin.

$$^-O{-}\overset{O}{\overset{||}{\underset{|}{P}}}_{O^-}{-}O{-}\overset{O}{\overset{||}{\underset{|}{P}}}_{O^-}{-}O{-}\overset{O}{\overset{||}{\underset{|}{P}}}_{O^-}{-}O{-}CH_2{-}CH_2{-}NH{-}\text{(nitroaryl azide)}$$

(LXXXIII)

Myosin is known to be a relatively nonspecific enzyme and cleaves the terminal phosphates of essentially all nucleoside triphosphate derivatives. In the analogue, the photolabile nitroaryl azide group is placed at approximately the same distance from the phosphate groups as the adenine ring is in ATP. The appropriate positioning of the groups allows the analogue to interact with the enzyme sites binding the anionic ATP moiety and the aromatic structure. Therefore, the analogue is an efficient affinity reagent for an ATP binding site in myosin.[210]

Similarly, 3-bromoacetyl pyridine obviously imitates the imidazole ring of histidine when it reacts specifically with histidine decarboxylase (EC 4.1.1.22).[211]

A number of aromatic compounds were found to imitate heterocyclic bases of nucleic acid components. This is the case for fluorescein isothiocyanate (LXXXIV) which modifies H,K-ATPase at the ATP binding site presumably mimicking the adenine moiety of the molecule.[212]

As was mentioned in Section II.A, triazine-based dyes such as Cibacron blue F3G-A (LXXXV) and some others are universal affinity reagents for nucleotide-dependent enzymes.[103,213] Part of the chromophores of the dyes were suggested to mimic the naturally occurring heterocycles. A number of examples where triazine-based dyes were applied for affinity modification of enzymes interacting with nucleotides and NAD can be found in Table 2 in the Appendix.

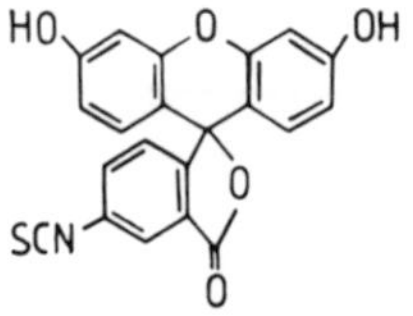

(LXXXIV)

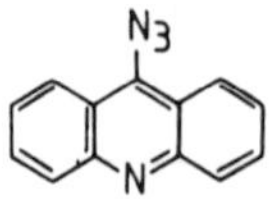

(LXXXVI)

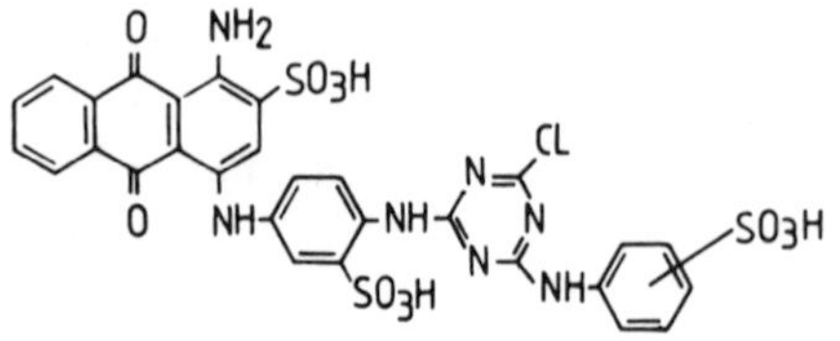

(LXXXV)

One more example of affinity reagent targeted to general nucleotide- and aromatic-binding sites in proteins is 9-azido-acridine (LXXXVI) which under irradiation efficiently modifies α-chymotrypsin, lactate dehydrogenase, alcohol dehydrogenase, ribonuclease, and several other enzymes dealing with various nucleotides, nucleotide coenzymes, and compounds.[214]

It was demonstrated that metal ions coordinated to enzymes can serve as specific groups for targeting the affinity reagents. Binding due to coordination with metal ions is sufficiently strong while specificity is provided by uniqueness of the metal ions in the protein structure. The reported case was the affinity modification of alcohol dehydrogenase with 2-bromo-3-(5-imidazolyl)propionic acid, the methyl ester of alcohol dehydrogenase, and similar chloroderivatives.[215] The enzyme is known to contain two zinc atoms. The first atom is a structural one and has all four ligands from the protein side chains. The other zinc atom has three protein ligands and one free coordinate. Since imidazole was known to bind to this free coordinate and inactivate the enzyme reversibly, the reagent containing both the imidazole moiety and alkylating group was prepared and was found to be a specific irreversible inhibitor of the enzyme.

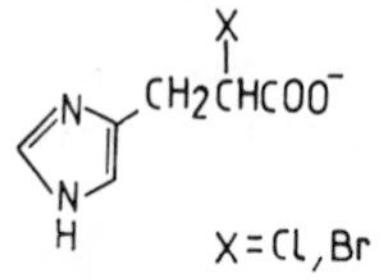

(LXXXVII)

X = Cl, Br

It should be noted that this metal-directed affinity modification was totally stereospecific. No modification was observed with (R)-enantiomers of the reagents while the (S)-enantiomers were all efficient. Such all-or-none effect could only be observed in the case of rigid fixation of the reactive group of the compound bound through the imidazole ring to the active-site zinc atom.

There are some cases where relatively unspecific affinity is used to guide chemical reagents to certain types of biopolymers or biopolymer complexes. These cases can be considered as affinity modification with reagents targeted by low-specific addressing groups. One example is represented by modification of nucleic acids with reactive derivatives of intercalating dyes. The dye groups, although being relatively low-specific to nucleotide sequences, intercalate into double helical regions of nucleic acids. Therefore, the reagents are targeted to double-stranded nucleic acids. The most widely used reagents of this type are photoreactive mono- and diazide derivatives of ethidium, 3-amino-8-azido-5-ethyl-6-phenylphenanthridinium chloride (LXXXVIIIa), and 3,8-diazido-5-ethyl-6-phenylphenanthridinium chloride (LXXXVIIIb).[216]

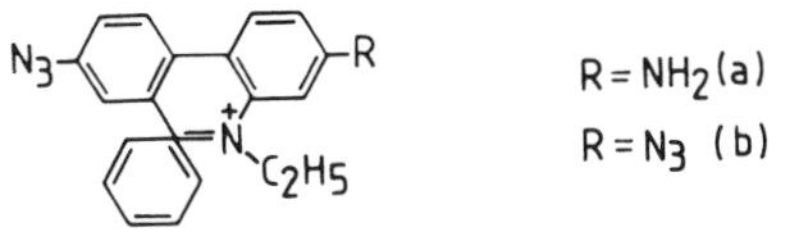

R = NH₂ (a)

R = N₃ (b)

(LXXXVIIIa,b)

Coupling of polyaromatic structures was used as an approach to modulate the binding specificity of the known DNA-modifying compounds. Thus, the anticancer drug *cis*-platinum was linked by a hexamethylene chain to acridine orange. The prepared compound (LXXXIX) exhibited enhanced binding to the DNA sequences CGG and AC as compared to the parent platinating reagent.[217]

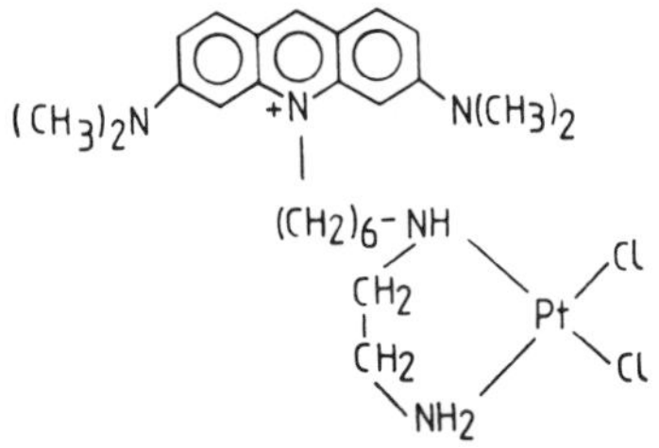

(LXXXIX)

Since reactive derivatives possessing polyaromatic structures intercalate between base pairs only in the helical nucleic acids, they can be used as affinity reagents directed to the double-stranded regions of RNAs. Thus, methidiumpropyl-EDTA·Fe(II) (LXXV) was shown to predominantly cleave double-stranded regions in tRNA and in an rRNA fragment.[218,219]

Still less-specific reagents which can be scarcely classified as affinity labels are the reactive hydrophobic compounds which are widely used in experiments with biological membranes. Due to the hydrophobic nature, the reagents concentrate within the lipid bilayer of membranes and modify any proteins or protein domains embedded into the lipid phase. Therefore, the type of specificity they exhibit is the recognition of hydrophobic vs. hydrophilic areas of membrane proteins. The application of hydrophobic labeling to the investigation of membrane proteins was extensively reviewed.[220-224]

The typical examples of hydrophobic labels are 5-iodonaphthyl-1-azide (XC) and a phospholipid derivative (XCI).[225]

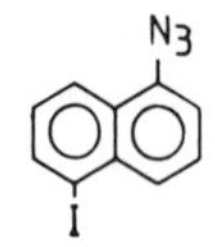

(XC)

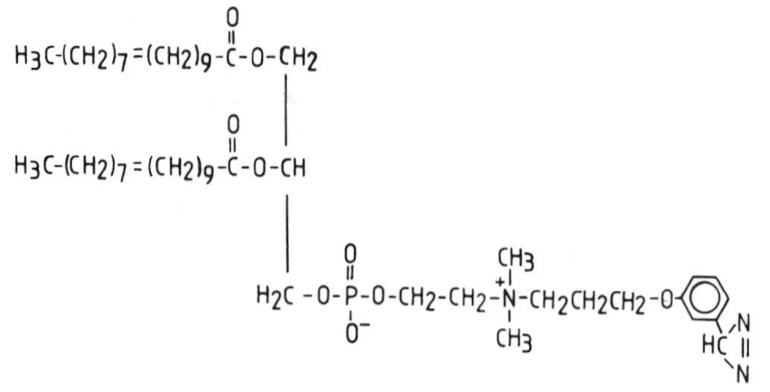

(XCI)

Phospholipid derivatives are the most specific among the hydrophobic labels since a number of membrane proteins have been shown to exhibit a specificity to a polar head group of phospholipids. Thus, phosphatidylcholine-exchange protein which catalyzes the transfer of the lipid between membranes forms a specific complex with this ligand. A photoreactive analogue of phosphatidylcholine (XCII) was capable of specific complex formation with the protein.

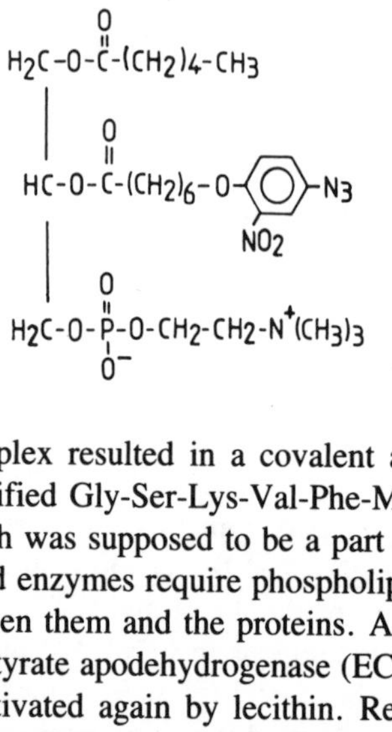

(XCII)

Irradiation of the complex resulted in a covalent attachment of the derivative and the peptide found to be modified Gly-Ser-Lys-Val-Phe-Met-Tyr-Tyr contained a pentapeptide hydrophobic cluster which was supposed to be a part of the lipid binding site.[226]

Some membrane-bound enzymes require phospholipids for activity which suggests some specific interaction between them and the proteins. A typical enzyme of this type is mitochondrial D-β-hydroxybutyrate apodehydrogenase (EC 1.1.1.30) which is inactive in a purified state but can be activated again by lecithin. Reconstitution of the enzyme with the photoreactive derivatives XCIIIa, b and following UV irradiation resulted in the covalent attachment of several lecithin analogue molecules to the protein.[227]

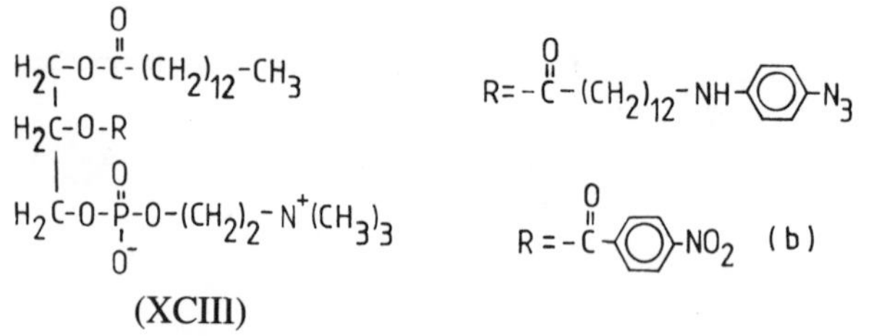

$R=-\overset{O}{\overset{||}{C}}-(CH_2)_{12}-NH-\bigcirc-N_3$ (a)

$R=-\overset{O}{\overset{||}{C}}-\bigcirc-NO_2$ (b)

(XCIII)

B. General Approaches to the Synthesis of Affinity Reagents

In this section general principles of the synthesis of affinity reagents will be considered. Detailed illustrations of these principles will be given in the next section on examples of an important family of affinity reagents — derivatives and components of nucleic acids.

When the addressing structure of an affinity reagent is chosen and the parts of this structure which are not essential for the recognition are established, one must find the way to build a reactive group in the molecule. There are several approaches in preparing affinity reagents which are based either on the synthesis of analogues of natural compounds bearing reactive groups near or even within the moiety which is recognized by the target biopolymer or on the coupling of a reactive grouping to the natural ligand via some spacer structure. According to classification introduced by Baker,[1] affinity reagents designed in these two ways are referred to as endo- and exoaffinity reagents. The first type of reagent is considered to accomplish modification directly within the binding sites on the target biopolymers. The reagents of the exo type are considered to modify the biopolymer residues in some area in the vicinity of the binding sites. The size of this area is determined by the nature of the spacer structure. As examples of endoaffinity reagents, one can consider the amino acid analogues aminoacyl methylene keto chlorides XX and ferrate anion which bind and react directly at the binding sites (Section A). A typical exoaffinity reagent is 4-azido-2-nitrophenyl phosphate which also imitates inorganic phosphate but possesses a photoreactive group remoted from the addressing phosphate moiety.

The most general, although usually the most laborious approach to the preparation of affinity reagents, is the total chemical synthesis of reactive derivatives. Using this approach, one can prepare affinity reagents which closely imitate the natural ligands and contain built-in reactive groupings in any part of the structure. Thus, the only structural difference between the pheromone of *Antheraea polyphemus* (XCIVa) and the affinity reagent designed for labeling the pheromone receptor (XCIVb) consists in the substitution of the diazoacetate group for the acetate one.[228]

$$CH_3(CH_2)_3CH=CH(CH_2)_3CH=CH(CH_2)_4-O-\overset{\overset{O}{\|}}{C}-R \qquad \begin{array}{ll} R=-CH_3 & (a) \\[6pt] R=-CHN_2 & (b) \end{array}$$

(XCIV)

This relatively simple affinity reagent is available by a complicated multistep synthesis.

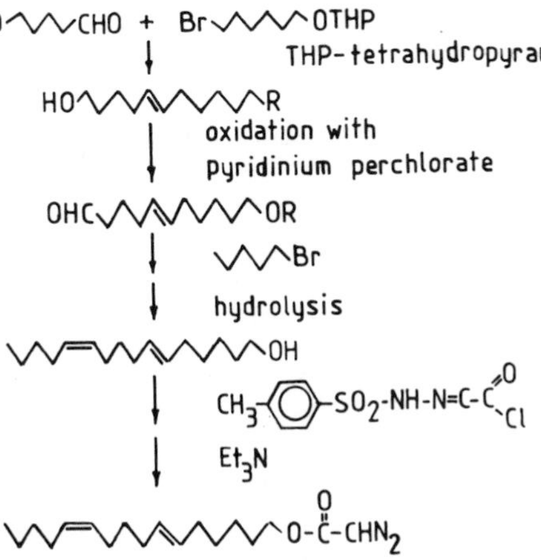

(72)

Similarly, using the total chemical synthesis, the affinity reagent bromoacetol phosphate XCV for the modification of one of the glycolysis enzymes, triose phosphate isomerase, was prepared.[229] The reagent structure is very similar to that of the natural substrate dioxyacetone phosphate, (XCVI) and Equation 73.

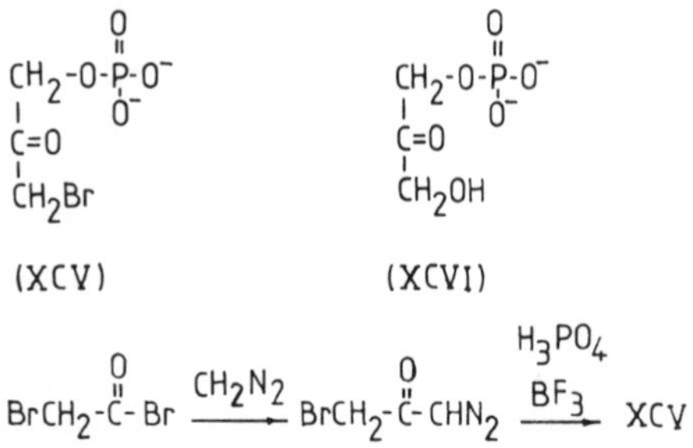

$$
\begin{array}{cc}
\underset{\text{(XCV)}}{\begin{array}{l} CH_2\text{-O-}\overset{\displaystyle O}{\overset{\|}{P}}\text{-O}^- \\ \quad\quad O^- \\ C=O \\ CH_2Br \end{array}} &
\underset{\text{(XCVI)}}{\begin{array}{l} CH_2\text{-O-}\overset{\displaystyle O}{\overset{\|}{P}}\text{-O}^- \\ \quad\quad O^- \\ C=O \\ CH_2OH \end{array}}
\end{array}
\tag{73}
$$

$$
BrCH_2\text{-}\overset{\displaystyle O}{\overset{\|}{C}}\text{- Br} \;\xrightarrow{CH_2N_2}\; BrCH_2\text{-}\overset{\displaystyle O}{\overset{\|}{C}}\text{-CHN}_2 \;\xrightarrow[BF_3]{H_3PO_4}\; XCV
$$

In some cases, the semisynthetic preparation of affinity reagents is possible, when functional groups of the natural ligand can be chemically transformed into reactive groupings. Thus, reactive analogues of deoxycorticosterone 21-diazo-21-deoxycorticosterone XCVIIIa and 21-chloro-21-deoxycortisosterone XCVIIIb for affinity modification of the corticosterone receptor were prepared from XCVII by the chemical conversion of carboxylic group to diazoacetate and chloromethylketo ones, respectively.[230,231]

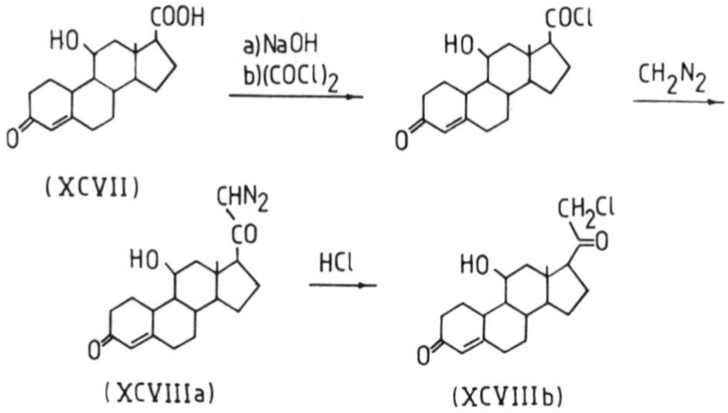

$$
\tag{74}
$$

(XCVII)   (XCVIIIa)   (XCVIIIb)

Sometimes natural compounds can be converted to the affinity reagents easily by specific chemical activation of the functional groups. An example of this approach is the conversion of folate to the corresponding anhydride derivative using the carbodiimide treatment (see Section II.B).

A commonly used technique for the preparation of exoaffinity reagents consists of the attachment of reactive structures to functional groups of natural ligands. An example is the synthesis of *p*-azidoanilide phenylalanine (XCIX) from phenylalanine and *p*-azidoanilide using carbodiimide as a condensing agent.[198]

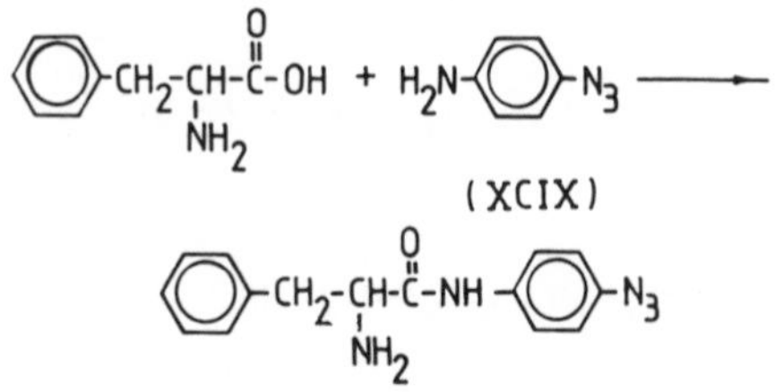

(XCIX)

$$
\tag{75}
$$

The popularity of this approach is related to the simplicity and the possibility of preparing affinity reagents from complex ligands and unstable compounds.

When the ligand molecule is large compared to the reactive group, the whole area of the ligand binding site can be investigated only by using several affinity reagents bearing reactive groups at different sites of the addressing structure. For example, one can consider the investigation of the steroid binding site of the estradiol-17 β-dehydrogenase (EC 1.1.1.62), the enzyme which transfers the 4-pro-S hydrogen atom of nicotinamide to 17-oxosteroids converting the latter into 17-hydroxysteroids. In this case, a series of alkylating derivatives bearing reactive groups in nine different positions was prepared by acylating amino or hydroxy groups in various steroid derivatives with bromoacetic or iodoacetic acid activated by dicyclohexylcarbodiimide (see Section II.B and Table 3). These derivatives were found to attack different residues in the enzyme active site. When an alkylating substituent was shifted from C3 to C16 of the steroid nucleus, modification of the active-site histidine residue decreased dramatically while modification of the active-site cysteine residue increased. Therefore, it was concluded that the active-site histidine is located in the vicinity of the steroid A ring and that the cysteine residue is located in the catalytic region of the active site near the D ring of the steroid molecule.[232]

The simplest method of synthesis of affinity reagents is based on the use of heterobifunctional reagents (see Chapter 1, Section II.E). In this case, one-step treatment of natural ligands with the reagents yields corresponding reactive derivatives. A great variety of heterobifunctional reagents was prepared by using combinations of different reactive groupings (Table 4) and linker structures (Table 5). Using these derivatives, reactive groups can be attached to various functional fragments present in biopolymers. Structures of the heterobifunctional reagents useful for the synthesis of affinity reagents are given in Table 1 in the Appendix. The structures are given in brief form using codes introduced in Tables 4 and 5. This system of codes allows the compact presentation of the reagent structures. Thus, Structures CII and CIII

can be presented by codes p2-(s38-s10-s2-c1-s2)-r1 and r1-(s26)-r7, respectively (reagent numbers 9 and 94 in the Appendix). The choice of a heterobifunctional reagent for the preparation of an affinity reagent is determined by several factors. The anchoring group of the reagent should react efficiently with the addressing structure to be derivatized under conditions nondamaging to this structure and to the second reactive group of the reagent. The choice of the group to be introduced is determined by the objective of the affinity modification experiment. When topography of the binding site is explored and the objective of the modification is to label the target structure, a high yield of reaction is not important. However, in this case the reactive group should be highly reactive and nonspecific, since the nature of the residues in the reagent binding site is unknown. Therefore, use of photoreactive groups such as azido is optimal in these types of experiments. On the contrary, when the efficiency of affinity modification is the primary concern and nucleophiles are known to be present in the reagent binding site, reagents with electrophilic reactive groups such as alkylating are advantageous.

The bridging structure of heterobifunctional reagents determines the distance between and

Table 3
REACTIVE DERIVATIVES OF ESTRADIOL-17

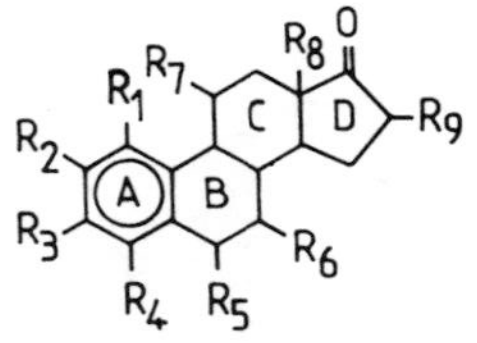

C(a–j)

	R1	R2	R3	R4	R5	R6	R7	R8	R9	R10
A	H	H	ICH₂COO–	H	H	H	H	–CH₃	H	H
B	H	H	BrCH₂-CH₂-COO–	H	H	H	H	–CH₃	H	H
C	H	BrCH₂CONH–	CH₃O–	ICH₂CONH–	H	H	H	–CH₃	H	H
D	H	H	CH₃O–	H	H	H	H	–CH₃	H	H
E	H	ICH₂COO–	CH₃O–	BrCH₂COO–	H	H	H	–CH₃	H	H
F	H	H	CH₃COO–	H	BrCH₂COO–	H	H	–CH₃	H	H
G	H	H	CH₃COO–	H	H	BrCH₂COO–	H	–CH₃	H	H
H	H	H	CH₃COO–	H	H	H	BrCH₂COO–	–CH₃	H	H
I	H	H	CH₃COO–	H	H	H	H	–CH₃	H	BrCH₂COO–
J	H	H	CH₃COO–	H	H	H	H	H	H	H

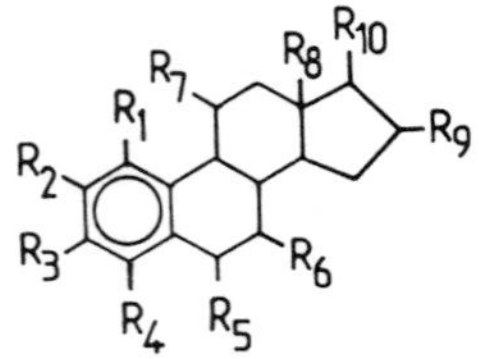

CI(a–e)

	R1	R2	R3	R4	R5	R6	R7	R8	R9	R10
A	H	H	ICH₂COO–	H	H	H	H	–CH₃	H	H
B	H	H	ICH₂CONH–	–CH₃	H	H	H	–CH₃	H	–OH
C	BrCH₂COO–	H	H	H	H	H	H	–CH₃	ICH₂COO–	–OH
D	H	H	CH₃O–	H	H	H	H	–CH₃	ICH₂COO–	H
E	H	H	CH₃O–	H	H	H	H	–CH₃	H	ICH₂COO–

From Pons, M., Nicolas, J.-C., Boussioux, A.-M., Descomps, B., and Crastes de Paulet, A., *Eur. J. Biochem.*, 68, 385, 1976.

Table 4
HETEROBIFUNCTIONAL REAGENTS: REACTIVE GROUPS

Code number	Formula	Code number	Formula

Chemically reactive ends (r)

Code number	Formula	Code number	Formula
r1	$-\overset{\overset{O}{\|\|}}{C}-O-N$ (succinimidyl)	r11	$-\overset{\overset{O}{\|\|}}{\underset{\underset{O}{\|\|}}{S}}-Cl$
r2	$-\overset{\overset{O}{\|\|}}{C}-O-N$ (sulfosuccinimidyl, $SO_3^{\ominus}$)	r12	$-S-\overset{\overset{O}{\|\|}}{\underset{\underset{O}{\|\|}}{S}}-$
		r13	$-SCl$
r3	$-\overset{\overset{NH_2^{\oplus}}{\|\|}}{C}-OCH_3$	r14	$-SH$
		r15	$-Hg^{\oplus}$
r4	$-\overset{\overset{NH_2^{\oplus}}{\|\|}}{C}-OC_2H_5$	r16	$-F$
r5	$-N=C=S$	r17	$-Cl$
r6	$-N=C=O$	r18	$-NHCH_2CH_2Br$
r7	$-N$ (maleimidyl)	r19	$-\underset{\underset{Br}{\|}}{C}H-\overset{\overset{O}{\|\|}}{C}-CH_3$
r8	$-S-S-$ (2-pyridyl)	r20	$-CH_2Br$
r9	$-S-S-$ (3-nitro-2-pyridyl, NO_2)	r21	$-\overset{\overset{O}{\|\|}}{C}CH_2Br$
r10	$-S-N$ (phthalimidyl)	r22	$-\overset{\overset{O}{\|\|}}{C}CH_2Cl$
		r23	$-\overset{\overset{O}{\|\|}}{C}-SeCH_2CH_2COO^{\ominus}$
		r24	$-\overset{\overset{O}{\|\|}}{C}-Cl$
r25	$-\overset{\overset{O}{\|\|}}{C}-$ (phenyl)	r28	$-CH-CH_2$ (glycidyl, carbonate)
r26	$-\overset{\overset{O}{\|\|}}{C}-O-$ (phenyl)$-NO_2$		

Table 4 (continued)
HETEROBIFUNCTIONAL REAGENTS: REACTIVE GROUPS

Code number	Formula	Code number	Formula
r27	$-\underset{\underset{O}{\parallel}}{C}-\underset{\underset{O}{\parallel}}{CH}$	r29	

Photoreactive ends (p)

p1	$-\overset{+}{N}\equiv N$	p4	$-\underset{\underset{O}{\parallel}}{C}-\underset{\underset{N_2}{}}{CH}$
p2	$-N=\overset{\oplus}{N}=\overset{\ominus}{N}$		
p3	$-\overset{N=N}{\underset{\diagdown\diagup}{C}}-CF_3$		

relative mobility of the reactive groups. Using longer spacer structures gives more freedom to the introduced reactive residues and increases the chances of achieving affinity modification, although the reaction may proceed outside the reagent binding site in this case. Using heterobifunctional reagents with cleavable spacer structures (Table 5) is helpful in the experiments where the identification of the modified biopolymers and residues is necessary. The possibility of cleaving the reagent moiety allows the removal of the addressing structure from the modified biopolymer which greatly simplifies the identification and analysis of the latter. Spacer structures and reactive groupings of heterobifunctional reagents can be labeled with radioisotopes. Thus, the heterobifunctional reagent CII contains the ^{35}S-labeled spacer arm (shown by the asterisk). The reagent can be coupled to an addressing structure using the acylating N-hydroxysuccinimide group. After affinity modification, the reagent moiety can be cleaved by dithiotrithol. The cleavage yields a free cold addressing structure and the biopolymer which carries the ^{35}S-labeled fragment of the reagent.

C. Reactive Derivatives of Biopolymers

In many cases both partners of specific interaction of some biological significance are biopolymers. To study this interaction by affinity modification, it is necessary to prepare reactive derivatives of one of these partners. Both proteins and nucleic acids contain many functional residues unessential for the interaction under investigation and usually no problem arises with the choice of the area for the introduction of reactive groups. As to the choice of reactive group, it should be emphasized that it is not convenient to introduce into the biopolymer molecule electrophilic reactive moieties due to the danger of numerous intra-molecular cross-linkings with various nucleophilic centers of addressing biopolymer. It is preferential to supply the biopolymer with photoreactive or some other activatable groups which are unreactive in the course of synthesis, purification, storage, and preparation of the reagent to affinity modification and may be switched on after the specific complex with the target partner is already formed.

Table 5
HETEROBIFUNCTIONAL REAGENTS: SPACER GROUPS

Code number	Formula	Code number	Formula
		Noncleavable spacer groups (s)	
s1	$-CH_2-$	s20	$-CH-CH_2-\langle\bigcirc\rangle-$ $O=C-OCH_3$
s2	$-CH_2CH_2-$		
s3	$-CH_2CH_2CH_2-$		
s4	$-CH_2CH_2CH_2CH_2-$	s21	$-\overset{O}{\overset{\|}{C}}-CH_3$
s5	$-CH_2CH_2CH_2CH_2CH_2-$		
s6	$-CH_2CH_2CH_2CH_2CH_2CH_2-$		
s7	$-O-CH_2-O-$	s22	$-\overset{O}{\overset{\|}{C}}-OH$
s8	$-O-$	s23	$-\overset{O}{\underset{O}{\overset{\|}{\underset{\|}{S}}}}-$
s9	$-\overset{O}{\overset{\|}{C}}-$		
s10	$-NH-$		
s11	$-\underset{CH_3}{\overset{\|}{N}}-$	s24	$\langle\bigcirc\rangle$
s12	$-\underset{O}{\overset{\|}{C}}-NH-$	s25	$\langle\bigcirc\rangle$
s13	$-NH-CH_2-\overset{O}{\overset{\|}{C}}-$	s26	$\langle\bigcirc\rangle$
s14	$-\underset{OH}{\overset{\|}{C}H}-$	s27	$^{125}I-\langle\bigcirc\rangle$
s15	$-\underset{COO^{\ominus}}{\overset{\|}{C}H}-$	s28	$\langle\bigcirc\rangle-OH$
s16	$-\overset{\|}{C}H-$ $\overset{N}{\diagup\diagdown}$ $CH_3\quad CH_3$	s29	$\langle\bigcirc\rangle-OH$
s17	$-\overset{\|}{C}H-$ $O=C-OC_2H_5$	s30	$\langle\bigcirc\rangle-OH$
s18	$-CH-CH_2-$ $NH-\underset{O}{\overset{\|}{C}}-O-C(CH_3)_3$		
s19	$-CH-$ CH_2 $\langle\bigcirc\rangle$ OH	s31	$\overset{CH_3}{\langle\bigcirc\rangle}$
		s32	$\langle\bigcirc\rangle-CH_3$

Table 5 (continued)
HETEROBIFUNCTIONAL REAGENTS: SPACER GROUPS

Code number	Formula	Code number	Formula
s33		s41	
s34		s42	
s35		s43	
s36		s44	
s37		s45	
s38		s46	
s39		s47	
s40			

Cleavable spacer groups (c)

Code number	Formula	Cleavage method
c1	$-S-S-$	H_2O_2 oxidation; $NaBH_4$ reduction; or RSH treatment
c2		$NaIO_4$ (periodate) oxidation
c3		Base hydrolysis

Table 5 (continued)
HETEROBIFUNCTIONAL REAGENTS: SPACER GROUPS

Code number	Formula	Code number	Formula
c4	$-N\overset{!}{\underset{!}{=}}N-$		$Na_2S_2O_4$ or $Na_2S_2O_6$ (dithionite) reduction
c5	$-S\overset{O}{\underset{O}{-}}S-$		RSH treatment
c6	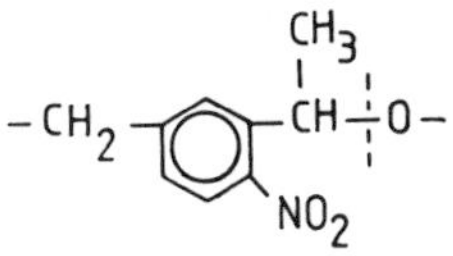 $-CH_2-\bigcirc\overset{CH_3}{\underset{NO_2}{-CH}}-O-$		Low energy UV light (365 nm)

Two main types of reactive derivatives of biopolymer should be distinguished. The first one is represented by derivatives bearing one reactive group attached to a definite point on the biopolymer. The other one is represented by derivatives with reactive groups scattered statistically over a number of similar residues of the addressing biopolymer.

Reagents of the first type are needed for a detailed investigation of some definite fragment of target biopolymer. The second type of reagent is used first to localize or damage the whole area of contact between interacting molecules.

In spite of impressive success in the chemical synthesis of proteins and nucleic acids, the direct chemical synthesis of biopolymers with reactive groups is poorly realized and the main approach to the preparation of such reagents is based upon the modification of biopolymers with bifunctional reagents.

The introduction of a reactive group into a definite residue of biopolymer is most easily performed at the points with unique reactivities. As such points, terminal residues, some minor components, and residues of major monomers of sole monomer present in a biopolymer may serve. It may be also carried out if some major monomer residue has enhanced reactivity compared with other similar residues due to a particular position in a macromolecular structure. For example, we may present the introduction of a photoreactive residue in the A chain of insulin. The N-terminal glycine residue of the A chain was eliminated by Edman degradation and instead N-(4-azido-2-nitrophenyl)glycine was attached by subsequent treatment with respective N-hydroxysuccinimide ester.[233] The B chain was temporarily protected with alkali-labile methylsulfonylethyloxycarbonyl residue.

Yeast iso-1-cytochrome *c* has a single strongly nucleophilic cysteine residue. 4-Azidophenyl residue was introduced at this point by treatment with N-(4-azidophenylthio)-phthalimide (CIV).[234]

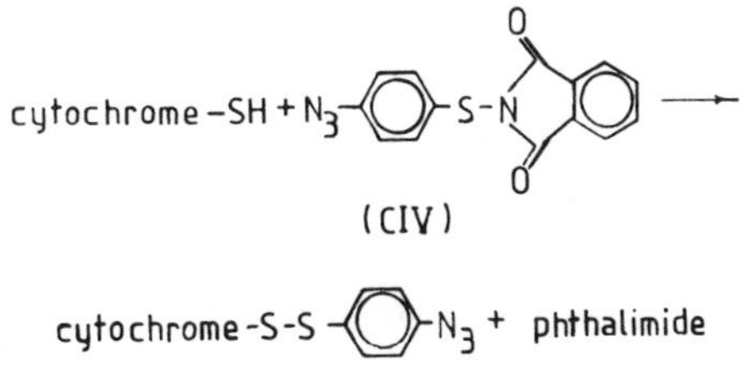

$$\text{(CIV)} \tag{76}$$

cytochrome-S-S-◯-N₃ + phthalimide

If the reaction of a biopolymer with a heterobifunctional reagent proceeds at a few number of points, the derivatives with a definite position of reactive groups may be obtained by the separation of the mixture of modified biopolymers produced by the treatment of the starting biopolymer with the reagent. Thus, horse heart cytochrome *c* contains two reactive lysine residues Lys-13 and Lys-22. Monomodified derivatives obtained by the treatment with 4-fluoro-3-nitrophenylazide (NAP-F), NAP-Lys(13)-cytochrome, and NAP-Lys(22)-cytochrome were separated by ion-exchange chromatography and used for the modification of beef heart cytochrome *c* oxidase, the terminal enzyme of the electron transport chain. It was found that UV irradiation of the enzyme mixture with the first derivative results in the covalent attachment of cytochrome *c* to oxidase whereas the second derivative was unreactive towards the enzyme in the same conditions.[235]

As an example of the use of a biopolymer modified statistically at a number of points, we may mention the S1 protein of the small 30S subunit of *Escherichia coli* ribosomes. In order to study the interaction of this protein with 16S rRNA, it was treated with ethyl 4-azidobenzoylaminoacetimidate (CV).

$$N_3-\text{\Large\textcircled{}}-CO-NH-CH_2-\overset{\overset{\displaystyle NH}{\|}}{C}-OC_2H_5 \qquad (CV)$$

This reagent attacks exposed NH_2 groups (lysine residues) of the protein. A modified S1 protein with statistically scattered photoreactive groups was associated with S1-depleted 30S subunits and reconstructed particles were irradiated by UV light. S1 was found to cross-link with the fragment 861-889 of 16S rRNA.[236]

In some cases biosynthetic methods may be used to introduce reactive groups in biopolymers. Thus, the enzyme mammalian transglutaminase (EC 2.3.2.13) catalyzes the substitution of amines and hydrazines for the amino group at glutamine residues of proteins. The enzyme is low-specific for amine substrates. Using this enzyme, a method of introduction of the above-mentioned NAP residues was elaborated by treatment proteins with *N*-(4-azido-2-nitrophenyl)-ethylene diamine (CVI) in the presence of transglutaminase.[237]

$$\text{Protein-}CH_2CH_2CONH_2 + NAP-NHCH_2CH_2NH_2 \longrightarrow$$

$$(CVI)$$

$$\text{Protein-}CH_2CH_2CO-NHCH_2CH_2NH-NAP + NH_3$$

(77)

Sometimes the systems of high specificity may be successfully used for the same purpose. Thus, the introduction of a photoreactive group in the growing polypeptide chain in the ribosomal protein biosynthesis system is described. It is known that only tRNA moieties of AA-tRNA are selected by messenger (m)RNA. Therefore, yeast Lys-tRNA^{Lys} derivative-bearing photoreactive residue was produced by the treatment of Lys-tRNA^{Lys} with *N*-hydroxysuccinimide ester of *p*-azidobenzoic acid in alkaline conditions favorable for the predominant acylation of the ε-amino group. This modified *N*ε-(N_3-Bz)-Lys-tRNA^{Lys} participated in the protein biosynthesis in a cell-free wheat germ system charged with pituitary RNA, thus substituting photoreactive N^ϵ-(N_3-benzoyl)-lysyl residues for lysyl ones in the growing peptidyl moiety of peptidyl-tRNA.[238]

D. Reactive Derivatives of Nucleic Acids and Nucleic Acid Components

In this section we shall illustrate the general approaches used to design the reagents for affinity modification by the description of the main groups of reactive derivatives of nucleic acids and their constituents — oligonucleotides, nucleotides, and nucleosides.

The enhanced interest in derivatives of nucleosides and especially nucleotides originates from the outstanding significances of nucleotides for living organisms. Almost all energy-consuming processes such as endergonic chemical reactions, transmembrane transfer of ions, muscle contraction, and unidirectional movement of template nucleic acids along the information-multiplying or -transforming systems are supplied with energy by ATP or sometimes GTP molecules. Besides this, ATP serves as a nearly universal donor of phosphate in numerous phosphorylation reactions. Nucleotide coenzymes NAD and NADP participate in the hundreds of oxidation-reduction transformations of hydroxy and carbonylic groups as well as the aldehyde and carboxylic groups. Adenosine moieties are found in coenzyme A, the main acyl groups carrier, and in S-adenosylmethionine, nearly a single source of methyl groups in a great number of biochemical methylation processes. Nucleotide sugars are intermediates in the oligo- and polysaccharides biosynthesis. Ribonucleoside and deoxyribonucleoside-5'-triphosphates are substrates of nucleic acid polymerases. Nucleotides play a great role in the regulation processes. 3',5'-Cyclophosphates of adenosine and guanosine (cAMP and cGMP) are participants of the systems transferring the signals produced by the interaction of hormones and neurotransmitters with cell surface receptors (see Chapter 1, Section I.F). Therefore, the reactive derivatives of nucleotides may be used for affinity modification of many hundreds of proteins.

The interaction of nucleic acids with proteins, nucleoproteins, and complementary polynucleotides plays a decisive role in the processes lying in the foundation of storage, multiplication, and expression of genetic information. Replication, reparation, and the recombination of DNA require a significant number of proteins interacting with DNA such as DNA polymerases, topoisomerases, DNA-unwinding proteins, damage-specific endonucleases and glycosidases, protein components of chromatin, etc. In the course of transcription of genetic information necessary for the production of m-, t-, r-, and other RNAs, DNA interacts with RNA polymerases and a number of regulatory proteins such as repressors in prokaryotes and transcription factors in eukaryotes. RNA-protein interactions are necessary for processing primary transcripts leading to the formation of the completed RNA molecules of the translation systems, as well as for specific aminoacylation of tRNAs, for formation of the complexes between ribosomes, mRNAs, and AA-tRNAs and many other processes. Reactive nucleic acid derivatives are some of the most powerful tools in the study of these numerous interactions. The use of reactive derivatives of oligonucleotides as the models of essential fragments of nucleic acids in many cases simplifies the interpretation of the experimental results.

At present, significant experience is accumulated concerning the introduction of the reactive groups in all three main moieties of nucleotides: heterocyclic bases, sugar fragments, and phosphate residues. The existence of many reagents is essential since various proteins differ significantly in the ability to react with the same nucleotide derivative. This may happen even within a group of similar proteins. Thus, all AA-tRNA-synthetases interact with ATP according to Equation 10 in Chapter 1. However, sensitivities of these enzymes toward the same analogue, ATP γ-p-azidoanilide, may be quite different. Thus, beef pancreas tryptophanyl-tRNA-synthetase is readily inactivated by 10^{-4} M reagent under UV irradiation whereas $E.$ $coli$ phenylalanyl-tRNA-synthetase remains active even under irradiation in the presence of $5 \cdot 10^{-3}$ M of ATP γ-p-azidoanilide.[239] Isoenzyme P1 of yeast hexakinase, the enzyme catalyzing the transfer of the phosphate group from ATP to hexoses, in monomeric form is readily inactivated with alkylating AMP derivative CVIIa and is insensitive to a similar ATP derivative CVIIb.

$$\mathrm{ClCH_2CH_2} \diagdown \mathrm{N} - \underset{\mathrm{CH_3}}{\diagup} \bigcirc -\mathrm{CH_2-NH-}\left(\overset{\overset{O}{\parallel}}{\underset{\underset{O^-}{\mid}}{P}}-O\right)_n - \mathrm{Ado} \qquad \begin{array}{l} n=1\,(a) \\ n=3\,(b) \end{array}$$

(CVII)

At the same time, a dimeric enzyme is inactivated by treatment with CVIIb and is insensitive to CVIIa.[240]

The introduction of a reactive group into a heterocyclic fragment opens the way for the preparation of universal derivatives of nucleosides, nucleosidemono-, di-, and triphosphates as well as of cyclic phosphates. This is the reason for the extreme popularity of 8-azido derivatives of adenine and guanine. Affinity modification of a significant number of enzymes was studied using derivatives CVIII and CIX.[241]

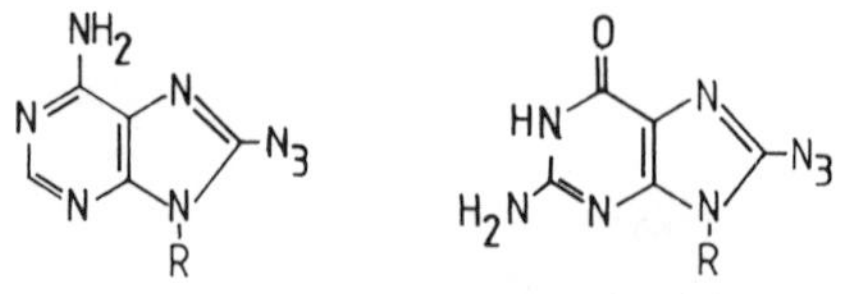

(CVIII)   (CIX)

R: ribose, ribose-5'-monophosphate, ribose-5'-diphosphate, ribose-5'-triphosphate, and ribose-3',5'-cyclophosphate.

The 8-azido group is converted under UV irradiation to nitrene and the latter is a very small substituent. Therefore, 8-azido-purine nucleotides should be considered as the reagents of an endo type.

However, it should be kept in mind that 8-N_3 substituent significantly influences the structure of a nucleoside making sin-conformation more favorable than native anticonformation. The same is true for more bulky substituents such as the alkylating groups Cl-$CH_2CONH(CH_2)_nNH-$ (n = 2,6) which also were introduced in position 8 of the purine ring.[242] Therefore, it seems preferential to use the recently introduced as a photoaffinity reagent 2-azido adenosine derivatives which have native anticonformation.[243]

A number of derivatives bearing a reactive group in position 6 of purine is described. Typical examples CX (a to e) are presented. Some additional reagents are given in Table 2 in the Appendix.

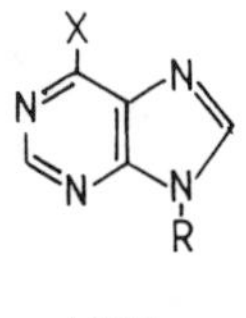

(CX)

X = Cl, (a);[244] X = SH, (b);[244] X = NH–CO–p–C_6H_4F, (c);[245] X = NH–CO–o–C_6H_4F, (d);[245] X = NH–$CH_2C_6H_4$NHCOCH$_2$Br, (e);[246] R = ribose-5'-mono-, di-, or triphosphate.

The conversion of an adenine to 1,N^6-ethenoadenine (ε Ade) moiety CXI was proposed to obtain photoactivatable derivatives.

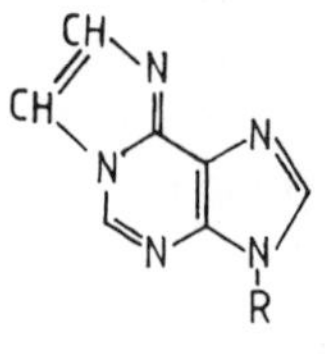

(CXI)

ε ATP was used as a photoaffinity reagent with *E. coli* phenylalanyl-tRNA-synthetase. It is interesting that εATP, in spite of a rather significant change of purine heterocycle, still remains a substrate of this enzyme with an enhanced affinity as compared with the parent ATP.[247]

Pyrimidine nucleotide analogues with a reactive group in a heterocyclic base are mainly obtained by derivatization of uracil at the 5 position since on the average it is least damaging to interact with specific proteins. Typical examples are given by the compounds CXII (a to d):

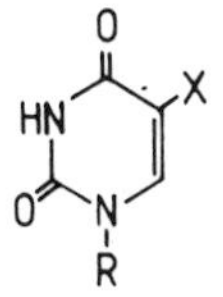

$X = COCH_2Br$ (dUMP) (a)[248]

$X = CH_2NHCOCH_2I$ (dUMP) (b)[248]

$X = CHO$ (UTP) (c)[249]

$X = N_3$ (dUTP) (d)[250]

(CXII)

The two first compounds are alkylating reagents. CX11c can form Schiff bases with amino groups of proteins. The latter compound is photoreactive.

Pyridineadeninedinucleotides (PyADs) with a derivatized pyridine ring may be used as reactive NAD or NADP analogues. Compounds CXIII (a to d) may serve as examples of such derivatives.

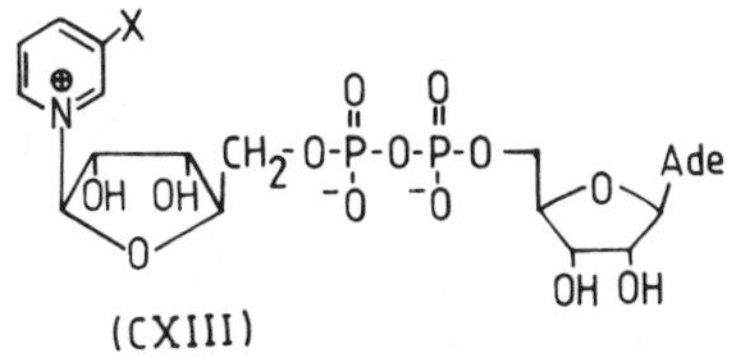

(CXIII)

$X = N_2^+$, (a);[251] $X = N_3$, (b);[252] $X = CHN_2$, (c);[253] and $X = COCH_2Cl$, (d).[254]

An oligomethylene fragment may be substituted for the ribose fragment of the pyridinenucleotide part of the reagent. Thus, the derivatives CXIV (a, b) are described which are irreversible inhibitors of some NAD-dependent proteins.[255]

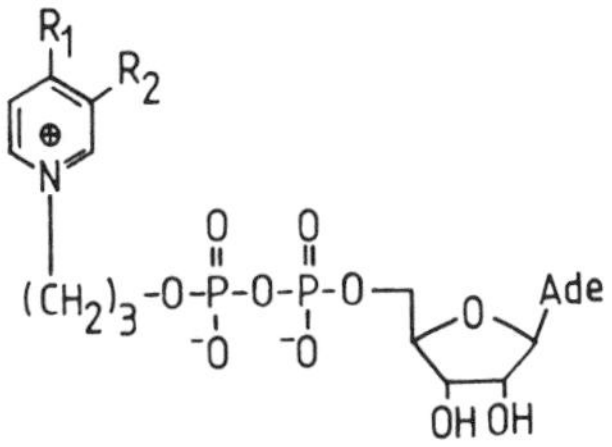

(CXIV)

The main approach to introducing reactive groups into ribose and deoxyribose moieties of nucleotides and nucleosides is the acylation of hydroxy groups. As two examples, an ATP derivative with an arylazido group (CXV)[256] and an AMP derivative bearing the diazomalonyl fragment (CXVI)[130] may be presented.

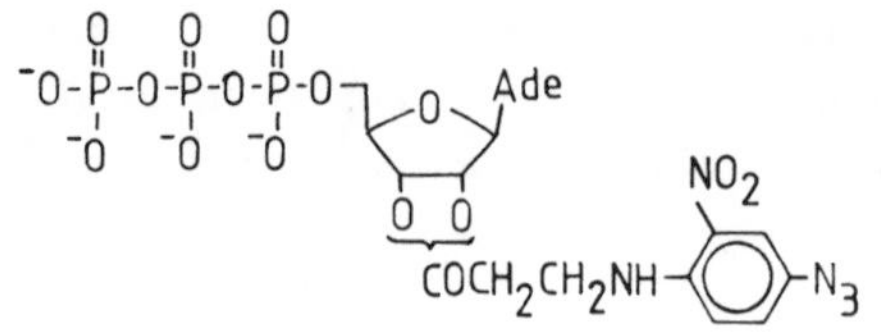

(CXV)

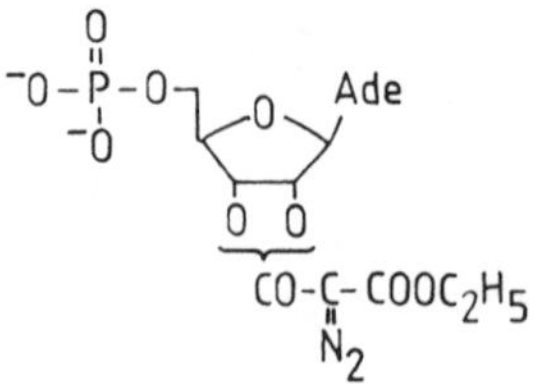

(CXVI)

Reactions specific for the *cis*-diol group of ribose may be used for the attachment of reactive moieties. Thus, acetals of ribonucleosides, ribonucleotides, and oligoribonucleotides (R) are described.[257]

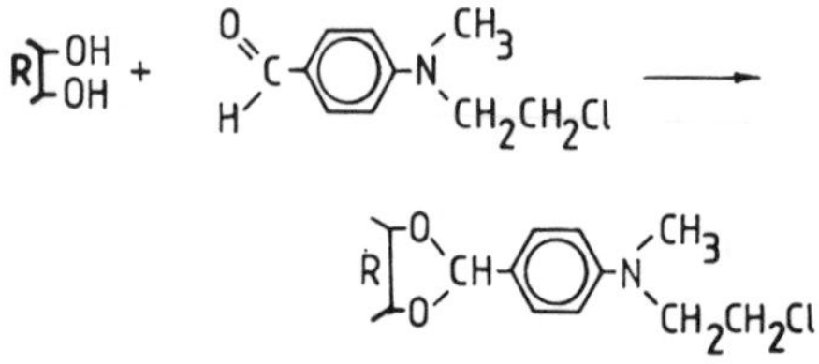

(78)

The examples of such compounds (IX in Chapter 1 and XVIII) were already presented.

As it was already considered in Section II.C, periodate oxidation of nucleosides, nucleotides, and oligo- or polymeric compounds with a 3′-terminal ribonucleotide is widely used to produce respective dialdehyde derivatives. These derivatives are reactive to NH_2-groups and may serve as affinity reagents. At the same time a dialdehyde fragment may be used to attach some other reactive group by treatment with respective hydrazides. As an example we present a photoreactive GTP derivative CXVIII produced in this way:[258]

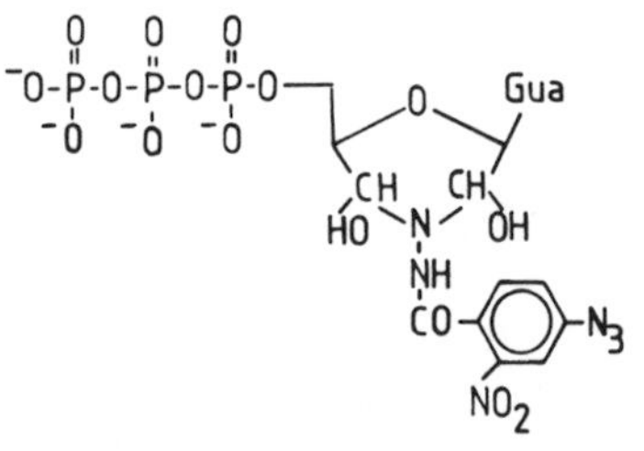

(CXVIII)

To obtain nucleotide derivatives with reactive centers in the region of a phosphate or oligophosphate group, the latter should be derivatized. Another way is to substitute some reactive fragment for some part of or the whole phosphate moiety. This latter approach is most widely represented by *p*-fluorosulfonylbenzoyl adenosine (XXX) and a similar guanosine derivative.[107] A series of other compounds is described with alkylating fragments substituted for phosphate moieties.[242,259]

Isosteric AMP and ATP analogues (CXIX) containing an oxidized adenosine fragment are acylating reagents with a reactive group in the phosphate region of nucleotides.[260]

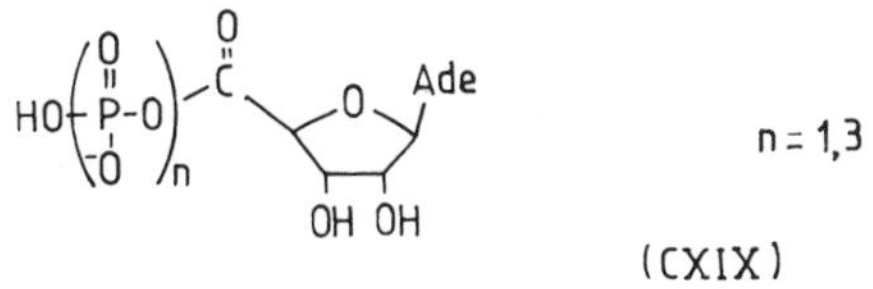

(CXIX)

n = 1,3

Derivatives of nucleosidemono-, di-, and triphosphates with small photoreactive azido groups directly linked to phosphate moieties are described.[261]

Derivatives with an activated mono- or oligophosphate group were described in Section II.B. Being used as affinity reagents, they form covalent adducts with the phosphate moiety directly attached to the nucleophilic center of the biopolymer and may be referred to as endo reagents. These derivatives, especially XXXII, XXXIII, and XXXIV, easily react with amines and may be used as intermediates in the synthesis of reagents of the exo type. The reaction with zwitterionic methylimidazolide phosphate is a general method for the synthesis of γ-derivatives of nucleosidetriphosphates. Most widely used are the derivatives CXX (a,b):

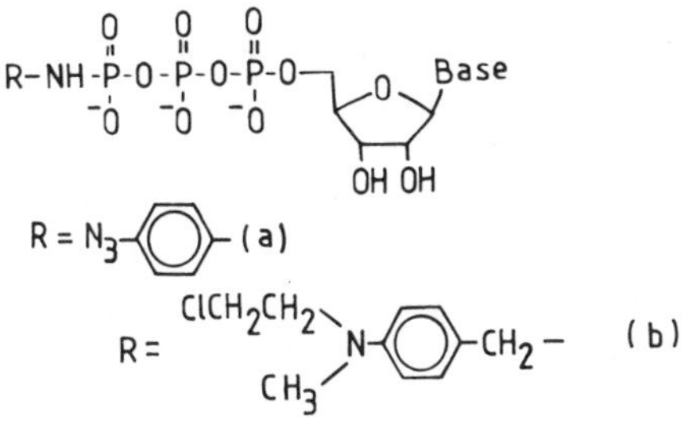

(CXX)

These derivatives and a number of other similar compounds are described in a special review.[112] They seem to be the best nucleosidetriphosphate analogues as they are readily obtained by a simple universal method for nucleotides of both ribo- and deoxyriboseries, contain undamaged base and ribose moieties and only slightly modified triphosphate fragments with the negative charge decreased by unity. Indeed, it was found that these derivatives efficiently modify most part of the investigated nucleosidetriphosphate-consuming enzymes. It should be emphasized that the majority of reactions of nucleosidetriphosphates occurs as either phosphate or nucleotidyl transfer and that triphosphate fragments reside in the catalytic centers of respective enzymes. Therefore, modification by reagents CXII proceeds in the active site of the enzyme close to the phosphate or nucleotidyl acceptor binding area. Mild procedures used in the preparation of derivatives of the CXII type permit the application of the approach to a synthesis of derivatives of rather labile nucleotides. Thus, the preparation of photoreactive γ-[4-(benzoylphenyl)methylamido]-7-methyl-GTP (CXXI) was described.[262]

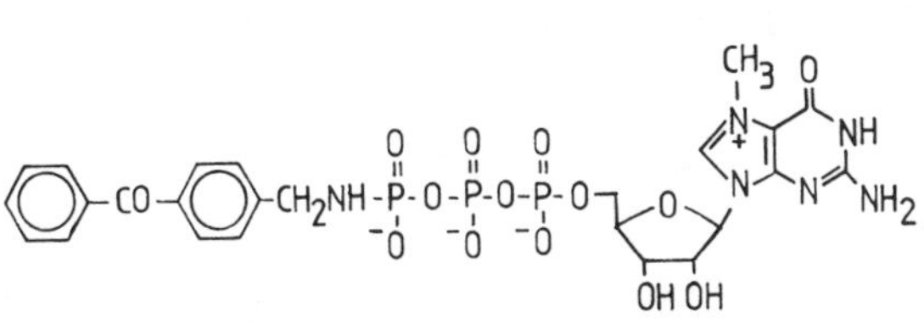

(CXXI)

The interest in this derivative is based upon the existence at the 5′ end of most eukaryotic mRNAs of so-called cap structure containing 7-methylguanosinetriphosphate bound via the triphosphate fragment to the 5′-hydroxy group of the first monomer of the polynucleotide chain of mRNA. The described reagent was successfully used for photoaffinity labeling of the influenza virus protein-recognizing cap structure.[262] Viral protein PR 2 (see Chapter 1, Table 3) was found to be responsible for this recognition.

It is possible to prepare nucleotide derivatives bearing two reactive groups for affinity cross-linking of two regions of the active center. As an example, the compound CXXII may be presented bearing reactive groups both in the base and sugar moieties.[263]

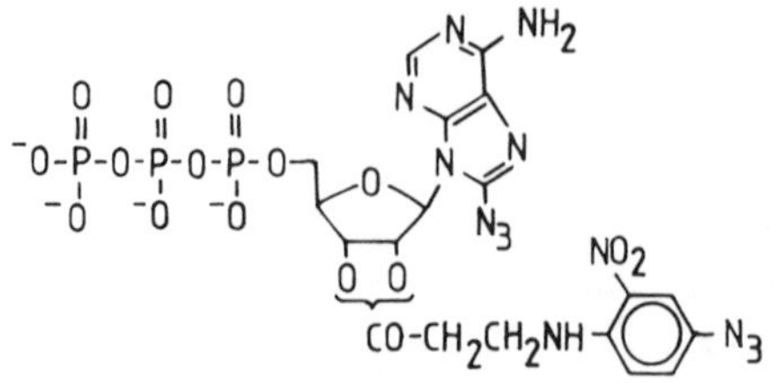

(CXXII)

Turning to reactive derivatives of oligonucleotides and nucleic acids we must emphasize that only part of the approaches elaborated for nucleotides may be easily transformed for oligo- and polynucleotides.

To prepare oligonucleotides with reactive bases in a definite position it is necessary to carry out oligonucleotide synthesis using a modified monomer at the respective step of the synthesis. This may be done only if the reactive base can survive in the course of the preparation of the oligonucleotide. Therefore, up to now only few examples of the use of this approach are described. In this way the alkylating oligonucleotide derivative XXI was prepared[99] as well as the derivative with the EDTA·Fe(II) group described in Section II.H (see Figure 1). To escape damage of the reactive group in the course of oligonucleotide synthesis it is preferable to introduce it in the already prepared oligonucleotide. This may be easily done if some specific auxiliary group is introduced in the oligonucleotide in the course of the synthesis. This approach was successfully used for the preparation of a number of codon analogues. 5-Aminouridine was used to prepare the initiator codon analog AUGU(5-NH_2). The amino group of uridine was acylated with bromoacetic anhydride giving rise to the compound CXXIII.[264]

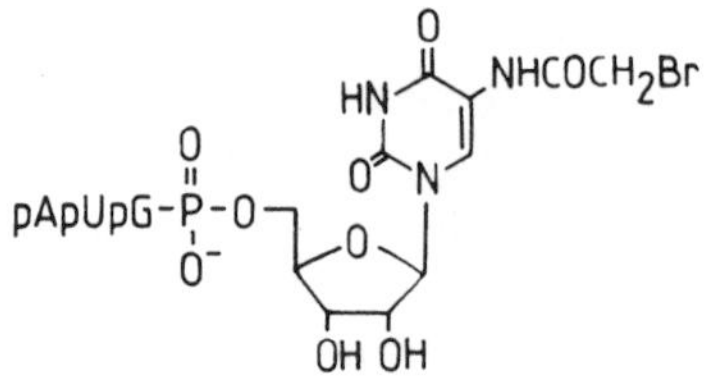

(CXXIII)

The similar analogue was prepared starting from *p*-nitrophenyl ester of oligonucleotide pApUpG. The nitro group was reduced to a amino group which was subsequently acylated with bromoacetic anhydride giving rise to the compound CXXIV.[265]

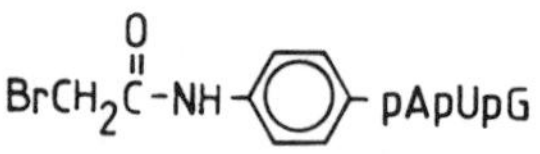

(CXXIV)

The derivatization of internal sugar fragments of oligo- and polynucleotides with reactive groups does not seem to be prospective. Deoxyribose moieties lack any reasonable point for derivatization; 2'-OH groups of ribose residues are sterically hindered and the modification should result in severe conformational changes of a polynucleotide fragment subjected to such modification. The derivatization of definite residues of internucleotide phosphates seems to be more promising. The modern state of oligonucleotide synthesis presents reasonable ways for such derivatization. However, no such attempt dealing with the introduction of reactive moieties is described up to now.

At the same time, 5′-terminal phosphomonoester group and 3′-terminal *cis*-diol group of RNAs, ribonucleotides, and oligodeoxyribonucleotides with a 3′-terminal ribonucleotide fragment are widely used for the preparation of reactive derivatives. Thus, oligonucleotides with an activated 5′-terminal phosphate were already described in Section II.B. Alkylating derivatives of oligonucleotides CXXV are easily obtained by the treatment of oligonucleotides with *p*-(*N*-2-chloroethyl-*N*-methylamino)benzylamine in the presence of triphenylphosphine and dipyridyldisulfide.[257]

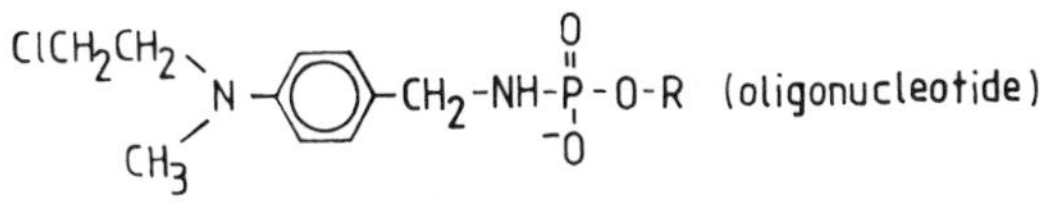

$$ClCH_2CH_2\diagdown N-\langle\bigcirc\rangle-CH_2-NH-\overset{\overset{O}{\|}}{\underset{\overline{O}}{P}}-O-R \quad (oligonucleotide)$$
$$CH_3\diagup$$

$$(CXXV)$$

Reaction 78 described for nucleosides and nucleotides was applied to a number of oligonucleotides and tRNA.[257]

Reactive derivatives of some tRNAs and oligonucleotides were obtained either by periodate oxidation[266] or by oxidation and subsequent treatment with hydrazides bearing photoreactive arylazido groups.[267,268]

Some minor components present in nucleic acids, especially in tRNA molecules, may be used for the attachment of reactive groups. Thus, derivatization of 3-(3-amino-3-carboxy-propyl)-uridine (CXXVI) present in the region of the extra loop of some tRNAs is described producing a photoreactive tRNA derivative.[269]

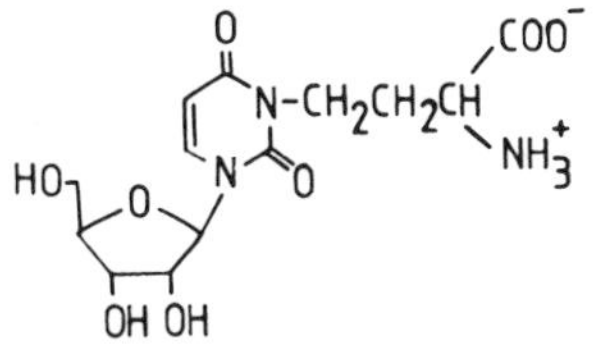

(CXXVI)

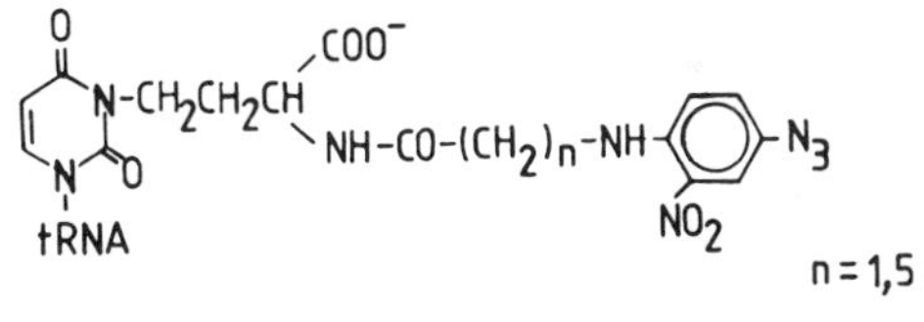

(CXXVII)

$$n = 1, 5$$

4-Thiouridine present in position 8 of several bacterial tRNAs was converted to photoreactive residue by treatment with bromacetyl-*p*-azidoanilide.[270]

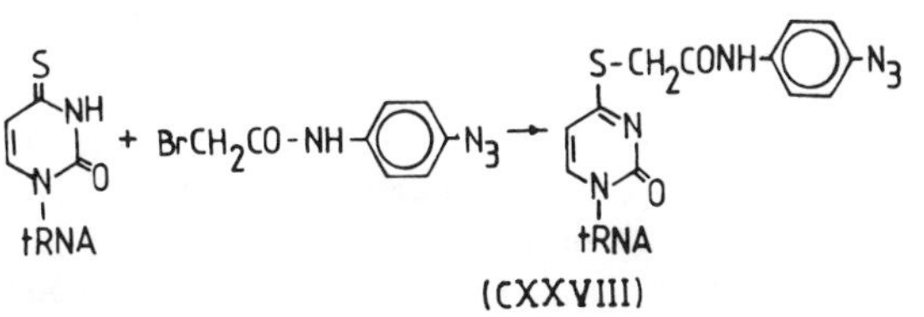

(79)

(CXXVIII)

Similar alkylation of *Escherichia coli* tRNA$_1^{Arg}$ containing 2-thiocytidine in an anticodon loop resulted in the preparation of several photoreactive derivatives CXXIX.[271]

R= $-CH_2CO-\langle C_6H_4 \rangle-N_3$

$-CH_2CO-OCH_2CO-\langle C_6H_4 \rangle-N_3$

$-CH_2CO-OCH_2CH_2-\underset{CH_3}{N}-\langle \underset{NO_2}{C_6H_3} \rangle-N_3$

(tRNA aminopyrimidine structure, S–R) (CXXIX)

Some other minor bases such as dihydrouracil present in the D loop of a great number of tRNAs and wybutin found in the anticodon loop of eukaryotic tRNA^{Phe} may be easily removed from tRNA in mild conditions. The aldehyde group is formed at the position of the removed base and the reactive group may be introduced instead of the eliminated base. Thus, consecutive treatment of tRNA^{Gly} with NaBH$_4$ and *N*-(4-azido-2-nitrophenyl)glycyl hydrazide resulted in the photoreactive tRNA derivative CXXX.[272]

$$\underset{tRNA}{NHNH_2}COCH_2NH-\underset{NO_2}{\langle C_6H_3 \rangle}-N_3 \qquad \text{(CXXX)}$$

A unique rather basic amino group is present in all AA-tRNAs. A number of reactive derivatives obtained by selective acylation (CXXXIa to g) is described.

$$X-CO-NHCHCO-tRNA^{(i)} \qquad \text{(CXXXI)}$$
$$\overset{|}{R_i}$$

$X = -(CH_2)_3-\langle C_6H_4 \rangle-N(CH_2CH_2Cl)_2$ (a)[273]

$X = BrCH_2-$ (b)[274]

$X = -O-\langle C_6H_4 \rangle-NO_2$ (c)[275]

$X = N_2 = \underset{|}{C}-COOC_2H_5$ (d)[276]

$X = -CH_2-\langle C_6H_4 \rangle-O-\langle \underset{NO_2}{C_6H_3} \rangle-N_3$ (e)[277]

$X = \langle \underset{NO_2}{C_6H_3} \rangle-N_3$ (f)[278]

$X = -CH_2CH_2-\langle C_6H_4 \rangle-CO-\langle C_6H_5 \rangle$ (g)[279]

The introduction of 5'-terminal thiophosphate residue which may be easily derivatized was proposed as a general approach to the synthesis of reactive derivatives of oligo- and polynucleotides. This residue may be introduced using γ-thio-ATP as a substrate of polynucleotide kinase. Alkylation with the aliphatic fragment of the heterobifunctional alkylating reagent

$$\underset{O}{\overset{H}{\underset{\diagdown}{C}}}-\langle C_6H_4 \rangle-\underset{(CH_2)_3-N}{N}\underset{\diagdown}{\overset{\diagup CH_2CH_2Cl}{\underset{\diagdown}{\underset{CH_2CH_2Cl}{\overset{\diagup CH_2CH_2Cl}{}}}}} \qquad \text{(CXXXII)}$$

was used to introduce low-reactive aromatic 2-chloroethylamine into the oligonucleotide derivative. The resulting derivative may be reduced by $NaBH_4$ which converts the electron acceptor formyl residue to a hydroxymethyl one, thus enhancing the reactivity of the aromatic nitrogen mustard.[280]

$$R-CH_2CH_2-N-(CH_2)_3-N \underset{CH_2CH_2OH}{\overset{CH_2CH_2Cl}{|}} \underset{}{\bigcirc} - C \overset{O}{\underset{H}{}} \longrightarrow \tag{80}$$

$$R-CH_2CH_2-N-(CH_2)_3-N \underset{CH_2CH_2OH}{\overset{CH_2CH_2Cl}{|}} \bigcirc - CH_2OH$$

R: oligonucleotide or polynucleotide.

As was mentioned in the previous section, a number of problems may be solved using biopolymer derivatives with reactive groups scattered over the polymer chain. To retain the functional properties of a biopolymer the total extent of the modification should be low.

Thus, treatment of RNAs with hydrogen sulfide results in the conversion of some cytosine residues to 4-thiouridines.

$$\tag{81}$$

Thiouridine residues formed may be used as photoactivatable groups.[281]

The biosynthetic approach was elaborated to produce rRNA containing 4-thiouridine residues.[282] *E. coli* was grown in the medium containing both uridine and 4-thiouridine. rRNA isolated from the produced bacteria contained some 4-thiouridine and could be used as a photoreactive biopolymer.

Copolymers of AMP and 8-N_3-AMP as well as IMP (inosine-monophosphate) and 8-N_3-IMP were produced by copolymerization of the respective ribonucleoside-5'-diphosphate with polynucleotide phosphorylase.[283] The introduction of photoreactive bromodeoxyuridine residues in DNA molecules by growing microorganisms in the presence of 5-bromodeoxyuridine is widely used in the preparation of photoreactive DNA. Photoactivatable iodocytosine residues may be produced by chemical iodination of DNA.[32]

The preparation of a photoreactive lac operator, a DNA fragment with a specific affinity to lac repressor (see Chapter 1, Section I.E), was carried out by incubating a synthetic oligonucleotide representing one strand of the operator with DNA polymerase from *E. coli* and a mixture of deoxynucleosidetriphosphates containing 5-azidodeoxyuridine-5'-triphosphate. A double-stranded DNA fragment obtained was cross-linked under UV irradiation with the lac-repressor.[250]

Statistically scattered reactive groups may be introduced into polynucleotide chains also by treatment with heterobifunctional reagents. Thus, the reagent CXXXII was used for treatment of the transcripts of definite genes of phage T7. A reaction proceeds in this case at N7 of guanine residues. By following the treatment of the modified RNA transcript with $NaBH_4$, aromatic 2-chloroethylamino groups were activated according to Reaction 81. The obtained modified transcripts were used to perform mutagenesis of T7 DNA at definite genes corresponding to these transcripts used as the carriers of alkylating functions.[284]

Similarly, the statistical derivatization of cytosines of T7 DNA was performed according to Scheme 82. The substitution of hydrazine derivatives (CXXXIII) for NH_2 groups of cytosine results in the introduction of latent alkylating fragments into the DNA molecule. An acid treatment leading to the conversion of a ketal to a keto group activates the C-Br bond and a reactive DNA derivative is obtained.[285]

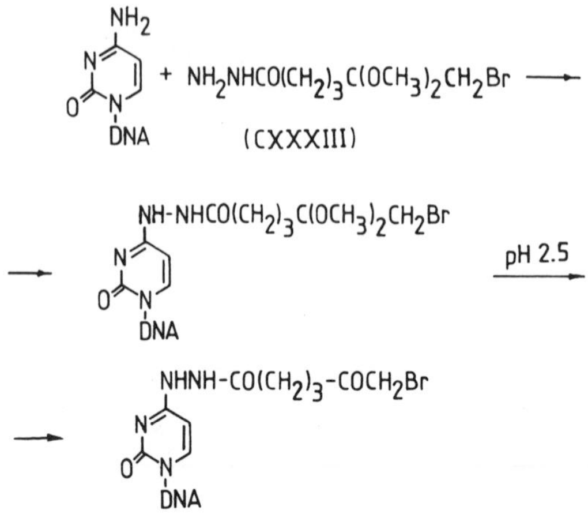

$$(82)$$

A useful tool for investigating tRNA interactions with ribosomes in a different functional state turned out to be tRNA derivatives bearing photoreactive groups scattered over guanine residues of tRNA. These derivatives are easily obtained by consecutive treatment of tRNA (*E. coli* tRNA[Phe] was mainly used in these investigations) with 4-(N-2-chloroethyl-N-methylamino)benzylamine and arylation of the introduced amino groups with 2,4-dinitro-5-fluorophenyl azide (Scheme 83).[286]

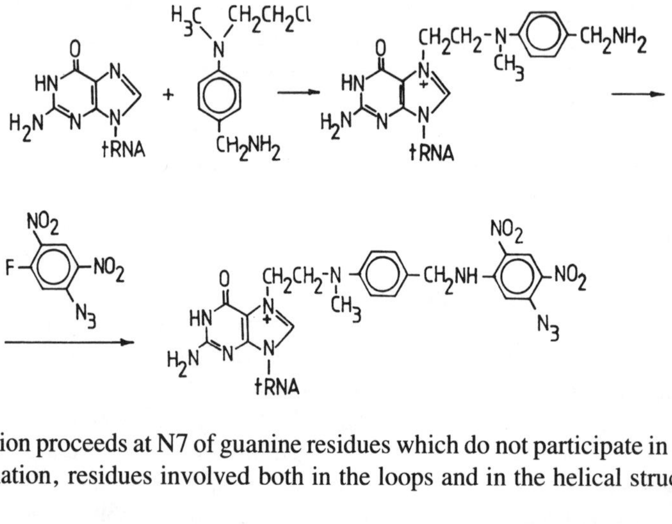

$$(83)$$

As the reaction proceeds at N7 of guanine residues which do not participate in the secondary structure formation, residues involved both in the loops and in the helical structures may be derivatized.

Reactive aldehyde groups scattered over the DNA molecule may be introduced by methylation of purines with dimethylsulfate and subsequent elimination of 7-methylguanines and 3-methyladenines. In the specific complexes with proteins, e.g., in chromatin, Schiff bases may be formed with a subsequent splitting of the DNA chain. The reduction of Schiff bases with $NaBH_4$ as shown in Scheme 84 results in a stable covalent adduct.[287]

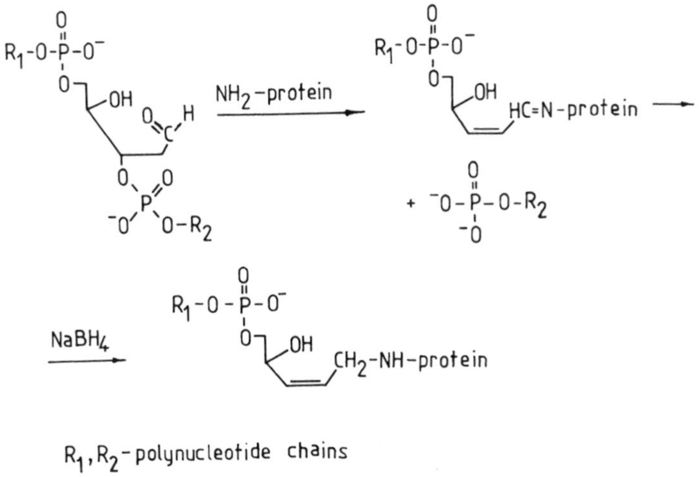

$$R_1, R_2 - \text{polynucleotide chains} \tag{84}$$

R_1, R_2 — polynuceotide chains.

Nucleic acids with reactive groups scattered over the whole chain are useful when it is desirable to outline the whole area of contact of nucleic acid under consideration with the specific partner. When it is done it is essential to get more precise information. For this goal it is convenient to produce similar derivatives with a restricted number of points of attachment of the reactive groups. As follows from the data presented in Chapter 1, Table 5, the first step of the process (Scheme 83) may result in a decreased number of alkylation points when carried out in the conditions stabilizing the tertiary structure of tRNA.

Indeed, *E. coli* tRNA[Phe] derivatives produced according to Scheme 83 by alkylation in conditions of labile and stable tertiary structure were shown to differ in the number of modified ribosomal proteins formed under irradiation of the ternary complex of the derivatives with ribosomes and poly(U).[288] In the first case, some additional modified proteins were found.

A more selective introduction of photoreactive groups was achieved by using oligonucleotide derivatives CXXV. Thus, the derivative

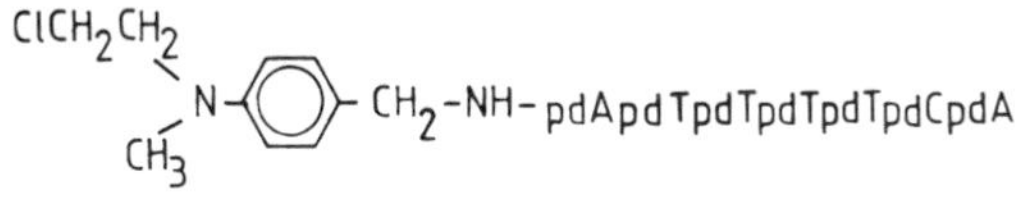

with an oligonucleotide moiety complementary to the part of the anticodon loop of tRNA[Phe] alkylated tRNA only at four guanine residues: G10, G24, G28, and G29.[289] As the phosphamide bond between the alkylating fragment and the oligonucleotide may be hydrolyzed by a mild acid treatment, only the benzylamide moiety remains attached to the abovementioned positions in tRNA and may be derivatized according to the second step of the process (Scheme 83).

Chapter 3

AFFINITY MODIFICATION — EXPERIMENTAL METHODS

I. INTRODUCTION

It is impossible within the framework of this book to present a detailed description of all the main methods used to perform affinity modification and to analyze the time course and final results. The goal of this chapter is to describe the general principles of these experiments and to attract attention to some problems arising in the course of affinity modification experiments. Necessary details may be found in original papers and special manuals concerning chemical modification and experimental procedures of affinity modification.[26,92]

II. PERFORMANCE OF AFFINITY MODIFICATION

A. General Procedures

The first question arising when one performs modification of a biopolymer with an affinity reagent is the choice of experimental conditions. This means the choice of concentrations of biopolymer and reagent, temperature, pH, buffer components, ionic strength, and auxiliary compounds for the stabilization of the biopolymer under consideration. This choice depends on the properties of the reacting components and on the main goal of the experiment. Thus, if one needs a high specificity of reaction, it is preferable to use both reacting components in stoichiometric amounts to escape nonspecific modification. However, this may be done only in the case of sufficiently high mutual affinity of the components. On the contrary, if one is interested in complete modification, e.g., for exhaustive inactivation of a biopolymer, sufficient excess of the reagent over the biopolymer must be used. The investigator should be aware of possible pitfalls related to the use of high reagent concentrations, in particular when very hot radiolabeled compounds are used. Sometimes small radioactive impurities which could be hardly detected in the starting reagent preparation bind to biopolymers and result in misleading radiolabeling patterns.

The excess of biopolymer over the reagent is convenient in kinetic investigations when the main goal is to measure with the highest accuracy available, and it is essential to have a simple experimental procedure for the separation of the modified biopolymer from the excess reagent since many samples are to be analyzed.

The choice of pH is determined by the pH range of stability of the reagent-biopolymer complex and the chemistry involved in the modification process. Reactions with nucleophilic groups protonated under physiological conditions, such as the lysine amino group, are facilitated in alkaline media. Therefore, when these groups are to be modified, slightly alkaline buffers are used when the reagent-biopolymer complex is stable in these conditions. Another type of limitation may result from lability of the reagent used in acidic or alkaline media. Thus, the acetal bond in the nucleotide and oligonucleotide derivatives IX in Chapter 1 and XVIII in Chapter 2 is acid-labile and therefore the reagents should be used at pH above 6.0.

Some reactive groups of affinity reagents are readily inactivated by buffer or other auxiliary components. A well-known example is the reduction of azido groups of arylazides and diazo groups by thiol compounds which are introduced in buffers to protect protein sulfhydryl (SH) groups from oxidation.[290,291] Dithiothreitol is the most efficient reductant of these photoreactive groups; 2-mercaptoethanol, cysteine, and thioglycolate are less reactive and the latter compound was proposed as the SH compound for use in buffers for photoaffinity modification. Another example of interferring reactions with buffer components is given by the 2′,3′-dialdehyde derivatives of nucleotides which react with the amino group of the widely used tris.[292]

Sometimes buffer components can interfere with the specific binding of an affinity reagent to the target biopolymer. Thus, the affinity reagents iodoacetic acid and 2-bromo-3-(5-imidazolyl)propionic acid bind at the active center of alcohol dehydrogenase mainly due to the interaction with the anion binding site of this center. Since phosphate and other anions can also efficiently interact with this site, they inhibit the reaction. Therefore, in this case, zwitterionic buffer compounds such as MES (4-morpholineethanesulfonic acid) were to be used.[293]

When affinity modification proceeds in crude tissue extracts or within cells, the reagents can be metabolized or degraded by the enzymes present in the medium. Thus, the suicide reagent 10-undecynoic acid efficiently inactivates hepatic cytochrome P-450 isozymes that catalyze ω- and (ω-1)-hydroxylation of lauric acid in vitro. However, it does not destroy the enzyme activity in vivo due to a rapid degradation of the inhibitor by the general mechanism of oxidative degradation of fatty acids. When analogues resistant to such degradation, CXXXIVa in which the carboxyl group is replaced by a sulfate moiety and CXXXIVb in which two methyls are placed vicinal to the carboxylic group, were prepared, they were found to be efficient inactivators of the hydroxylases both in vitro and in vivo.[294]

$$^-O-\overset{\overset{O}{\|}}{\underset{\underset{O}{\|}}{S}}-O-(CH_2)_9^- C\equiv CH \qquad\qquad HO-\overset{\overset{O}{\|}}{C}-\overset{\overset{CH_3}{|}}{\underset{\underset{CH_3}{|}}{C}}-(CH_2)_8^-C\equiv CH$$

$$(a) \qquad\qquad\qquad\qquad (b)$$

$$(CXXXIV)$$

Some experimental procedures include the generation of reactive species directly in solution. ''Electroaffinity modification'' was described, which was affinity modification of lipoxygenase with reactive species produced by oxidative electrolysis of 1,5-dihydroxynaphthalene.

The reactive species were suggested to be intermediate oxidation products such as semiquinone or phenoxy radicals since the addition of electrochemically oxidized solution of 1,5-dihydroxynaphthalene to the enzyme did not inactivate the protein.[295] Another example of reactive species formation in solution is given by affinity modification of the enzyme catechol *O*-methyltransferase (EC 2.1.1.6) which catalyzes the transfer of a methyl group from *S*-adenosylmethionine to a catechol substrate. Meta- and para-*O*-methylated products are produced by this enzyme which is responsible for the extraneuronal inactivation of catecholamines and the detoxification of various xenobiotic catechols. The enzyme is inactivated by 5,6-dihydroxyindole derivatives in the presence of air. Air oxidation of these compounds yields respective aminochromes.

$$(1)$$

$$R_1, R_2, R_3 \text{ are either H or CH}_3$$

These quinonoid products alkylate active site residues of the enzyme.[296]

B. Photoaffinity Modification

In this section chemical aspects of photoaffinity modification experiments will be considered. Techniques of photoaffinity experiments including conditions of photolysis, choice of apparatus for photolysis, source of light, filters, and recommendations concerning commercially available equipment can be found in *Photogenerated Reagents in Biochemistry and Molecular Biology.*[92]

Photoaffinity modification experiments are characterized by specific features since the reactions are carried out in the transparent media only and UV-excited groups of reagents and biopolymers can be involved in many specific and nonspecific reactions. Thus, irradiation of biopolymers with UV light in the presence of oxygen can result in photooxidative damage of the molecules. Although in many cases successful photoaffinity modification was achieved without any attempts to remove oxygen, it is generally recommended to perform the experiments in an inert atmosphere of argon or nitrogen.

To avoid UV inactivation of biopolymers, using photoreactive groups activatable by long wavelength UV light nondamaging to biopolymers is usually recommended. It should be noted, however, that excitation with short wavelength UV light can be used as well if the photoactivatable group of the reagent has a high extinction coefficient and the quantum yield of photomodification is large compared to that of by-reactions. In this case, a target process can be accomplished rapidly before substantial damage of a biopolymer occurs. Indeed, investigation of stability of the biopolymer under irradiation conditions in the absence of an affinity reagent is the standard control of the affinity modification experiments. Another usual control is to check the absence of the reaction between the biopolymer and the reagent in the dark.

It was mentioned that sometimes the highly reactive primary species produced by photolysis of photoaffinity reagents are converted to less reactive electrophilic species which interact further with biopolymers and can be considered as true affinity reagent molecules. Such affinity modification processes are known as "pseudophotoaffinity modification".[92] In this case, the main advantages of the photoaffinity modification, the possibility to perform nearly instant modification with relatively low specificity towards attacked residues, are lost. To ensure the absence of intermediate low-reactive species formation, it is recommended to check whether the treatment of a biopolymer with the prephotolyzed reagent solution results in modification. Thus, it was found that long-living reactive species are formed upon irradiation of ATP γ-p-azidoanilide solution and these species inactivate creatin kinase.[297]

In some cases, reactive species produced by photolysis may dissociate prior to a reaction with a biopolymer. At the same time in parallel with biopolymer modification by the species formed at the reagent binding site, nonspecific modification by the species formed in solution may occur.

Nonspecific modification of biopolymers by reactive species formed in solution or dissociated from the specific complex can be prevented by using scavengers which trap free reactive species. As a scavenger, any compound which efficiently reacts with photolysis products and does not interfere with the interaction of an affinity reagent and biopolymer can be used. Typical scavengers are thiols, hydroquinone, amines such as p-aminebenzoic acid, and water-soluble proteins which can interact with various reactive particles. The most efficient scavengers are thiol compounds such as thioglycolate since they are powerful nucleophiles and readily react with both electrophilic species and free radicals. However, one should keep in mind the possibility of the reaction with azido and diazo groups of the starting compound and optimize the scavenger concentration and the incubation times. Use of scavengers can be illustrated by experiments on modification of serotonin-S_2 and histamine-H_1 brain receptors with two azide analogues of ketanserin which is a ligand of these receptors, namely 6- and 7-azidoketanserin (CXXXV a and b, respectively).

$R_1=H \quad R_2=H \quad$ Ketanserin
$R_1=H \quad R_2=N_3 \quad$ (a)
$R_1=N_3 \quad R_2=H \quad$ (b)

(CXXXV)

Both derivatives showed high affinity for both receptor types. 7-Azidoketanserin was found to be a true photoaffinity reagent irreversibly blocking both serotonin and histamine receptors upon irradiation. Scavenger *p*-aminobenzoic acid, which traps the biradicals formed in solution and suppresses the nonspecific labeling by the species formed outside the complex, did not influence the modification. In contrast, modification with 6-azidoketanserin inhibited only histamine receptors. This blockade was abolished by the addition of the scavenger indicating that the inactivation was not due to a real photoreaction and suggesting the formation of intermediate reactive species from the nitrenes produced by the photolysis of 6-azidoketanserin.[298]

The performance of photoaffinity modification at very low temperatures was proposed to suppress the dissociation of a reagent from the complex with biopolymer molecules and to prevent the thermoinactivation of a biopolymer in the course of irradiation. Irradiation at 80°K followed by slow warming was used in experiments on photoaffinity modification of bacterial photosynthetic centers with 2-azidoanthraquinone. Under these conditions, efficient covalent modification occurred. In experiments performed at 25°C, protein was denatured before covalent coupling was accomplished.[299]

It was found that direct photocross-linking of adenosine 3',5'-cyclic phosphate and phosphofructokinase in frozen solution at $-77°C$ proceeds with the rate more than tenfold above that achieved in aqueous solution at ambient temperatures. The nature of this effect remains to be elucidated.[300]

Indeed, the most interesting applications of photoaffinity modification are related to the possibility of studying fast processes such as macromolecular complex formation. Several experimental studies of rapid cross-linking of biopolymers have been reported. An example of such experiments is given by the investigation of binding of RNA polymerase to DNA.[301] Results of this experiment will be described in Chapter 5, Section II.C.

Affinity modification with photoactivatable reagents usually yields a low extent of biopolymer modification. Higher modification extents can be achieved by repeating treatments of the biopolymer with fresh samples of reagents. More efficient approaches were developed in model experiments on photoaffinity modification of phosphofructokinase with $O^{2'}$-(ethyl-2-diazomalonyl)adenosine 3':5'-cyclic monophosphate. One of the methods proposed consisted in the performance of a photoaffinity modification reaction in a cell which was separated by the dialysis membrane from another cell. In the course of photolysis, fresh solution of the reagent was pumped through this second cell. Therefore, fresh reagent was introduced in the reaction cell and the photolyzed reagent was removed permanently in the course of the experiment. In the other case, the enzyme was embedded into diethylamino-ethyl-Sephadex® A-50 granules and the latter were packed into a quartz column. The column was irradiated with UV light while the reagent solution passed through the gel, delivering fresh reagent molecules and removing photolysis products. Using these two approaches achieved 70 to 75% modification extent as compared to ~20% observed in experiments performed in the usual way.[302]

III. EXPERIMENTAL PROCEDURES USED FOR QUANTITATIVE INVESTIGATION OF MODIFICATION OF BIOPOLYMERS

A. Extent of Modification

To study the physical chemistry of a process of biopolymer modification, it is necessary to obtain quantitative characteristics of the process. Two main sets of data are used for this purpose. The first is the extent of modification of a biopolymer expressed usually as the number of moles of reagent covalently bound per mole of biopolymer. The second is the biological activity A of a biopolymer remaining after modification, e.g., the enzymatic activity of enzyme. This value is usually expressed as a fraction of the initial activity A_o. The extent of modification and the level of activity are studied as functions of time and reaction conditions at various concentrations of the components. Further, we shall describe briefly the main experimental approaches to measuring these values. Certainly, these approaches are equally valid for the direct modification of a biopolymer and for the process proceeding within the specific complex of a reagent with a biopolymer.

In the most simple cases the remaining activity coincides with the fraction of nonmodified biopolymer

$$A/A_o = 1 - Z \tag{2}$$

where Z is the modification extent. This is the case if the reaction proceeds only within the complex and the modified biopolymer is completely inactive. The same takes place if the direct modification proceeds only at one definite point on the biopolymer molecule and this point is absolutely necessary for biological activity. Expression 2 is not fulfilled when some reaction outside the active site proceeds in parallel with the reaction within the specific complex. This is also the case when modification within the complex does not result in the complete loss of biological activity.

In the case of low molecular weight reagents, a modified biopolymer may be easily separated from the excess reagent and the products of by-conversions by precipitation or gel filtration. Therefore, any physical property of the reagent which is not inherent to the biopolymer may be used to determine the modification extent, provided this property does not disappear in the course of modification.

Radioactivity is best as it does not undergo any change after the covalent attachment of a reagent. As an experimental example we present in Table 1 the results of alkylation of transfer (t)RNA with derivatives of oligonucleotides bearing N-2-chloroethyl-N-methylam-inophenyl residue attached via the acetal bond to the 3'-terminal ribose. The extensive washing of precipitate is a prerequisite for correct measuring of the modification extent.

The optical density measurements may be used when the UV spectrum of the reagent does not overlap with that of the biopolymer. Reagents with absorbance in visible and near UV ranges are preferential. Thus, for the above-mentioned reaction, mild acid treatment of modified tRNA results in the hydrolysis of the acetal bond and the formation of the fragment of p-aminobenzaldehyde absorbing at 350 nm. Nucleic acids do not absorb in this wavelength range and the number of the bound reagent residues may be easily calculated using A_{350} measurements. As it is seen in the table, the results coincide with those obtained by counting ^{14}C-label incorporation.

If the absorption ranges of biopolymer and reagent nearly overlap, optical density measurements may be used only if the molar extinction of the reagent is at least of the same order of magnitude as that of the biopolymer or exceeds it. Thus, the attachment of nucleotide or oligonucleotide derivatives to proteins may be followed spectrophotometrically if the

Table 1
ALKYLATION OF UNFRACTIONATED YEAST tRNA WITH 2′,3′-*O*-[4-(*N*-2-CHLOROETHYL-*N*-METHYLAMINO)-BENZYLIDENE]THYMIDYLYL(3′ → 5′)THYMIDYLYL(3′ → 5′) URIDINE, dTpdTpUCHRCl

Time (min)	Radioactivity bound to tRNA aliquote, (c/min)	Modification extent × 10^2 (radiochemical data)	A_{350} (o.d.u.) of tRNA aliquot	Modification extent × 10^2 calculated from A_{350}
40	1325	0.76	0.40	0.84
90	2315	1.33	0.68	1.42
150	3360	1.93	0.98	2.07
230	4270	2.45	1.02	2.16

Note: Modification was performed in 0.04 *M* tris-HNO$_3$, pH 7.6 at 40°C, specific radioactivity of dTpdTpUCHRCl was 8.7 · 10^4 c/min/mol, concentrations of tRNA and dTpdTpUCHRCl were 4 · 10^{-3} *M* (in moles of nucleotide residues) and 3.2 · 10^{-3} *M*, respectively. Aliquots of 0.5 mℓ were taken and tRNA precipitated with ethanol was either counted or subjected to treatment with 0.04 *M* acetate buffer, pH 4.0 at 40°C for 1 hr and A$_{350}$ was measured.

From Vlassov, V. V., Grineva, N. I., Zarytova, V. F., and Knorre, D. G., *Mol. Biol. (USSR)*, 4, 201, 1970.

protein is poor in tryptophane or lacks it at all. For example, modification of pancreatic RNAase with dideoxyribonucleotide derivative[115]

which possesses a phosphorylating ability may be followed by measuring A$_{260}$ and A$_{280}$; as for RNAase E$_{260}$ = 5000 M^{-1} cm^{-1} and E$_{280}$ = 9400 M^{-2} cm^{-1}, and for dinuleotide moiety of the reagent E$_{260}$ = 24,000 M^{-1} cm $^{-1}$ and E$_{280}$ = 9000 M^{-1} cm^{-1}. Assuming the additivity of molar extinction one obtains for the modification extent x:

$$x = \frac{5(A_{280}/A_{260}) - 9.4}{24(A_{280}/A_{260}) - 9}$$

The ratio A$_{280}$/A$_{260}$ falls from 1.88 at x = 0 (nonmodified enzyme) to 0.63 at x = 1 and to 0.52 at x = 2. Therefore, one may estimate modification extent by measuring this ratio at least up to the binding of two residues per protein molecule. However, these estimates should be considered only as semiquantitative since some changes of molar extinctions may occur due to interactions of the covalently bound chromophores with a protein molecule.

Another physical property which may be introduced into a biopolymer in the course of chemical modification fluorescence should be mentioned due to high sensitivity. However, analogues bearing some fluorescent group should be designed. Among such groups, ethenoadenine residues are widely used. Thus, 2-[(4-bromo-2,3-dioxobutyl)thio]-1-N^6-ethenoadenosine 2′,5′-biphosphate (CXXXVI) was proposed recently as NSDP$^+$ analogue.

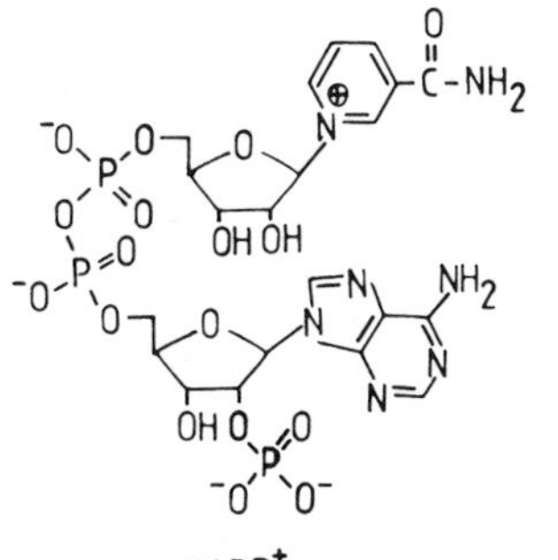

NADP⁺ (shown) — structure labeled **NADP⁺**

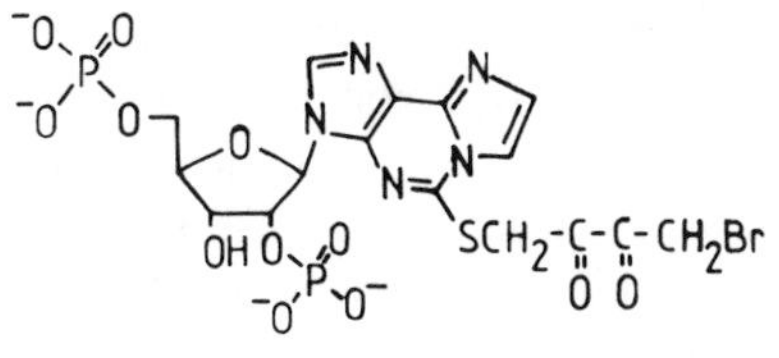

(CXXXVI)

The affinity labeling of isocitrate dehydrogenase (EC 1.1.1.40) catalyzing the oxidation of isocitrate to α-ketoglutarate

$$
\begin{array}{l}
CH_2-COO^- \\
CH-COO^- \\
CHOH \\
COO^-
\end{array}
\; + \; NADP^+ \longrightarrow
\begin{array}{l}
CH_2-COO^- \\
CH_2 \\
CO-COO^-
\end{array}
\; + \; CO_2 + NADPH \tag{3}
$$

was performed. To measure the extent of modification, the enzyme subjected to treatment with the reagent was denatured by the addition of guanidine hydrochloride, and the modified enzyme was separated from the excess of reagent by centrifugation. The intensity of fluorescence as compared with that of the reagent was taken as a measure of the amount of the reagent attached to the enzyme.[304]

However, it should be kept in mind that fluorescence intensity is extremely sensitive to the surrounding of the fluorescent group. Therefore, a simple comparison of the intensities of free reagent and modified biopolymer may lead to an essential under- or overestimate of the extent of modification. In the above-mentioned investigation, special measures were undertaken to escape this source of errors by using denatured modified protein to determine the intensity of fluorescence. The best way is to isolate homogeneous modified protein and use it as a standard.

As an example of a significant change of the fluorescence intensity due to binding to a biopolymer we present the interaction of the ATP analogue γ-etheno-ATP-*p*-azidoanilidate with rabbit muscle creatine kinase (EC 2.7.3.2) catalyzing the transfer of the phosphate moiety between ADP and creatine.

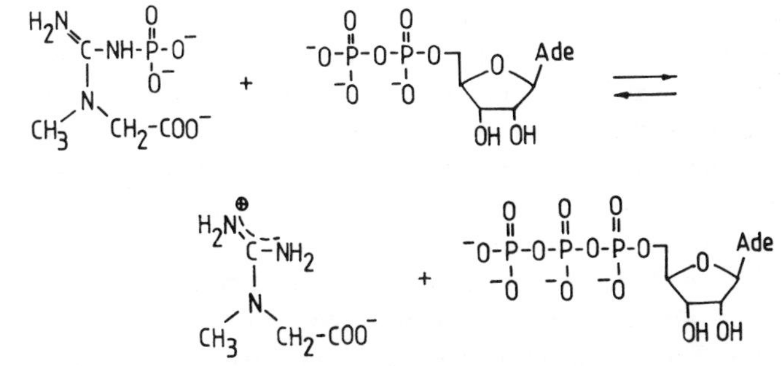

$$\tag{4}$$

It was found that UV irradiation of the enzyme in the presence of the reagent results in the parallel loss of the activity and the increase of fluorescence. The experiments demonstrate that binding results in a four- to fivefold rise of the fluorescence intensity.[305]

B. Inactivation of Biopolymer

The second approach widely used to quantify the modification extent is the measurement of the activity inherent to the biopolymer in the course of modification. The obvious advantage of this approach is the possibility of dealing with nonhomogeneous species, thus starting the investigation of a biopolymer prior to complete purification is achieved. Although the results should be taken with caution, they may accelerate the course of the investigation. The second advantage is that no label should be introduced into the biopolymer with reagent, thus sometimes making the preparation of affinity reagents much easier.

Two main types of approaches are used. The first is the measurement of the response of a biopolymer or more complicated system on the modification. This approach is certainly most widely used for enzymes. Incubation of the enzyme with reagent is followed by taking the aliquots and putting them in the medium containing substrate (or substrates) in conditions favorable for the reaction. Thus, in the case considered in the preceding section, the sample of isocitrate dehydrogenase incubated with the $NADP^+$ analogue was added to the solution containing 30 mM triethanolamine buffer pH 7.4, 0.1 mM $NADP^+$, 4 mM D,L-isocitrate, and 2 mM $MnSO_4$, and the rate of the appearance of NADPH was measured from the absorbance at 340 nm.[304] According to the other approach, modification of the enzyme is carried out in the presence of the substrate. If the enzymatic reaction results in the change of spectral properties of the reaction mixture, the course of this reaction can be monitored continuously and kinetic characteristics of the inactivation process can be obtained.

Respective equations and the representations suitable for the calculation of kinetic parameters of modification were given by several authors. Thus, Hart and O'Brien[306] used this approach to study the inhibition of acetylcholinesterase from bovine erythrocytes with organophosphates by measuring the hydrolysis of *p*-nitrophenyl acetate. Tian and Tsou[307] have treated a similar system in their study of the kinetics of inactivation of chymotropsin with diisopropyl fluorophosphate or phenylmethanesulfonyl fluoride followed by hydrolysis of either acetyl- or benzoyl-L-tyrosine ethyl ester. Walker and Elmore[308] have studied by this approach the inactivation of urokinase, kidney-cell plasminogen activator, plasmin, and β-trypsin with 1-(*N*-6-amino-*n*-hexyl)carbamoylimidazole in the presence of *N*-methylpyrrolid-2-one as a substrate. Tudela et al.[309] have used this procedure to follow the inactivation of trypsin with *N*-tosyl-L-lysylmethylene chloride with benzoyl-L-arginine *p*-nitroanilide as a substrate. Walker et al.[310] have followed the inactivation of thrombin by a series of peptides containing C-terminal arginyl chlormethane in the presence of Boc-Val-Pro-Arg-Amc (7-amino-4-methylcoumarin).

Respective equations used for the calculation of kinetic parameters from these data are presented in Chapter 4, Section I.B.

Enzymatic activity is the most suitable but not the only type of biopolymer property which can be studied in order to follow the course of affinity modification. Irreversible occupation of the ligand binding site by a covalently bound reagent molecule destroys the biopolymer ability to bind specific ligands. This consequence of the reaction can be used to follow the time course of the process and to measure the extent of modification. As an example one can consider the experiment on affinity modification of serotonin-S_2 receptors with photoreactive ketanserin derivatives (Section II.B). The time course of the receptor inactivation was studied by measuring the binding of the known specific ligand of the serotonin-S_2 receptors [^{3}H]-spiperone to the modified membrane preparations (Figure 1).[298]

Some specific physiological responses to affinity modification can also be used to study the reaction process. Thus, covalent binding of specific ligands to the receptors can activate irreversibly or deactivate specific biochemical processes. As an example we can consider experiments on affinity modification of a sodium channel system (see Chapter 1, Section I.F) with photoactivatable derivatives of tetrodotoxin (CXXXVII). This extremely efficient natural neurotoxin binds tightly to the selectivity filter of the channel which selects Na^+ ions.

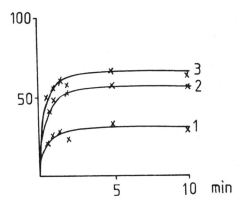

FIGURE 1. Kinetics of photolabeling of serotonin-S_2 receptors from membrane preparations of rat prefrontal cortex (%) with 7-azidoketanserin at concentrations of the latter 10^{-9} (1), 10^{-8} (2), and 10^{-7} M (3) as revealed by binding of [^{3}H] spiperone. (From Wouters, W., Dun, J. V., Leysen, J. E., and Laduron, P. M., *J. Biol. Chem.*, 260, 8423, 1985.)

Irreversible blocking of the channel was studied electrophysiologically. The model was the giant axon isolated from the circumesophageal nerve of the crab *Carcinus maenas*. After UV irradiation of the nerve preparation in the presence of the tetrodotoxin derivatives CXXXVII a and b and washing to remove the noncovalently bound reagent molecules, electrical stimulation was applied at one end of the nerve and resting and action potentials were recorded using an inserted glass capillary microelectrode. Inactivation of the sodium channels was measured by following the decrease of the rate of the rising phase of the action potential.[311]

In most cases affinity modification results in the covalent attachment of reagents to the binding sites on the biopolymer surface and the latter are inactivated. Therefore, usually the extent of covalent modification of biopolymers coincides with the inactivation extent observed. For example, the inactivation of creatin kinase exactly parallels the covalent binding of dialdehyde ADP derivative to this enzyme.[312] However, the perfect correlation between modification and inactivation data is not the general rule. When an enzyme possesses several different binding sites which can be attacked by the same reagent, the covalent modification extent may exceed the inactivation level observed. Thus, complete inactivation of tryptophanyl tRNA-synthetase which has two ATP binding sites with ATP γ-*p*-azidoanilide was achieved when, on the average, six molecules of the reagent were incorporated in the enzyme molecule. It was suggested that in addition to the two reagent molecules bound at the ATP binding site, two more molecules reacted at the effector nucleotide binding centers of the enzyme and two reagent molecules were attached to the protein due to nonspecific reaction.[313]

It should be kept in mind that the inactivation of a biopolymer is not necessarily related to the covalent attachment of the reagent to the biopolymer. Thus, in the mentioned experiments on photoaffinity modification of 3-ketosteroid Δ^5-Δ^4-isomerase with 3-oxo-4-estren-17β-yl acetate (Chapter 2, Section II.E), the main inactivating event was the conversion of aspartate to alanine in the enzyme active site.[135]

In conclusion, it should be mentioned that in some cases inactivation of enzymes in the presence of reactive derivatives of ligands is not related to chemical modification at all. Thus, haloketone derivatives of D,L-leucine and D,L-alanine were found to be efficient yet reversible inhibitors of enzyme leucine aminopeptidase (EC 3.4.11.1) which catalyzes consecutive degradation of peptides from the N-termini.[314] Similarly, several steroid derivatives bearing the alkylating group were found to be reversible inhibitors of 3-ketosteroid Δ^5-Δ^4-isomerase.[315] On the other hand, there are examples where enzymes were inactivated in the presence of derivatives which do not contain reactive groups at all. Thus, S-adenosyl-L-homocysteine hydrolase (EC 3.3.1.1) is inactivated by adenosine analogues including 2′,3′-dideoxyadenosine apparently lacking any reactive group.[316] One of the possible explanations in such cases may be the increased thermolability of proteins induced by the binding of ligands.

IV. DETERMINATION OF THE MODIFICATION POINTS

There are two main experimental levels in the investigation of products of affinity modification. The first one is associated with the identification of the macromolecules which are attacked by the affinity reagent. Typical cases where this experimental level is sufficient are the experiments on the identification of receptor proteins among the biopolymers present in tissue homogenates and the localization of ligand binding sites on the components of macromolecular complexes. Another level is required when the topography of the ligand binding sites and enzyme active centers are studied. In this case, one needs to identify biopolymer residues which are attacked by affinity reagents and to establish the chemical nature.

Analysis of modified biopolymers starts from the isolation from the reaction mixture. Before fine fractionation and structural study have started, one should check whether the label bound to the biopolymer is covalently linked to the target and that the modified residues survive procedures which will be used in the course of the investigation. It is recommended to remove all the noncovalently bound radioactive material from the biopolymer preparation since radioactive by-products of the reagent transformation may interfere with the identification of the modified biopolymers.

A. Distribution of the Label Among Subunits in Multisubunit Systems

Modified proteins are identified usually according to chromatographic or electrophoretic mobility in standard conditions. Sodiumdodecyl sulfate (SDS) polyacrylamide gel electrophoresis, which resolves proteins according to molecular weight, is most widely used for the investigation of moderately complex protein mixtures. After separation, protein bands are stained and those possessing radioactive proteins are localized by means of autoradiography or by cutting out pieces of gel and counting in a scintillation counter. Indeed, this approach can be used only in the cases when modification does not change significantly electrophoretic mobility of biopolymers. This is a problem when the affinity reagent is a macromolecular derivative. In this case the modified biopolymers can also be identified according to electrophoretic mobility if the addressing moiety of the bound reagent can be removed prior to separation. This can be achieved easily with macromolecular reagents bearing cleavable reactive groupings. Addressing structures can also be degraded by specific enzymes. As a typical example one can consider experiments on identification of *Escherichia*

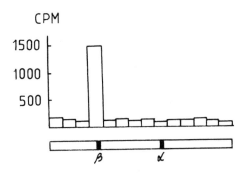

FIGURE 2. Distribution of radioactivity in the SDS polyacrylamide gel after electrophoretic separation of subunits of *E. coli* phenylalanyl-tRNA-synthetase modified with photoreactive tRNA^{Phe} derivative, obtained according to Scheme 83 in Chapter 2. Below positions of the protein bands in the gel revealed by Coomassie staining are shown. (From Vlassov, V. V., Lavrik, O. I., Khodyreva, S. N., Chiszikov, V. E., Shvalie, A. F., and Mamaev, S. V., *Mol. Biol. (USSR)*, 14, 531, 1980.)

coli phenylalanine tRNA synthetase subunits which bind tRNA. The enzyme was modified with the tRNA^{Phe} derivative, bearing arylazido groups randomly distributed over the molecule surface, which were introduced according to reactions (Chapter 2, Scheme 83).[286] The alkylation reagent used for the tRNA derivatization was labeled with a ^{14}C radioisotope. After modification, the cross-linked enzyme and tRNA were isolated by gel chromatography and the tRNA component of the complex was digested with pancreatic ribonuclease. SDS polyacrylamide gel electrophoretic separation of the enzyme subunits followed by radioactivity counting in the slices of the gel (Figure 2) showed that of the two types of the enzyme subunits only the heavy β-subunits were modified.[286]

For the separation of complex protein mixtures, two-dimensional electrophoretic systems are used. Thus, proteins of *E. coli* ribosomes modified with the above-mentioned photoreactive tRNA derivatives were analyzed using the two-dimensional electrophoretic system combining polyacrylamide gel electrophoresis in 7 *M* urea, pH 8.7 in the first dimension and polyacrylamide gel electrophoresis at pH 4.5 in the second dimension (Figure 3).[317-319] This system is characterized by a high resolving power and modified proteins could be separated from the corresponding unmodified ones. In this case identification of the modified proteins would be impossible since it is based entirely on the determination of the protein position of the gel. Therefore, in this case exhaustive digestion of the tRNA moiety with the mixture of ribonucleases A and T1 and phosphomonoesterase was used in order to reduce the modifying moieties present on the protein molecules to a minimum.[317]

Using cleavable reactive groupings greatly facilitates the analysis of modified biopolymers and allows the modified molecules to be identified without using radiolabeled reagents. In this case, first electrophoretic separation of the mixture treated with affinity reagent is performed in order to isolate protein bands or spots showing abnormal mobility which are candidates for the modified proteins. After isolation, these products are subjected to treatments cleaving the reactive groupings, and the produced material is electrophoresed again under standard conditions in order to identify the modified components. A simplified experimental version of such analysis which allows easy identification of spots containing modified biopolymer molecules is the diagonal electrophoresis method. According to this method, a modified protein mixture is gel electrophoresed in one dimension and the gel is subjected to the treatment cleaving the reactive grouping. Second-dimension electrophoresis

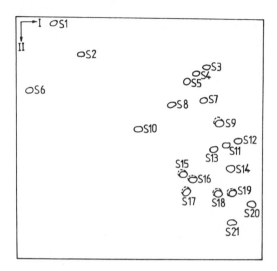

FIGURE 3. Two-dimensional electrophoregram of ribosomal proteins of 30S subunit of *E. coli* ribosomes modified
with photoreactive Phe-tRNA[Phe] derivative, obtained according to Scheme 83 in Chapter 2. Proteins were developed
by staining with Coomassie. Dashed lines indicate radioactive spots. (From Babkina, G. T., Karpova, G. G., and
Matassova, N. B., *Mol. Biol. (USSR)*, 5, 1287, 1984.)

performed under conditions identical to those of the first-dimension separation places all the
nonmodified protein molecules at the diagonal of the gel, while the modified molecules
which had different properties in the two electrophoretic runs are shifted off the diagonal.
Indeed, specific probes interacting with certain biopolymers, such as antibodies specific to
certain proteins, greatly facilitate the identification of the modified biopolymers. Isolation
of covalently bound reagent and biopolymer can be achieved using antibodies specific to
the reagent moiety.[320] When proteins covalently bound to a certain nucleic acid sequence
are to be identified, the cross-linked RNA-protein complex can be separated from the
nonbound proteins using hybridization to DNA complementary to the nucleic acid sequence
under consideration.[321,322]

B. Determination of the Monomer Residues Modified

The identification of modified residues in the biopolymer structure is the final and usually
the most laborious step of the affinity modification experiment. Success of this step is
determined primarily by the availability of sufficient amounts of pure modified biopolymer
and by the chemical nature of the modified residues. Very often affinity reagents can attack
several residues in the binding sites giving rise to the formation of several forms of the
modified biopolymers. In these cases structural analysis can be greatly facilitated by the
separation of the modified preparations to homogeneous products. This can be achieved by
using efficient separation techniques. Thus, reversed-phase HPLC resolves at least four
different monomodified forms of pancreatic ribonuclease produced by a reaction with 4-(*N*-
2-chloroethyl-*N*-methylamino)benzylamide of dinucleotide d(pTpA) (Figure 4).[115]

Nonspecific modification can introduce considerable difficulties in the analysis of the
affinity-modified biopolymers. In the case of photoaffinity modification, using scavengers
can help to suppress the nonspecific reactions (see Section II.B). To discriminate between
the specific and nonspecific modification, protection experiments are carried out. Biopolymers are modified in the presence of specific ligands which protect specific binding sites
from the reagents. In further analysis, only the revealed specifically modified areas are taken

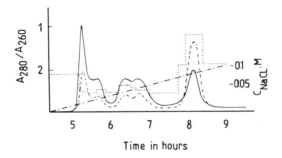

FIGURE 4. Chromatography on carboxymethylcellulose column of the products of affinity modification of pancreatic RNAase with 5'-[p-(N-2-chloroethyl-N-methylamino)benzyl] amide of dinucleotide pdTpdA. Elution was carried out in 5 mM Tris-HCl, with a linear NaCl gradient. (—) A_{260}, (---) A_{280}, and (⋯) A_{280}/A_{260}. (From Knorre, D. G., Buneva, V. N., Baram, G. I., Godovikova, T. S., and Zarytova, V. F., FEBS Lett., 194, 64, 1986.)

into account. For the localization of the specific reaction area, a differential modification approach developed originally for chemical modification studies of biopolymer-ligand interaction can also be used. According to this approach, a biopolymer is modified in two steps. First, it is premodified with a cold affinity reagent in the presence of a specific ligand protecting the active site. At this step nonspecific sites are modified while the specific ones are protected. After removal of the reagent and the ligand, the biopolymer is treated with the radioactive affinity reagent and specific modification of the active site is achieved.[323]

To choose optimal methods for analysis of modified biopolymers, it is useful to have preliminary information about the chemical nature of the modified residues. The chemical nature of the modified groups is obvious when the affinity reagent contains a highly specific reactive group such as a SH group interacting with cysteine residues only; the glyoxal group interacts with the arginine or dialdehyde group of oxidized nucleotide derivatives, which react exclusively with primary amino groups. In other cases, useful information about the nature of the modified residues can be obtained by testing the chemical stability under various treatments. Thus, acylated hydroxy and amino groups are easily discriminated since the former are easily hydrolyzed by alkaline treatment while the latter are stable under conditions of the alkaline hydrolysis. In experiments on affinity modification of pyruvate kinase with 5'-p-fluorosulfonylbenzoyl-1,N^6-ethenoadenosine mentioned in Chapter 2, Section II.B, the formation of an intramolecular disulfide bond was suggested since the enzyme inactivation could be reversed by the addition of dithiothreitol.[108] One more example is given by one of the experiments on affinity modification of E. coli RNA polymerase with derivatives of nucleoside triphosphates.[324] In one of the experiments, RNA polymerase was modified with guanosine-5'-trimetaphosphate in the presence of a promotor containing template and UTP. The reagent used was a phosphorylating one and hence it could attack Lys, His, Arg, Ser, Thr, Tyr, Asp, Glu, and Cys residues. The label introduced was found to be stable at natural pH and to be removed by mild acidic treatment; therefore, Ser, Thr, Tyr, and Cys could be excluded from the list since the phosphorylated derivatives are acid-stable. Arg, Asp, and Glu also could not be the modified residues since in model experiments it was found that the phosphorylated guanidine group is very unstable even in a neutral medium, and phosphorylated Asp and Glu are known to be highly reactive anhydrides. Therefore, the targets of the affinity modification could be His and Lys residues only. In order to discriminate between these two possibilities, the kinetics of the labeled RNA polymerase demodification was compared with those of hydrolysis of the phosphoamide bonds in ATP γ-methylamide

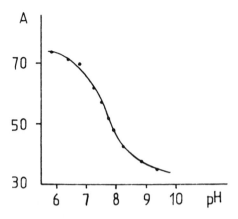

FIGURE 5.   Relative activity (%) of rabbit muscle
creatine kinase incubated at 30°C for 10 min with 5
m*M* ATP γ-[p-(*N*-2-chloroethyl-*N*-methylam-
ino)benzyl] amide at different pH values. (From
Nevinsky, G. A., Gazaryants, M. G., and Lavrik,
O. I., *Bioorg. Khim. (USSR)*, 10, 656, 1984.)

and ATP γ-imidazolide which were models of phosphorylated Lys and His, respectively. It
was found that the rate of the enzyme demodification at pH 4.9, 20°C was the same as that
of the hydrolysis of ATP γ-imidazolide (k_{app} = 0.016 min^{-1}) and considerably faster as
compared to the hydrolysis of ATP γ-methylamide (k_{app} = 0.014 min^{-1} at 50°C). Therefore,
the data suggested that the modified enzyme residue was His.

Investigation of the affinity modification kinetics under different conditions can sometimes
give helpful ideas about the residues attacked. Thus, in experiments on the modification of
creatine kinase monomer subunits by the alkylating reagent, an ATP analogue (CXXb), it
was observed that the extent of the modification process depends upon the pH considerably
suggesting that the residue attacked has pK_a of ~7.7 (Figure 5) which corresponds to the
pK_a value of one of the active site histidine residues determined by NMR.[325]

The usual way to localize the modified regions in protein molecules consists in specific
enzymatic or chemical cleavage of the biopolymer to smaller peptide fragments, separation
of the peptides by electrophoresis or chromatography, and identification of the modified
peptide. Peptides can be identified according to mobility in standard separation systems as
was described in the previous section for the large peptides. Precise identification is accomplished by amino acid analysis and sequencing. Identification of the modified sites in the
protein molecules on the peptide resolution level is achieved easily if a small number of
biopolymer residues are derivatized. Thus, the affinity reagent folate analogue bearing a
glyoxal group modifies a single residue in the active site of dihydrofolate reductase (EC
1.5.1.3), which is obviously arginine residue taking into account narrow specificity of the
reactive group and stability of the product formed.

(CXXXVIII)

The modified protein was cleaved with CNBr to three peptides and the peptides were
separated by consecutive gel chromatography and reversed-phase HPLC. The only peptide
which contained 340-nm absorbing pteridine chromophore of the affinity reagent was isolated
and subjected to sequenator analysis. The sequence determined allowed the peptide to be
identified as the N-terminal one.[326]

A more complicated case where peptide analysis helped to study the specificity of affinity modification is represented by experiments on affinity modification of a steroid-binding site of human placental estradiol 17β-dehydrogenase with 3-bromo-[2'-^{14}C]acetoxy-1,3,5(10)estratrien-17-one.

(CXXXIX)

Modified enzyme was digested with trypsin and the radioactive peptides were isolated by consecutive gel chromatography and ion exchange column chromatography. Analysis of the protein samples modified with 60-μM affinity reagent revealed four peptides containing radioactive 3-carboxymethylhistidine residues while only one labeled peptide was found in the tryptic digest of the enzyme alkylated with 6-μM reagent CXXXIX. The results suggested that at high reagent concentration nonspecific hydrophobic sites on the enzyme were occupied and labeled.[327]   Some specific approaches simplifying the isolation of the affinity modified peptides from complex peptide mixtures were proposed. A simple procedure was described for the isolation from protein hydrolyzates of peptides containing covalently bound residues with *cis*-diol groups on columns with dihydroxyborylaminoethyl cellulose which absorbs *cis*-diol-bearing molecules due to a complex formation with the dihydroxyborylic groups. The method is useful for the isolation of peptides modified with derivatives or components of RNAs.[328]

The use of affinity reagents covalently linked to a solid phase was proposed for the isolation of modified peptides. This efficient approach is useful in the cases where the modification extent is low. Using this method, the product of the minor photoreaction between 3-oxosteroid Δ^5-Δ^4 isomerase and the reagent XXXVII in Chapter 2 was investigated. This minor reaction, in contrast to the major one leading to photochemical decarboxylation of aspartate 38 of the enzyme (Chapter 2, Section II.E) results in the covalent attachment of the reagent to the protein. The minor reaction proceeds with very low yield. Therefore, the solid-phase photoaffinity reagent, Δ^6-testosterone-agarose, was prepared. After photomodification of the enzyme with this reagent, the covalently bound enzyme was chymotrypsin-digested. Washing of the agarose removed the digestion products except for the covalently bound peptide which could be detached from the solid phase by cleaving the reagent-agarose linkage.[329]

For the fast localization of peptides containing modified radiolabeled residues in the protein structure, a special approach was proposed which is similar to the Maxam-Gilbert DNA sequencing methodology (see Chapter 1, Section II.B). According to this approach, modified protein is subjected to limited specific chemical or enzymatic cleavage (less than one cut per protein molecule, on the average) and the peptides produced are separated by SDS gel electrophoresis according to the molecular weights. Single hit cleavage at n specific sites in the protein structure produces two series of peptides, n peptides containing the C end and n peptides containing the N end. Due to the uneven distribution of the cleavage sites in the protein, electrophoretic separation of the peptide mixture results in a specific pattern of bands which can, in principle, be assigned to corresponding peptides when the protein primary structure is known. If the radioactive reagent is attached between cleavage sites i and i + 1, the label will be present in the peptides of the N series produced by cleavage at sites from i + 1 to n and absent from the shorter peptides. In the peptides of the C series the label will be in all peptides longer than those obtained by cleavage at i site, i.e., in peptides terminated at the cleavage points from i to 1. Analysis of the radioactive band pattern allows the shortest labeled peptide of the N series produced by cleavage at the site

Table 2
**EDMAN DEGRADATION OF THE LABELED PEPTIDE
FROM SPINACH RIBULOSEBISPHOSPHATE
CARBOXYLASE/OXYGENASE MODIFIED WITH 3-
BROMO-1,4-DIHYDROXY-2-BUTANONE 1,4-
BISPHOSPHATE**

Amino acid	L	S	G	G	D	H	I	H	S	G
R	1.6	1.7	1.9	1.5	1.8	1.7	1.7	1.6	1.3	1.2

Amino acid	T	V	V	G	K*	L	E	G	E	R
R	1.1	1.0	1.1	3.1	10.1	6.2	4.4	3.4	2.7	1.9

Note: R: radioactivity, percent of the total radioactivity of the peptide.

From Stringer, C. D. and Hartman, F. C., *Biochem. Biophys. Res. Commun.*, 80, 1043, 1978.

i + 1 and the shortest labeled peptide of the C series produced by cleavage at the site i to be identified and hence to identify the protein fragment between the neighboring cleavage points i and i + 1 which contains the modified residue.[330] This approach was used for localization of the modified histidine residue in *E. coli* RNA polymerase treated with the reactive substrate analogue GDP β-imidazolide. CNBr cleavage was used as a specific fragmentation procedure (see Chapter 1, Section II.B). It was found that the modified residue is located between the cleavage sites 29 and 30, i.e., between methionine residues 1230 and 1243. Since there is only one histidine residue in this sequence, namely His 1237, it was concluded that this residue is attacked by the affinity reagent.

Localization of the modified amino acid residues within peptides is achieved by sequencing. Clear-cut results can be obtained when the peptides contain single modified residues. In this case, in the course of stepwise Edman degradation, no normal phenylthiohydantoin appears at the cycle corresponding to cleaving off of the modified amino acid residue. If the reagent was radiolabeled, the label will be released at this step. Thus, automated Edman degradation of the labeled tryptic peptide from the plant enzyme required for the photosynthetic fixation of CO_2 ribulosebisphosphate carboxylase/oxygenase (EC 4.1.1.39) modified with affinity reagent 3-bromo-1,4-dihydroxy-2-butanone 1,4-bisphosphate has revealed no normal amino acid phenylthiohydantoin in sequentation cycle 15. It was also found that the fraction obtained in this cycle contained radioactivity which evidenced that the liberation of the residue-carrying radiolabeled reagent has occurred (Table 2). It was concluded that modified amino acid residue was located in this position.[331]

Some affinity reagents, in particular photoreactive ones, modify several amino acid residues within one peptide. An example can be the experiment on photoaffinity modification of mitochondrial F_1-ATPase with 2-azido [α-^{32}P]adenosine diphosphate. The modified enzyme was cleaved to small peptides and the peptide containing radioactivity was isolated, identified on the basis of amino acid composition and N-terminal amino acid, and subjected to Edman degradation. It was found that the label was localized on four amino acids (Figure 6).[332] In the cases cited above, Edman degradation allowed the modified amino acid residues to be identified precisely. It should be noted that it is not always possible. Some cases have been reported where sequencing the modified peptides using Edman degradation was complicated. Thus, when modified peptides isolated from myosin treated with *N*-(4-azido-2-nitrophenyl)-2-aminoethyl diphosphate were sequenced, some radioactivity was released at each cycle. The most reasonable explanation was that the linkage of the [^{3}H]-labeled reagent with the modified residue was acid-labile and the reagent grouping was released each cycle during the acid cleavage steps of the Edman procedure.[333] Analysis of the peptides isolated

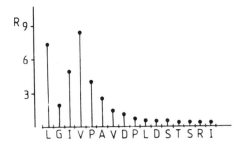

FIGURE 6. Radioactivity (R, c/min × 10^{-3}) of phenylthiohydantoins obtained by sequential Edman degradation of labeled peptide from beef heart mitochondrial F$_1$-ATPase modified with 2-azido [α-^{32}P] adenosine diphosphate. (From Garin, J., Boulay, F., Issartel, J. P., Lunardi, J., and Vignais, P. V., *Biochemistry*, 25, 4431, 1986.)

from *E. coli* ADP-glucose synthetase modified by 8-azido-ADP[^{14}C] glucose could not be accomplished due to the lability of the glucosidic linkage at acidic pH.[334]

The identification of the modified sites in nucleic acids can also be carried out using fragmentation of the biopolymers with specific enzymes. Thus, the first step of localization of modified residues in double-stranded DNAs consists in the digestion of the nucleic acid with restriction enzymes which cleave DNA in sequence-specific way, separation of the restriction fragments by gel electrophoresis, and identification of the fragments which carry the modified residues. Similarly, modified RNAs are hydrolyzed by specific ribonucleases and identification of the modified oligonucleotides is very often the final level of resolution achieved. For example, in experiments on direct UV cross-linking of tRNAs to aminoacyl-tRNA-synthetases the cross-linked proteins and tRNAs were treated with ribonuclease T1 which cleaves RNA at positions of guanosine residues and the released oligonucleotides were separated and quantitized. Comparison of the data with results of similar oligonucleotide analysis of the free tRNA allowed the oligonucleotides which were lacking from the digests to be revealed. These oligonucleotides were considered as the protein-cross-linked ones.[335]

The most advanced experiments on the localization of the modified residues in RNA by the traditional technique based on specific enzymatic degradation were related to cross-linking studies of the ribosome structure. An example can be the analysis of the UV-cross-linked ribosomal protein L4 and ^{32}P-labeled 23S RNA.[336] To identify the modified residue in the RNA, the complex was subjected to mild digestion with a small amount of pyrimidine-specific ribonuclease A under conditions where cleavage is incomplete and large RNA fragments are formed. The products were separated by two-dimensional gel electrophoresis and the cross-linked L4-RNA fragments were isolated. The cross-linked product obtained was treated with proteinase K to remove the protein. At the next step of analysis, the obtained RNA fragment was digested with ribonuclease T1 and the oligonucleotides produced were separated in a two-dimensional thin-layer chromatographic system. The secondary digestion with ribonuclease A and analysis of the cleavage products permitted the identification of oligonucleotides obtained according to mobilities in a standard chromatographic system. T1 oligonucleotides found evidenced that the RNA fragment was originated from the 23S RNA sequence in positions 580 to 669. It was also noticed that the oligonucleotide AAUAGp from the central part of this sequence was not found. Instead, an anomalous oligonucleotide was present. At the next step of analysis the resolution was extended to the nucleotide level. Exhaustive digestion of the anomalous oligonucleotide with ribonuclease A was performed. The digestion yielded AGp together with an anomalous product. Treatment of this product with nonspecific ribonuclease T2 released Ap. Measurement of the radioactivities of Ap and of the peptide-containing spot gave the ratio of about 2. It was concluded therefore that the oligonucleotide studied was AAUAGp cross-linked to protein via central uridine which is uridine U615 of the 23S RNA.

The most efficient approaches for the analysis of modified nucleic acids are based on Maxam-Gilbert and Sanger methodologies developed for sequencing of nucleic acids.[337] The first methodology can be used when nucleic acid with a ^{32}P-labeled end can be cleaved at positions of the modified residues. The produced specific labeled fragments can be electrophoresed in parallel with the products of Maxam-Gilbert sequencing reactions of the nucleic acid under study, and in this way positions of cleavage and the modification points can be identified. As an example experiments on affinity modification of a single-stranded DNA fragment with complementary oligonucleotide derivatives bearing alkylating groups can be considered (Figure 7). The fragment was reacted with the derivatives, cleaved at positions of the alkylated purines by piperidine treatment, and subjected to gel electrophoretic analysis which has revealed the modification sites.[338]

According to the other methodology, modified nucleic acid is used as a template for enzymatic synthesis of the complementary nucleic acid strand with DNA polymerases. An oligonucleotide complementary to a nucleic acid sequence ahead of the modified residue is annealed to the DNA and used as a primer. In the presence of labeled nucleoside triphosphates the enzyme extends the primer and synthesizes labeled single-stranded DNA until it meets the damaged residue in the template. Due to the high specificity of the polymerases toward nucleotides in the templates, the enzyme stops at the damaged residue and terminates the synthesis, thus producing fragments of specific length. Gel electrophoretic analysis of the fragments reveals positions of the modified residues in the template. An example of application of such methodology to the localization of modified residues in RNA is represented by experiments on investigation of 23S RNA isolated from *E. coli* ribosomes modified with the affinity reagent 3-(4'-benzoylphenyl)propionyl[^{3}H]-Phe-tRNA (CXXXIg in Chapter 2).[339] Preliminary localization of the modified region of the RNA was performed by hybridization of the RNA bearing labeled modified residue to the known rDNA fragments, hydrolysis of the nonhybridized single-stranded RNA, and electrophoretic analysis of the produced heteroduplexes. In these experiments it was found that the modified residue lies within a 183-nucleotide region between positions 2442 and 2625 of the 23S RNA. To identify the precise position of the modified nucleotide, primers complementary to the 3' side of this region were annealed to the RNA and were extended with reverse transcriptase, RNA-dependent DNA polymerase which is mainly found in the oncogenic viruses. The enzyme stops exactly at the nucleotide immediately preceding the modified base from the 3' side. Analysis of the elongation products showed that the synthesis was terminated at positions U2584 and U2585 of the 23S RNA which were identified as the modified bases.

Various experimental versions of the methodologies described which are helpful for the identification of modified residues in the structure of various nucleic acids including double-stranded polynucleotides can be found in a special review.[337]

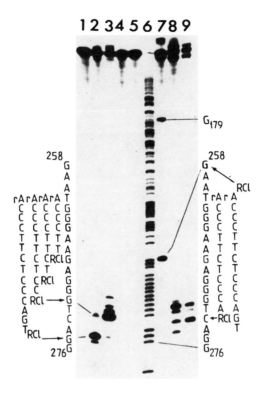

FIGURE 7. Results of gel electrophoretic analysis of single-stranded DNA fragment alkylated by oligonucleotide derivatives bearing *N*-2-chloroethyl-*N*-methylaminophenyl groups at 5'-phosphate or 3'-terminal rA residue. Lane 1 is a control, noncleaved fragment. Lane 6 contains products of the (A + G) sequencing reaction (Chapter 1, Table 4). Lanes 2, 3, 4, 5, 7, and 8 are cleavage products resulted from alkylation of fragment with reagents Cl-RpTGACCCTCTTCCCrA, ClRpCCTCTTCCCrA, Cl-RpTCTT CCCrA, ClRpTTCCCrA, pTGACCCTCTTCC-CrARCl, and ClRACCCTCTTCCCrA, respectively. Lane 9 is the same as lane 8 except for an additional treatment with hydrazine resulting in cleavage at alkylated cytidine residues was introduced. On the sides of the gel the target sequence G258 to G276 of the DNA fragment is shown. Positions of the corresponding bonds on the gel are shown with solid lines. Arrows indicate nucleotides attacked by the reagents. (From Vlassov, V. V., Zarytova, V. F., Kutiavin, I. V., Mamaev, S. V., and Podyminogin, M. A., *Nucleic Acids Res.*, 14, 4065, 1986.)

Chapter 4

AFFINITY MODIFICATION: PHYSICAL CHEMISTRY

I. KINETICS OF AFFINITY MODIFICATION

A. The Simplest Scheme of Affinity Modification and the Range of Validity

The simplest scheme of affinity modification may be represented as follows

$$P + X \rightleftarrows (k_1, k_{-1})$$

$$PX \rightarrow PZ(k_2) \tag{1}$$

where P, biopolymer; X, reagent; PX, specific complex; PZ, covalent adduct of X or of some fragments to P; and k_1, k_{-1}, and k_2, rate constants of respective reactions. The scheme was considered by Baker et al.[340] and the kinetic equation corresponding to this scheme was presented by the same group of authors.[95] However, the commonly used treatment of the scheme is that proposed by Kitz and Wilson.[341] Below, we shall present this treatment performed in quasiequilibrium approximation.

According to this approximation, equilibrium retains between complex PX and constituents in spite of consumption of the former. Therefore,

$$\frac{[PX]}{[P][X]} = K_x = k_1/k_{-1} \tag{2}$$

where K_x is the association constant, and square brackets notify concentrations of the particles taken in the brackets. Hereafter in this chapter we shall use preferentially association constants. This should be kept in mind when comparing expressions presented in this chapter with those given in the original papers where dissociation constants are widely used.

The total concentration of biopolymer p_o according to this scheme is given by

$$p_o = [P] + [PX] + [PZ] \tag{3}$$

and, therefore

$$[PX] = K_x[X](p_o - [PZ])/(1 + K_x[X])$$

Consequently

$$d[PZ]/dt = k_2[PX] = k_2K_x[X](p_o - [PZ])/(1 + K_x[X]) \tag{4}$$

If the reagent is present in sufficient molar excess over the biopolymer, the initial concentration of the reagent x_o may be substituted for [X] and Equation 4 may be integrated under initial conditions [PZ] = 0 at t = 0. Integration results in the expression

$$[PZ] = p_o[1 - \exp(-k_{app}t)] \tag{5}$$

with the apparent rate constant

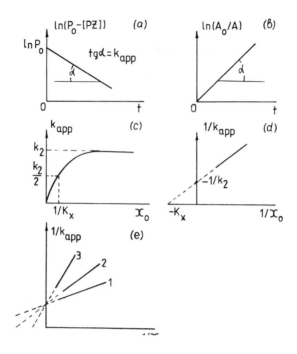

FIGURE 1.   Semilogarithmic plots of kinetic curves of affinity
modification in the Kitz and Wilson approximation. (a) Kinetics
of accumulation of modification product [PZ], p_o is the initial
concentration of biopolymer; (b) kinetics of inactivation of biopolymer, A is the activity, and A_o is the initial activity. Dependence of the apparent rate constant k_{app} on the initial reagent
concentration: (c) k_{app} vs. x_o, graphic determination of k_2 and K_x
values of Equation 6 is shown; (d) $1/k_{app}$ vs. $1/x_o$; and (e) $1/k_{app}$
vs. $1/x_o$ in the absence of unreactive analogue (1) and in the
presence of two different concentrations of such an analogue, (2)
and (3).

$$k_{app} = k_2 K_x x_o/(1 + K_x x_o) \qquad (6)$$

Thus, the reaction behaves as a pseudo first-order process. The k_{app} value may be easily
found as

$$k_{app} = (1/t) \ln\{p_o/(p_o - [PZ])\} \qquad (7)$$

It may be also estimated as a slope of the linear semilogarithmic plot $\ln(p_o - [PZ])$ vs.
t (Figure 1). The dependence of k_{app} on x_o is hyperbolic. At high concentrations of reagent
at $K_x x_o \gg 1$ it reaches limit value k_2. Half of this value is reached at $K_x x_o = 1$.

If modification proceeds at one definite point of biopolymer and results in the complete
loss of biological activity of biopolymer A (e.g., enzymatic activity), the relative residual
amount of unmodified biopolymer is equal to relative residual activity A/A_o (A_o, activity in
the beginning of the process). Therefore, Equation 7 may be written as

$$k_{app} = (1/t) \ln(A_o/A) \qquad (8)$$

Measuring k_{app} at several different concentrations x_o and using Equation 6 it is easy to
calculate k_2 and K_x. The equation may be used for a direct least square minimization procedure
as well as in linearized form

$$1/k_{app} = 1/k_2 + (1/k_2 K_x)(1/x_o) \qquad (9)$$

The rough estimates may be obtained by plotting $1/k_{app}$ vs. $1/x_o$ as an intersection of the experimental straight line with ordinates $(1/k_2)$ and abscisses axes $(-K_x)$ (Figure 1).

If a nonreactive specific ligand Y is present in sufficient molar excess over a biopolymer, a similar derivation results in Equation 5 with

$$k_{app} = k_2 K_x x_o/(1 + K_x x_o + K_y y_o) \qquad (10)$$

Y may be a natural ligand recognized by a biopolymer in the course of functioning, otherwise it may be a competitive reversible inhibitor of the activity of the biopolymer. Plotting $1/k_{app}$ vs. $1/x_o$ for each set of experiments carried out at a definite Y concentration, one should obtain straight lines with common intersection at the ordinate axis (Figure 1). Comparing k_{app} values obtained in the absence of Y and in the presence of a definite Y concentration, it is easy to calculate K_y provided K_x is already found as described above.

As it is obvious from Equation 6, to separately estimate k_2 and K_x values it is essential to carry out the experiments at concentrations on the order of magnitude of $1/K_x$. If $K_x x_o \gg 1$, k_{app} does not change with X concentration and is simply equal to k_2. If $K_x x_o \ll 1$, the expression transforms to

$$k_{app} = k_2 K_x x_o \qquad (11)$$

The reaction obeys the law of the second-order reactions with the rate constant $k_2 K_x$ and only this product may be determined from the experimental data.

Quite similar equations may be obtained for Equation 1 in the steady-state approximation with a single difference that in this case[342,343]

$$K_x = k_1/(k_{-1} + k_2) \qquad (12)$$

In some special situations it may be convenient to deal with the excess of biopolymer over reagent. In this case, the mass conservation law for the reagent

$$x_o = [X] + [PX] + [PZ] \qquad (13)$$

should be taken into account whereas [P] may be considered as a constant value equal to p_o. Therefore,

$$d[PZ]/dt = k_2 K_x p_o(x_o - [PZ])/(1 + K_x p_o) \qquad (14)$$

$$[PZ] = x_o[1 - \exp(-k_{app}t)] \qquad (15)$$

$$k_{app} = k_2 K_x p_o/(1 + K_x p_o) \qquad (16)$$

Sometimes due to complications arising in the course of modification (consumption of the reagent in some by-processes, spontaneous inactivation of a biopolymer due to thermal denaturation, etc.), it is desirable to restrict kinetic measurements to the initial period of the reaction. In this case, the equation describing the dependence of initial rate v_o on the initial concentration of the reagent x_o may be useful. Under the same assumptions of quasiequilibrium and excess of reagent over biopolymer, this equation is easily obtained from Equation 4 by substituting x_o for [X] and v_o for $d[PZ]/dt$ as well as by putting $[PZ] = 0$.

$$v_o = k_2 K_x x_o p_o / (1 + K_x x_o) \tag{17}$$

In the presence of the excess of nonreactive ligand Y this expression transforms to

$$v_o = k_2 K_x x_o p_o / (1 + K_x x_o + K_y y_o) \tag{18}$$

These equations were at first proposed by Baker et al.[95] It is seen that k_2, K_x, and K_y values may be easily calculated from the experimental dependences of v_o on x_o and y_o.

The simple expressions, either Equations 5 and 6 or 15 and 16 do not hold if p_o and x_o are of the same order of magnitude. The more general case suitable for any x_o/p_o ratio was analyzed by Gorshkova and Chimitova.[344] Using Equations 2, 3, and 13, one easily obtains

$$K_x [P]^2 + [1 + K_x(x_o - p_o)][P] = p_o - [PZ] \tag{19}$$

The differentiation of Equation 19 and substitution of $k_2[PX]$ for $d[PZ]/dt$ and $K_x[P][1 + K_x(x_o - p_o)]$ for $[PX]$ results in the differential equation for $[P]$

$$\{2K_x[P] + [1 + K_x(x_o - p_o)]\} d[P]/dt = k_2 K_x[P][1 + K_x(x_o - p_o)]$$

which is easily integrated

$$\left[1 + \frac{1}{(x_o - p_o)K_x}\right] \ln \frac{[P]_o}{[P]} + \left[1 - \frac{1}{(x_o - p_o)K_x}\right] \ln \frac{[P]_o + x_o - p_o}{[P] + x_o - p_o} = k_2 t \tag{20}$$

where $[P]$ and $[P]_o$ are positive roots of Equations 19 and 21.

$$K_x[P]_o^2 + [1 + K_x(x_o - p_o)][P]_o = p_o \tag{21}$$

Expression 20 does not hold if $x = p_o$. For this case a similar consideration leads to Equations 22 to 24.

$$2\ln \frac{[P]_o}{[P]} + \frac{1}{K_x}\left(\frac{1}{[P]} - \frac{1}{[P]_o}\right) = k_2 t \tag{22}$$

$$K_x[P]^2 + [P] = p_o - [PZ] \tag{23}$$

$$K_x[P]_o^2 + [P]_o = p_o \tag{24}$$

All the above-considered equations were derived in quasiequilibrium or steady-state approximation valid for most affinity modification processes. An exact solution may be easily obtained if one of the components is present in great excess over the other. We shall restrict ourselves by considering the case $x_o \gg p_o$.[345] The kinetics of the process may be described by two differential equations:

$$d[PX]/dt = k_1[P][X] - (k_{-1} + k_2)[PX] \tag{25}$$

$$d[PZ]/dt = k_2[PX] \tag{26}$$

and since

$$[P] = p_o - [PX] - [PZ], \qquad [X] \approx x_o$$

Equation 25 turns to

$$d[PX]/dt = k_1 p_o x_o - (k_{-1} + k_2 + k_1 x_o)[PX] - k_1 x_o[PZ] \qquad (27)$$

Equations 26 and 27 form the system of two first-order differential equations with constant coefficients. Integration results in

$$[PZ] = \frac{k_1 k_2 x_o p_o}{q_1(q_2 - q_1)} [1 - \exp(-q_1 t)] + \frac{k_1 k_2 x_o p_o}{q_2(q_2 - q_1)} [1 - \exp(-q_2 t)] \qquad (28)$$

where

$$q_1, q_2 = \frac{1}{2} [k_{-1} + k_2 + k_1 x_o \mp \sqrt{(k_{-1} + k_2 + k_1 x_o)^2 - 4k_1 k_2 x_o}] \qquad (29)$$

As

$$(k_{-1} + k_2 + k_1 x_o)^2 - 4k_1 k_2 x_o = (k_{-1} - k_2 + k_1 x_o)^2 + 4k_{-1} k_1 x_o$$

q_1 and q_2 are always real positive values.

If

$$4k_1 k_2 x_o \ll (k_{-1} + k_2 + k_1 x_o)^2$$

$$q_1 \approx k_1 k_2 x_o/(k_{-1} + k_2 + k_1 x_o)$$

$$= k_2 K_x x_o/(1 + K_x x_o) \qquad (30)$$

where K_x is given by Equation 12. Under the same conditions

$$q_2 \approx k_{-1} + k_2 + k_1 x_o \gg q_1$$

Therefore, after a short transient period $t \sim 1/q_2$, Equation 28 transforms to Equations 5 and 6. Thus, the steady-state approximation is valid if the inequality (Equation 30) holds. This inequality is easily transformed to

$$4k_2/k_1 x_o \ll [1 + (k_{-1} + k_2)/k_1 x_o]^2$$

which is fulfilled both if $k_2/k_1 x_o \ll 1$ and $k_2/k_1 x_o \gg 1$. Consequently, the approximate treatment of Equation 1 is not valid only if k_2 and $k_1 x_o$ are of the same order of magnitude.[346]
A quasiequilibrium approximation may be used instead of a steady-state one if $k_2 \ll k_{-1}$. In this case, K_x defined by Equation 12 transforms to an association constant. However, rate constants necessary for the estimation of the validity of this condition are scarcely accessible. It was proposed to make a respective estimation using an experimental k_{app} value and assuming that k_1 is close to the diffusion-controlled rate constant.[347] The assumption means that association of X with P proceeds without an activation barrier.
The rearrangement of Equation 6 with K_x defined by Equation 12 results in the expression:

$$k_{app}/k_1 x_o = [1 + (k_{-1}/k_2) + (k_1 x_o/k_2)]^{-1}$$

Therefore, if $k_{-1}/k_2 \gg 1$, $k_{app}/k_1 x_o \ll 1$ and this inequality may be used for the estimation of the validity of a quasiequilibrium condition.

Some complications of Equation 1 and kinetic treatment were considered by Childs and Bardsley.[348] The authors have analyzed the scheme

$$P + X \rightleftarrows PX \rightleftarrows PZ$$

with reversible covalent attachment of affinity reagent. An exact solution was derived for the case $x_o \gg p_o$. Without this condition, the third-order nonlinear differential equation should be integrated and only an approximate solution can be obtained, e.g., by a reversion method. The cumbersome expressions derived by the authors may be found in their original paper. In the same paper, an approach is briefly considered to treat the scheme of affinity modification proceeding via several reversible intermediate steps

$$P + X \rightleftarrows PX \rightleftarrows PZ_1 \rightleftarrows PZ_2 \dots \rightleftarrows PZ_n$$

However, the use of the expressions presented should be done with caution as some errors are involved, e.g., in Chapter 1, Equation 8 of the paper as indicated by Cornish-Bowden.[346]

B. Simultaneous Enzymatic and Affinity Modification Processes

As it was already mentioned in Chapter 3, Section III.B, the kinetic study of affinity modification of enzymes may be carried out by following the accumulation of the product of enzymatic process in the presence of an irreversible inhibitor. Kinetic treatment of the simplest scheme describing simultaneous enzymatic reaction and affinity modification was described in several papers.[306-310] This scheme may be presented as follows:

$$E + S \rightleftarrows ES(k_1, k_{-1})$$

enzymatic process

$$ES \rightarrow E + Q(k_2)$$

$$E + X \rightleftarrows EX(k_1^*, k_{-1}^*)$$

affinity modification

$$EX \rightarrow EZ(k_2^*) \tag{31}$$

where E, enzyme; S, substrate; Q, product of enzymatic process; X, reagent; EZ, modification product; and ES and EX, respective reversible complexes. To describe the kinetic behavior of seven components we need seven equations. Under steady-state approximation, three mass conservation laws, two steady-state conditions for EX and ES, and two differential equations may be written

$$[E] + [ES] + [EX] + [EZ] = e_o \tag{32}$$

$$[X] + [EZ] = x_o([EX] \text{ neglected}) \tag{33}$$

$$[S] + [Q] = s_o([ES] \text{ neglected}) \tag{34}$$

$$d[ES]/dt = k_1[E][S] - (k_{-1} + k_2)[ES] \approx 0 \tag{35}$$

$$d[EX]/dt = k_1^*[E][X] - (k_{-1}^* + k_2^*)[EX] \approx 0 \tag{36}$$

$$d[Q]/dt = k_2[ES] \tag{37}$$

$$d[EZ]/dt \; = \; k_2{}^*[EX] \tag{38}$$

where e_o, x_o, s_o are the concentrations of enzyme, reagent, and substrate introduced into the reaction mixture.

We shall restrict ourselves by the process carried out with significant excess of X and S over the enzyme and assume that only a small part of substrate is converted to a product when complete inactivation of enzyme is achieved. This means that we may neglect [EZ] in Equation 33 and [Q] in Equation 34, and consequently x_o may be substituted for [X] and s_o for [S]. Designating

$$k_1/(k_{-1} + k_2) = \overline{K}_m \quad \text{(reversed Michaelis constant)}$$

$$k_1{}^*/(k_{-1}{}^* + k_2{}^*) = K_x$$

and using Equations 35 and 36 it is easy to transform Equation 32 to

$$[E] \; = \; \frac{e_o \, - \, [EZ]}{1 \, + \, \overline{K}_m s_o \, + \, K_x x_o}$$

Therefore

$$[EX] \; = \; \frac{K_x x_o(e_o \, - \, [EZ])}{1 \, + \, \overline{K}_m s_o \, + \, K_x x_o} \tag{39}$$

$$[ES] \; = \; \frac{\overline{K}_m s_o(e_o \, - \, [EZ])}{1 \, + \, \overline{K}_m s_o \, + \, K_x x_o} \tag{40}$$

Dividing Equation 38 by Equation 37 with the help of Equations 39 and 40 we easily come to

$$d[EZ]/d[Q] \; = \; k_2{}^* K_x x_o/k_2 \overline{K}_m s_o$$

and, therefore,

$$[EZ] \; = \; (k_2{}^* K_x x_o/k_2 \overline{K}_m s_o)[Q]$$

Thus, we come to a single differential equation

$$d[Q]/dt \; = \; \frac{k_2 \overline{K}_m s_o \, e_o \, - \, (k_2{}^* K_x x_o/k_2 \overline{K}_m s_o)[Q]}{1 \, + \, \overline{K}_m s_o \, + \, K_x x_o}$$

which is easily integrated at initial conditions $Q = 0$ at $t = 0$ giving

$$[Q] \; = \; [Q]_\infty\{1 \, - \, \exp(-\lambda t)\} \tag{41}$$

where

$$[Q]_\infty \; = \; (k_2 \overline{K}_m s_o/k_2{}^* K_x x_o)e_o \tag{42}$$

$$\lambda \; = \; k_2{}^* K_x x_o/(1 \, + \, \overline{K}_m s_o \, + \, K_x x_o) \tag{43}$$

Values k_2 and $\overline{K}_m$ are easily found from the kinetic measurements in the absence of X by traditional enzymological methods. Determining $[Q]_\infty$ and λ for a set of kinetic curves recorded for reactions performed at different x_o values one easily calculates k_2^* and K_x. For example,

$$1/\lambda = 1/k_2^* + [(1 + \overline{K}_m s_o)/k_2^* K_x](1/x_o) \tag{44}$$

and, therefore, plotting $1/\lambda$ vs. $1/x_o$ we may calculate k_2^* and K_x from the set of experiments carried out at definite s_o. If x_o and e_o are varied simultaneously without changing the x_o/e_o ratio, a set of kinetic curves with different λ with common $[Q]_\infty$ may be obtained according to Equation 44.[309]

As the method was used for proteases operating via an intermediate acyl enzyme formation, the adaptation of the above equations to this mechanism was done by Leytus et al.[349] The scheme treated by the authors may be written as follows

$$
\begin{array}{c}
E + S \underset{}{\overset{K_s}{\rightleftharpoons}} ES \xrightarrow{k_2} ES' \xrightarrow{k_3} E + Q_2 \\[4pt]
+ \qquad\qquad\qquad + \\[4pt]
X \qquad\qquad\qquad Q_1 \\[4pt]
K_x \Big\uparrow\Big\downarrow \; k_2^* \\[4pt]
EX \longrightarrow EZ
\end{array}
\tag{45}
$$

where Q_1 and Q_2 are the hydrolysis products, one formed in parallel with the acyl enzyme ES' and the other by hydrolysis of ES'. In quasiequilibrium, approximation (K_s as the association constant of E and S) with the sufficient excess of both substrate S and reagent X over enzyme and assuming that k_3 is sufficiently small, the following expression for the accumulation of Q_1 may be derived

$$[Q_1] = [Q_1]_\infty[1 - \exp(-\lambda t)]$$

where

$$[Q_1]_\infty = k_2 K_s e_o s_o/(k_2 K_s s_o + k_2^* K_x x_o)$$

$$\lambda = (k_2 K_s s_o + k_2^* K_x x_o)/(1 + K_s s_o + K_x x_o)$$

C. Simultaneous Modification of Several Independent Centers by the Same Reagent

It may happen that several different sites are present in a biopolymer with the affinity to the same reagent. If no cooperative interactions take place between these sites, the kinetics of modification measured as a total extent of modification in the presence of sufficient excess of the reagent is described by the obvious equation

$$\zeta = [PZ]/p_o = \sum_{i=1}^{N} [1 - \exp(-k_{app}^{(i)}t)] \tag{46}$$

where [PZ] is the concentration of the reagent residues attached covalently to the biopolymer, $k_{app}^{(i)}$ is the pseudo first-order rate constants of modification of i-th sites, and N is the total number of these sites. It may be easily shown that the second derivative of the function (Equation 46) is negative at any positive t value and therefore the kinetic curve is upward

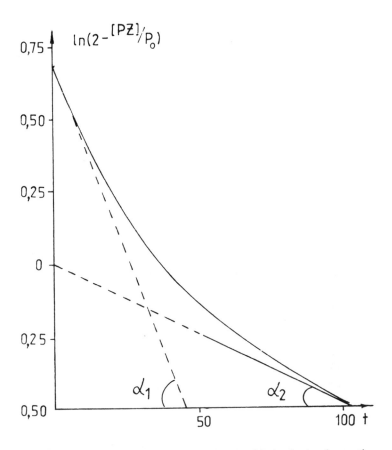

FIGURE 2. Semilogarithmic plot of kinetic curve calculated using the equation $[PZ]/p_o = \exp(-k_1 t) + \exp(-k_2 t)$ with $k_1 = 0.05$ and $k_2 = 0.005$. Initial and asymptotic slopes $\mathrm{tg}\alpha_1 = (k_1 + k_2)/2$ and $\mathrm{tg}\alpha_2 = k_2$.

convex. The expression should be tried to describe the kinetic curve of modification if the extent of modification exceeds unity or the number of identical subunits of the biopolymer. The presentation of the experimental curve as a sum of exponential terms may be done by the least-square fitting of the data. In the case of two exponential terms a rough estimate of the apparent rate constants may be performed by plotting $\ln(2 - [PZ]/p_o)$ vs. t. The initial slope of the curve gives $(k_{app}^{(1)} + k_{app}^{(2)})/2$, and the slope of asymptote gives the minimal k_{app} value (Figure 2).

In general the possibility and reliability of the presentation of experimental data in the form (Equation 46) depends on the accuracy of the measurements and the number of experimental points.

The problem is simplified if $k_{app}^{(i)}$ values for various sites differ by several orders of magnitude. In this case, the parameters of exponential terms may be estimated from experimental data obtained at different time intervals.

As an example the kinetics of alkylation of acetylcholine esterase from cobra venom (*Naja naja oxiana*) with ³H labeled *N,N*-dimethyl-2-chloro-2-[4'-³H]phenylethanolamine may be presented (Figure 3).[350] It is easily seen that two different processes take place. A calculation of $k_{app}^{(i)}$ (i = 1,2) values gives $k_{app}^{(1)} = 2.4 \times 10^{-3}$ sec^{-1}, $k_{app}^{(2)} = 3.2 \times 10^{-5}$ sec^{-1}.

The discrimination of the different types of centers becomes essentially easier if the properties have some sharp qualitative differences. This may be different depending on the affinity or reaction rate on pH, temperature, ionic strength, different sensitivity to some inhibitors, or different influence of modification on the functional properties of a biopolymer.

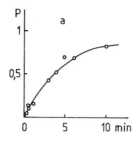
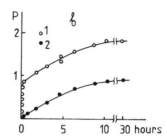

FIGURE 3.   Kinetics of affinity modification of acetylcholinesterase with *N,N*-dimethyl-2-phenylaziridinium: (a) initial period of reaction and (b) final period of reaction. Open circles represent the modification extent found by incorporation measurements using ^{3}H-labeled reagent. Full circles represent the results obtained by measuring the residual activity $(1 - A/A_o)$.

Thus, in the above-mentioned example, the first center of acetylcholine esterase being alkylated does not lose catalytic activity. Therefore, when following catalytic properties (Figure 3), the authors simply observed one-step reaction, curve 2, Figure 3b. The authors suggest that the first site subjected to alkylation resides rather far from the catalytic site.

The fulfillment of Equation 46 does not mean that we deal with affinity modification, the same is valid for a direct process without the intermediate formation of a specific complex. To discriminate between these two possibilities it is necessary to check the validity of the main criteria of affinity modification discussed in Section II.B. At least it is necessary to study the dependences of the respective $k_{app}^{(i)}$ on the reagent concentration which should obey Equations 6 or 9.

The possibility to resolve the kinetic curve of modification or inactivation of a biopolymer into several exponential terms does not prove that we really deal with several parallel independent modification processes. A number of other cases may lead to the similar appearance of the kinetic curves. For example, it was demonstrated by Fritzsch and Koepsell[351] that the bend on the semilogarithmic plot of affinity modification of several $(Na^+ + K^+)$ ATPases is better described by a two-states model

$$P \rightleftarrows P^*$$

$$P + X \rightleftarrows PX \rightarrow PZ \tag{47}$$

than by a two-sites one. According to their model, the enzyme exists in two slowly interconverted forms, P and P*, the former being able to react with X.

Another case which leads to a kinetic curve of inactivation represented by two positive exponential terms was considered by Tomich and Colman.[108] This case is derived from a two-step inactivation process proceeding via rapid formation of a partly active intermediate followed by slower transformation of the latter to a completely inactive form

$$\underset{\text{(active)}}{P} \xrightarrow{k_1} \underset{\substack{\text{(partly} \\ \text{active)}}}{P_1} \xrightarrow{k_2} \underset{\text{(inactive)}}{P_2} \quad (k_1 > k_2) \tag{48}$$

where k_1, k_2 are the apparent rate constants of two steps. Relative activity A/A_o is described as

$$A/A_o = ([P] + F[P_1])/p_o$$

where F is the ratio of activities of P_1 and P. With the excess of reagent X we deal with two consecutive first-order reactions and, therefore

$$[P] = p_o \exp(-k_1 t)$$

$$[P_1] = [p_o k_1/(k_1 - k_2)][\exp(-k_2 t) - \exp(-k_1 t)]$$

and, therefore

$$A/A_o = [1 - Fk_1/(k_1 - k_2)] \exp(-k_1 t) + [Fk_1/(k_1 - k_2)] \exp(-k_2 t) \qquad (49)$$

Obviously, both terms are positive if $Fk_1/(k_1 - k_2) < 1$ and consequently

$$F < \frac{k_1 - k_2}{k_1}$$

With $k_1 \gg k_2$, Equation 49 transforms to

$$A/A_o = (1 - F) \exp(-k_1 t) + F \exp(-k_2 t) \qquad (50)$$

This is the form used in the above-mentioned paper.

Equation 48 is a special case of a more general and reliable one considered by Ray and Koshland,[352]

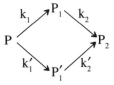

where P_1 and P_1' are two biopolymer derivatives modified at different points and P_2 is the derivative with both modified points. Assuming that activities of P_1, P_1', and P_2 are the parts F_1, F_1', and F_2, respectively, of activity of P, the authors come to the expression

$$A/A_o = F_2 + (1 - F_2) \exp[-(k_1 + k_1')t]$$
$$+ [k_1(F_1 - F_2)/(k_2 - k_1 - k_1')]\{\exp[-(k_1 + k_1')t]$$
$$- \exp(-k_2 t)\} + [k_1'(F_1 - F_2)/(k_2' - k_1 - k_1')]$$
$$\times \{\exp[-(k_1 + k_1')t] - \exp(-k_2' t)\} \qquad (51)$$

Equation 49 is easily obtained from Equation 51 by putting $k_1' = O$, $F_2 = O$, $F_1 = F$.

D. Main Sources of Kinetic Complications in Affinity Modification

The scheme put into foundation of the Kitz and Wilson equation (Section A) is only the most simple version of numerous schemes really met in various affinity modification processes. In this section we shall briefly consider the main arising complications. In the next sections the typical cases will be presented and subjected to mathematical treatment. In conclusion the general approach to the treatment of the complicated affinity modification schemes will be presented.

Equation 1 suggests that in the system under consideration there exists only one chemical conversion, namely a one-step reaction within the specific complex of a biopolymer with reagent PX. No transformation of the reagent outside the complex is taken into account. This is not the case for a great number of reagents used for modification of proteins and nucleic acids. As was emphasized in Chapter 2, Section I, both types of biopolymer contain a number of nucleophilic centers and therefore electrophilic reagents are mainly used for modification. As the process is usually performed in aqueous solution in the presence of buffers, reagents should attack water molecules as well as nucleophilic (basic) buffer components. These reactions should lead simultaneously to two complications.

The first one is the consumption of the reagent. The other complication is the transformation of reagent X in this process to nonreactive product R which retains the affinity towards the same polymer, as the address usually is not touched by this transformation. Equation 1 should be supplemented by at least two additional processes and may be written as

$$P + X \rightleftarrows PX$$

$$PX \rightarrow PZ$$

$$X \rightarrow R$$

$$P + R \rightleftarrows PR \tag{52}$$

The kinetic treatment of this scheme will be given in the next section.

Equation 52 as well as the simplest scheme (Equation 1) suggests that chemical conversion proceeds as a simple one-step reaction. As considered in Chapter 2 this is not the case for many reactions applied for the chemical modification of biopolymers. Aromatic 2-chloroethylamines primarily convert to ethyleneimmonium cation (Chapter 2, Section II.A), photoreactive compounds convert to molecules in excited states or to biradicals such as carbenes and nitrenes (Chapter 2, Section II.E), and chelating reagents in the presence of transition metals and oxidizers produce free radicals (Chapter 2, Section II.H). Different by-processes may occur with reactive intermediates.

As an example we shall consider the alkylation with aromatic 2-chloroethylamines suggesting that an intermediate ethyleneimmonium cation I either reacts with water or attacks some definite group of biopolymer within the specific complex. Even in this rather simple case several additional processes should be taken into consideration.

Intermediate I may be formed both in solution and within the complex. Being formed in solution, it may be recognized by the active site of the biopolymer producing the specific complex PI. Being produced within the complex PX, it may either attack the nucleophilic group of a biopolymer converting to PZ or dissociate off the complex. Moreover, the intermediate is not necessarily screened from the reaction with water. This process leads to the direct transformation of I to R within the complex. All these processes may be combined in the scheme.

$$
\begin{array}{ccc}
P + X & \rightleftarrows & PX \\
\downarrow & & \downarrow \\
I & \underset{-P}{\overset{+P}{\rightleftarrows}} PI & \longrightarrow PZ \\
\downarrow & & \downarrow \\
R & \underset{-P}{\overset{+P}{\rightleftarrows}} PR &
\end{array}
\tag{53}
$$

This scheme will be treated in Section H.

The second group of complications originates from cooperative processes proceeding within biopolymers and complexes subjected to affinity modification. Again, numerous cases are met in different biochemical systems.

Many biopolymers interact with several ligands, e.g., two or three substrates, substrate and effector, substrate and template in nucleic acids polymerases, etc. The presence of a ligand at one site may influence the affinity modification of the other site. The simplest scheme takes into account that the reagent binds to the biopolymer either containing or lacking the other ligand L. In general, affinity modification may proceed both within the complex containing or lacking this ligand. Assuming that the same product PZ is formed in both cases the scheme may be presented as follows:

$$
\begin{array}{ccc}
P & \underset{-L}{\overset{+L}{\rightleftharpoons}} & PL \\
+X \big\updownarrow \text{-X} & \text{-X} \big\updownarrow +X & \\
PX & \underset{-L}{\overset{+L}{\rightleftharpoons}} & PLX \\
\big\downarrow & & \big\downarrow \\
PZ & \underset{-L}{\overset{+L}{\rightleftharpoons}} & PZL
\end{array}
\tag{54}
$$

The affinity towards X may differ for P and PL. This means that some cooperative interaction exists between the binding sites for X and L. The quantitative measure of this cooperativity is the ratio of association constants of X with PL and P, K_{PLX}/K_{PX}. This type of cooperativity will be further referred to as the thermodynamic one. If $K_{PLX}/K_{PX} > 1$, we deal with positive thermodynamic cooperativity, if $K_{PLX}/K_{PX} < 1$ there is a negative cooperativity.

The rate constant within the complexes PX and PLX may differ also. The ratio of these rate constants k_{PLX}/k_{PX} will be further referred to as kinetic cooperativity, a positive one if $k_{PLX}/k_{PX} > 1$ and negative one if $k_{PLX}/k_{PX} < 1$.

The final result of modification may differ for the conversion of PX and PLX, as due to different conformation of these complexes different groups of a biopolymer may be available for the attack by the reactive group of the reagent. By analogy, we may refer to this case as structural cooperativity. The treatment of Scheme 54 will be presented in Section G.

Another type of cooperativity is observed in biopolymers consisting of several identical subunits. Cooperative interactions may occur between identical active sites residing at different subunits. The simplest scheme may be presented as follows:

$$
\begin{array}{ccccc}
P + X & \rightleftharpoons & PX & \longrightarrow & PZ \\
& +X \big\updownarrow \text{-X} & \text{-X} \big\updownarrow +X & & \\
& PX_2 & \longrightarrow & PZX & \longrightarrow & PZ_2
\end{array}
\tag{55}
$$

As in the preceding case, the affinity of X to P, PX, and PZ may differ due to thermodynamic cooperativity. Kinetic cooperativity may be revealed as a difference between rate constants of transformation within the complexes PX, PX_2, and PXZ. Certainly, the results of modification may be not identical in these three complexes. This complication is not shown in the scheme. The treatment of Scheme 55 will be carried out in Section G.

As an example of a more complicated scheme we shall present the process theoretically considered by Rakitzis.[353] The author has discussed an enzyme E with two modifiable sites A and B, the former being an active site. The modification process is presented by the scheme:

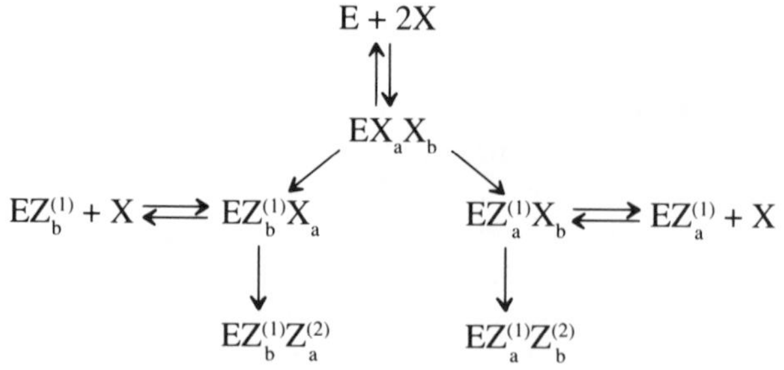

$$E + 2X$$

$$\Updownarrow$$

$$EX_aX_b$$

$$EZ_b^{(1)} + X \rightleftarrows EZ_b^{(1)}X_a \qquad EZ_a^{(1)}X_b \rightleftarrows EZ_a^{(1)} + X \qquad (56)$$

$$\downarrow \qquad\qquad\qquad \downarrow$$

$$EZ_b^{(1)}Z_a^{(2)} \qquad\qquad EZ_a^{(1)}Z_b^{(2)}$$

where X_a and X_b notify the reagent bound noncovalently to site A or B; $Z_a^{(1)}$ and $Z_a^{(2)}$ notify the covalently bound reagent in the site A within two different complexes, EX_aX_b and $EZ_b^{(1)}X_a$; and similarly $Z_b^{(1)}$ and $Z_b^{(2)}$ notify the covalently bound reagent in the site B within complexes EX_aX_b and $EZ_a^{(1)}X_b$. Mathematical treatment of this scheme as well as another similar one may be found in the original papers.[353,354]

E. Affinity Modification with Unstable Reagents

The kinetic treatment of the affinity modification with an unstable reagent was performed at first by Purdie and Heggie.[355] The same derivation was later repeated by Jacobson and Colman.[356] In both papers the latter process of Scheme 52 was not taken into account. The more complete treatment was carried out by Knorre and Chimitova.[357] In the latter paper, Scheme 52 was supplemented by the presence of some unreactive competitor Y. The complete scheme with kinetic parameters and association constants used may be written as follows:

$$P + X \rightleftarrows PX(K_x) \qquad PX \rightarrow PZ(k_1)$$

$$X \rightarrow R(k_2) \qquad P + R \rightleftarrows PR(K_r)$$

$$P + Y \rightleftarrows PY(K_y) \qquad (57)$$

The derivation of the kinetic equation was performed in quasiequilibrium approximation under the assumption that X and Y are taken in concentrations x_o and y_o significantly exceeding the initial concentration of biopolymer p_o. In this case, the main way of consumption of X is hydrolysis in the pseudo first-order process and concentration x as a function of time t is described as

$$x = x_o \exp(-k_2 t)$$

The mass conservation equation for biopolymer

$$[P] + [PX] + [PY] + [PR] + [PZ] = p_o$$

is easily transformed to

$$[P](1 + K_x x + K_y y_o + K_r r) = p_o - [PZ]$$

where r is the concentration of R

$$r = x_o - x$$

As

$$d[PZ]/dt = k_1[PX] = k_1 K_x [P]x$$

the final kinetic equation may be written as follows:

$$\frac{d[PZ]}{dt} = k_1 K_x x_o \exp(-k_2 t) \frac{P_o - [PZ]}{1 + K_r x_o + K_y y_o + (K_x - K_r)x_o \exp(-k_2 t)} \tag{58}$$

Integration of this differential equation permits to express the kinetics of accumulation of the modification product PZ as follows:

$$\frac{[PZ]}{p_o} = 1 - \left[\frac{1 + K_x x_o + K_y y_o}{1 + K_r x_o + K_y y_o + (K_x - K_r)x_o \exp(-k_2 t)}\right]^{(k_1/k_2)/(K_r/K_x - 1)} \tag{59}$$

The expression is not valid at $K_x = K_r$ and in this case turns to

$$[PZ]/p_o = 1 - \exp\left[-\frac{k_1 K_x x_o}{k_2} \frac{1 - \exp(-k_2 t)}{1 + K_x x_o + K_y y_o}\right] \tag{60}$$

An expression similar to Equation 59 in the absence of Y was derived by Palumaa et al.[207] Their equation may be obtained from Equation 59 putting $y_o = 0$. The equation was applied to calculate the kinetic parameters and association constants of alkylation of the horse serum acylcholine acylhydrolase (EC 3.1.1.8) with N,N-dimethyl-2-phenylaziridinium ion.[207] The set of $[PZ]_{ij}$ values corresponding to different initial concentrations of the reagent $(x_o)_i$ at times t_j was treated by nonlinear least-square regression program using the function:

$$\ln \frac{p_o}{p_o - [PZ]_{ij}} = B_o \ln \frac{B_1/(x_o)_i + 1}{B_1/(x_o)_i + B_2 + (1 - B_2)B_3^{t_j}} \tag{61}$$

where $B_o = (k_1/k_2)(1 - K_r/K_x)$, $B_1 = 1/K_x$, $B_2 = K_r/K_x$, and $B_3 = \exp(-k_2)$. Parameters B_i found by a minimization procedure permitted to estimate k_1, k_2, K_x, and K_r. At 25°C in a phosphate buffer pH 7.5 and ionic strength 0.15 M with $(x_o)_i$ taken in the range 0.104 to 0.766 M, the following values were obtained: $K_x = 1.01 \times 10^4$ M^{-1}, $K_r = 2.94 \times 10^3$ M^{-1}, $k_1 = 1.02 \times 10^{-4}$ sec^{-1}, and $k_2 = 2.01 \times 10^{-4}$ sec^{-1}. The accuracy of the data (SDs within 1 to 10%) proves the validity of the equation applied. The same authors have applied the full equation (Equation 59) to estimate the affinity of the cationic inhibitors to acylcholine acylhydrolase.[358] The kinetic curves of alkylation of the enzyme with N,N-dimethyl-2-phenylaziridinium ion in the presence of different competitors were obtained. Using the function shown in Equation 61 with the same significance of B_o, B_2, B_3, and $B_1 = (K_r + y_o)/K_x K_r$, the authors have calculated from these data by nonlinear least-square regression program K_y values for several competitors. These values are in good agreement with the K_i values indicated in the brackets for the same compounds found in the study of competitive inhibition of acetylthiocholine hydrolysis by the enzyme. The $K_y(K_i)$ values were found to be for tetramethylammonium chloride: 4.28×10^{-3} (4.36×10^{-3}) M^{-1}, for tetraethylammonium chloride: 2.87×10^{-3} (2.84×10^{-3}) M^{-1}, and for N-methylacridinium iodide 1.85×10^{-7} (1.15×10^{-7}) M^{-1}.

According to Equation 1, the modification extent of a biopolymer reaches 100% at any concentration of reagent equal or exceeding the molar concentration of the biopolymer. On the contrary, Scheme 57 leads to incomplete final modification. The extent of modification at $t = \infty$, $[PZ]_\infty/p_o$ is

$$\frac{[PZ]_\infty}{p_o} = 1 - \left[\frac{1 + K_x x_o + K_y y_o}{1 + K_r x_o + K_y y_o}\right]^{(k_1/k_2)/(K_r/K_x - 1)} \tag{62}$$

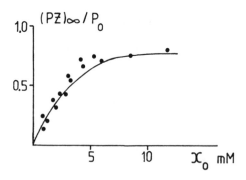

FIGURE 4. The dependence of the limit modification extent measured via an inactivation level in the enzymatic aminoacylation ($[PZ]_\infty/p_o = 1 - A/A_o$) on the initial concentration of reagent x_o for the reaction of *E. coli* phenylalanine-tRNA-synthetase with adenosine-5′-trimetaphosphate.

if $K_x \neq K_r$ and

$$\frac{[PZ]_\infty}{p_o} = 1 - \exp[-(k_1/k_2)K_x x_o/(1 + K_x x_o + K_y y_o)] \tag{63}$$

if $K_x = K_r$.

Equations 62 and 63 permit calculation of K_x, K_r, K_y, and k_1/k_2 from the experimentally found dependence of $[PZ]_\infty$ or residual activity A_∞ on x_o and y_o. This dependence may be obtained without kinetic experiments simply by measuring the final result of modification. Therefore, the above expressions may be especially useful if the time course of modification cannot be followed, e.g., due to a high reactivity of X.

As an example of an application of Equation 62 we present the data obtained for the inactivation of phenylalanyl-tRNA-synthetase from *Escherichia coli* with the ATP analog adenosine-5′-trimetaphosphate (Chapter 2, XXXII, B = Ade). The experimental data are presented in Figure 4. The solid line is calculated using Equation 62 with the parameters $K_x = 10\,M^{-1}$, $K_r = 10^3\,M^{-1}$, $k_1/k_2 = 61.2$, obtained from experimental data by a nonlinear least-square minimization procedure. As the reagent is hydrolyzed to ATP, K_r represents the affinity of ATP to the enzyme which is found to be significantly greater than the affinity of adenosine-5′-trimetaphosphate.[359]

The equation of Purdie and Heggie[355] neglecting the affinity of the product of hydrolysis of the reagent to the biopolymer is easily obtained from Equation 59 putting $K_r = 0$, $y_o = 0$

$$[PZ]/p_o = 1 - \{(1 + K_x x_o)/[1 + K_x x_o \exp(-k_2 t)]\}^{-k_1/k_2} \tag{64}$$

If the residual activity A/A_o of the biopolymer is measured instead of [PZ], Equation 64 turns to

$$A/A_o = \{(1 + K_x x_o)/[1 + K_x x_o \exp(-k_2 t)]\}^{-k_1/k_2} \tag{65}$$

Still more simplified forms of the Equation 65 were presented by Rakitzis.[360] For the activity A_∞ remaining after completion of the reaction in the case $K_x x_o \gg 1$ it may be written

$$\ln(A_o/A_\infty) = (k_1/k_2) \ln x_o + (k_1/k_2) \ln K_x \tag{66}$$

Table 1
PARTITION RATIOS k₃/k₄ FOR SOME
SUICIDE AFFINITY REAGENTS[156-158,169]

Reagent	Target enzyme	k_3/k_4
Propargylglycine	Methionine-γ-lyase	4—8
Vinylglycine	Aspartate aminotransferase	<1
	D-Amino acid transaminase	800
	L-Amino acid oxidase	2000
β-D-Fluoroalanine	Alanine racemase	800
Clavulanic acid	β-Lactamase	150
Penicillanic acid sul-fone	β-Lactamase	4500

If, on the contrary, $K_x x_o \ll 1$, Equation 65 turns to

$$\ln(A_o/A) = (k_1 K_x x_o/k_2)[1 - \exp(-k_2 t)] \tag{67}$$

In these conditions, as mentioned in Section A, the reaction proceeds as a second-order process with an apparent second-order rate constant $k_{app} = k_1 K_x$ (Equation 11). The plot $\ln(A/A_o)$ vs. $[1 - \exp(-k_2 t)]/k_2$ was proposed by Topham[361] to measure $k_1 K_x$. At infinite time[353,362]

$$\ln(A_o/A_\infty) = (k_{app}/k_2)x_o \tag{68}$$

A specific case of Equation 59 was considered in connection with the study of inactivation of acetylcholine esterase by phenylmethane sulfonylfluoride. The affinity of the reagent is very low if at all and the second-order rate constant $k'_1 = k_1 K_x$ should be introduced in accordance with Equation 11. However, a fluoride anion formed is a strong competitive inhibitor and, therefore, $K_r \neq 0$. Neglecting $K_x x_o$ as compared with unity and putting $y_o = 0$ (any additional ligand is absent) we may easily transform Equation 59 to the expression:[363]

$$\ln(1 - [PZ]/p_o) = - (k'_1 K_r/k_2) \ln\{1 + K_r x_o[1 - \exp(-k_2 t)]\} \tag{69}$$

F. Kinetics of Suicide Inhibition
The simplest scheme of inhibition by suicide substrates was treated by Waley.[364] This scheme may be presented as follows:

$$E + X \underset{k_{-1}}{\overset{k_1}{\rightleftarrows}} EX \overset{k_2}{\longrightarrow} EI \overset{k_3}{\longrightarrow} E + Q \tag{70}$$
$$\downarrow k_4$$
$$EZ$$

where I, reactive intermediate formed by enzymatic transformation of suicide substrate X; Q, the final product of enzymatic conversion of X; and EZ, inactivated enzyme. As both the product Q and the inactivated enzyme form from the common precursor, the ratio k_3/k_4 characterizes the average number of catalytic acts carried out by the enzyme prior to inactivation. This value may change from zero (inactivation without any catalytic conversion of suicide substrate) to significant values (Table 1).

Assuming that steady-state approximation is valid for the intermediate complexes PX and PI, we easily obtain the system of two differential and four algebraic equations describing the kinetic behavior of the system

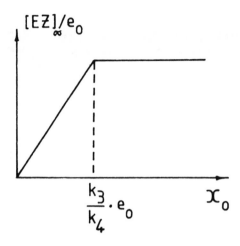

FIGURE 5.   Dependence of the relative limit yield of affinity modification on the suicide reagent concentration.

$$[E] + [EX] + [EI] + [EZ] = e_o$$

$$[X] + [EX] + [EI] + [EZ] + [Q] = x_o$$

$$d[EX]/dt = k_1[E][X] - (k_{-1} + k_2)[EX] = 0$$

$$d[EI]/dt = k_2[EX] - (k_3 + k_4)[EI] = 0$$

$$d[Q]/dt = k_3[EI]$$

$$d[EZ]/dt = k_4[EI] \tag{71}$$

Dividing the last equation by penultimate one and integrating at initial conditions [Q] = 0 at [PZ] = 0 we easily find that

$$[EZ] = (k_4/k_3)[Q] \tag{72}$$

In particular at t = ∞

$$[EZ]_\infty = (k_4/k_3)[Q]_\infty$$

It is evident that the reaction with a suicide substrate continues up to complete exhaustion of either enzyme or substrate. If the substrate is completely consumed at the end of the reaction $[EZ]_\infty = (k_4/k_3)x_o$. If the enzyme is thoroughly inactivated prior to exhaustion of the substrate $[EZ]_\infty = e_o$, $[Q]_\infty = (k_3/k_4)e_o$. Therefore, studying the inactivation of the enzyme as a function of x_o we obtain two straight lines.

$$[EZ]_\infty = (k_4/k_3)x_o \qquad (x_o < k_3e_o/k_4)$$

$$[EZ]_\infty = e_o \qquad (x_o > k_3e_o/k_4) \tag{73}$$

The two straight lines are connected by a sharp bend at $x_o = k_3e_o/k_4$ (Figure 5). This appearance of the dependence is typical for suicide substrates and differs strongly from similar dependences for other mechanisms (compare, e.g., the dependence presented in Figure 4).

Using a steady-state approximation under the assumption that steady-state concentrations [EX] and [EI] may be neglected in the mass conservation equations, we come to the kinetic equation for the accumulation of an inactivated enzyme

$$\frac{d[EZ]}{dt} = k_{app}K'(e_o - [PZ]) \frac{x_o - (1 + r)[EZ]}{1 + K'\{x_o - (1 + r)[EZ]\}} \tag{74}$$

where

$$k_{app} = k_2k_4/(k_2 + k_3 + k_4)$$

$$K' = [k_1/(k_{-1} + k_2)] \times [(k_2 + k_3 + k_4)/(k_3 + k_4)]$$

$$r = k_3/k_4$$

Integration under the initial condition $[PZ] = 0$ at $t = 0$ results in the expression

$$\frac{1}{k_{app}K'[x_o - (1 + r)e_o]} \ln \left\{ 1 - \frac{(1 + r)[EZ]}{x_o} \right\}$$

$$- \left\{ \frac{1}{k_{app}K'[x_o - (1 + r)e_o]} + \frac{1}{k_{app}} \right\} \ln \left(1 - \frac{[EZ]}{e_o} \right) = t \tag{75}$$

The time course of the reaction described by Equation 75 permits the independent parameters k_{app}, K', and r to be calculated by nonlinear regression analysis. More accurate data may be obtained using an approach recently proposed by Waley.[364]

Substituting in Equation 75 $t_{0.5}$ (half time of the enzyme inactivation) for t and $e_o/2$ for [EZ] it is easy to transform Equation 75 to the expression:

$$\frac{1}{k_{app}K'[1 - (1 + r)e_o/x_o]} \ln[2 - (1 + r)e_o/x_o] + \frac{\ln 2}{k_{app}} x_o = t_{0.5}x_o \tag{76}$$

If $t_{0.5}$ is measured as a function of x_o in a set of experiments performed at the constant e_o/x_o, $t_{0.5}x_o$ is a linear function of $t_{0.5}x_o$. Provided r is determined in the separate experiment as a ratio of the amounts of the enzymatic reaction product and inactivated enzyme, it is easy to calculate k_{app} and K' from the linear plot of $t_{0.5}x_o$ vs. x_o.

Using Equation 72, an equation similar to Equation 75 may be written for the accumulation of the product of enzymatic reaction Q.

$$\frac{1}{k_{app}K'[x_o - (1 + r)e_o]} \ln \left\{ 1 - \frac{1 + r}{r} \frac{[Q]}{x_o} \right\}$$

$$- \left\{ \frac{1}{k_{app}K'[x_o - (1 + r)e_o]} + \frac{1}{k_{app}} \right\} \ln \left(1 - \frac{[Q]}{re_o} \right) = t \tag{77}$$

Computer simulation of the system of differential equations for [X], [PX], [PI], [PZ], and [Q] without steady-state approximation with some reasonable sets of parameters has demonstrated rather good agreement of approximate solutions given by either Equations 75 or 77 for the time course of the suicide inactivation of the enzyme with exact solutions obtained by numerical integration.[365]

G. Cooperative Effects in Affinity Modification

The influence of some ligand L on the affinity modification of a biopolymer P with a reagent X is described by Scheme 54. If the final result of modification does not depend on the presence of the ligand, the scheme may be presented as:

$$
\begin{array}{ccc}
& K_x & k \\
P \rightleftharpoons & PX & \longrightarrow PZ \\
K_l \updownarrow & \quad \updownarrow \alpha K_l & \\
PL \rightleftharpoons & PXL \longrightarrow PZ \\
\alpha K_x & k\beta &
\end{array}
\tag{78}
$$

where K_x, K_l are the association constants for X and L, k is the rate constant in the complex PX, α is the factor of thermodynamic cooperativity, β is the factor of kinetic cooperativity. The same scheme holds if different points of the polymer are modified in the PX and PXL complexes but only the total extent of modification is followed. In this scheme, as compared with Scheme 54, we omit the existence of PZL complexes which are in rapid equilibrium with PZ. Quasiequilibrium treatment of the scheme results in the kinetic equation:[344,366]

$$
v = \frac{d[PZ]}{dt} = \frac{kK_x[X](1 + \alpha\beta K_l[L])(p_o - [PZ])}{1 + K_x[X] + K_l[L] + \alpha K_x K_l[X][L]}
\tag{79}
$$

Under the assumption that X and L are present in the sufficient excess over P, [X] and [L] may be considered as nearly constant and equal to initial concentrations (x_o, l) and the equation of pseudo first-order reaction holds

$$
\frac{d[PZ]}{dt} = k_{app}(p_o - [PZ])
$$

where

$$
k_{app} = \frac{kK_x x_o(1 + \alpha\beta K_l l)}{1 + K_x x_o + K_l l + \alpha K_x K_l x_o l}
\tag{80}
$$

(Comparing the appearance of this equation with that given by Dahl and McKinley-McKee[366] one should take into account that these authors used dissociation constants whereas association constants are used in this book.)

As an example of a process proceeding in accordance with this scheme, affinity modification of horse liver alcohol dehydrogenase either with iodoacetate or with 3-(5-imidazolyl)-2-bromopropionate may be presented. Binding of these reagents takes place at an anion-specific center of the enzyme formed by arginine residue Arg-47. Cys-46 coordinated to Zn^{2+} is alkylated by the reagents. A great number of nitrogen- and sulfur-containing compounds are found to influence the reaction due to coordination to the same Zn^{2+}.[367,368] Kinetic measurements and treatment of the results by Equation (80) permitted many compounds by α and β values to be characterized. The same was carried out for a set of sulfonamides used against alcohol poisoning due to the ability to penetrate the blood-brain barrier.[369]

If modification proceeds at a saturating concentration of the reagent, Equation 80 turns to

$$
k_{app} = \frac{k(1 + \alpha\beta K_l l)}{1 + \alpha K_l l}
\tag{81}
$$

Therefore, in this case k_{app} changes with the concentration of effector L from $k_{app} = k$ at $l = 0$ to $k_{app} = k\beta$ at $l = \infty$. The concentration of L corresponding to a half of the change, i.e., $k_{app}/k = (1 + \beta)/2$, is easily found as

$$l_{1/2} = 1/\alpha K_l$$

This means that the affinity of the ligand estimated by conventional procedure as a reversed concentration corresponding to a half of the complete change of k_{app} as a function of the ligand concentration is the affinity in the presence of the reagent and should differ from the affinity measured in the absence of this reagent.[344]

Thus, it was found that the affinity modification of *E. coli* phenylalanyl-tRNA-synthetase with the *E. coli* Phe-tRNA^{Phe} derivative chlorambucilyl-Phe-tRNA^{Phe} is suppressed in the presence of phenylalanine. However, the concentration range of the effect is two or three orders of magnitude higher than the K_m value for phenylalanine in the enzymatic aminoacylation of tRNA^{Phe} as well as the dissociation constant of the enzyme-phenylalanine complex measured by equilibrium dialysis. This means that the presence of chlorambucilyl-Phe-tRNA significantly decreases the affinity of the amino acid to the enzyme.[370]

Scheme 55 was treated in connection with the study of the alkylation reaction of *E. coli* phenylalanyl-tRNA-synthetase with another analogue, *N*-bromoacetyl-Phe-tRNA^{Phe}, taking into account only thermodynamic and kinetic cooperativity.[371] We shall present below a more complicated scheme considered by the same authors, keeping in mind that cooperativity may result in the change of the modification point.[372]

The above reaction proceeds with β subunits of the $\alpha_2\beta_2$ enzyme and therefore a polymer with two sets of identical active sites should be considered. Therefore, statistical factors must be introduced in kinetic and thermodynamic constants. Thus, dimeric P may bind reagent X at two equal points and, consequently, the association constant must be written as $2K_x$. Similarly, the intracomplex reaction in EX_2 may proceed within each of the two subunits and it is reasonable to designate a respective rate constant as $2\beta k$. Dissociation of PX_2 may occur at each of two subunits and, therefore, the association constant must be presented as $\alpha K_x/2$. The derivation was performed under the assumption that the reagent molecule X noncovalently bound to the subunit and reagent moiety Z covalently attached to a subunit influence all three types of cooperativity in a similar way. By $Z^{(1)}$ and $Z^{(2)}$ the reagent moieties are designated formed, respectively, in the course of the reaction within complex EX with the other subunit being vacant and within complexes EX_2, $EZ^{(1)}X$, and $EZ^{(2)}X$ with the second subunit occupied by either covalently or noncovalently bound reagent. The scheme is written as follows:

$$
P \underset{2K_x}{\rightleftarrows} PX
\begin{cases}
\overset{k}{\nearrow} PZ^{(1)} \overset{\alpha K_x}{\rightleftarrows} PZ^{(1)}X \overset{\beta k}{\longrightarrow} PZ^{(1)}Z^{(2)} \\
\underset{\alpha K_x/2}{\searrow} PX_2 \overset{2\beta k}{\longrightarrow} PZ^{(2)}X \overset{\beta k}{\longrightarrow} PZ^{(2)}Z^{(2)} \\
\qquad\qquad\quad \updownarrow \alpha K_x \\
\qquad\qquad\quad PZ^{(2)}
\end{cases}
\tag{82}
$$

The scheme was treated in quasiequilibrium approximation. This means that consumption of either PX or PX_2 is followed by rapid redistribution among P, PX, and PX_2 and only the expression for the consumption rate of total concentration $P_t = [P] + [PX] + [PX_2]$ can be written. The same holds for $PZ_t^{(1)} = [PZ^{(1)}] + [PZ^{(1)}X]$ and for $PZ_t^{(2)} = [PZ^{(2)}] +$

$[PZ^{(2)}X]$. At the same time, the fractions f of the reacting forms should be taken into account in kinetic equations, for example:

$$-\frac{dP_t}{dt} = kf_{PX}P_t + 2\,\beta kf_{PX_2}P_t$$

These fractions are readily found from the equilibrium conditions. For example:

$$\frac{[PZ^{(1)}X]}{[PZ^{(1)}][X]} = \alpha K_x \qquad [PZ^{(1)}] + [PZ^{(1)}X] = [PZ_t^{(1)}]$$

and, therefore,

$$f_{PZ^{(1)}X} = \alpha K_x[X]/(1 + \alpha K_x[X])$$

whereupon

$$\frac{d[PZ^{(1)}Z^{(2)}]}{dt} = \alpha\beta kK_x[X]/(1 + \alpha K_x[X])$$

In a similar way three other kinetic equations may be derived. The remaining set of differential equations describing the kinetic behavior of all enzymatic forms may be written as

$$\frac{d[PZ_t^{(1)}]}{dt} = \frac{2K_x k[X]}{1 + 2K_x[X] + \alpha K_x^2[X]^2}[P_t] - \frac{2\alpha\beta kK^2[X]^2}{1 + 2K_x[X] + \alpha K_x^2[X]^2}[PZ_t^{(1)}]$$

$$\frac{d[PZ_t^{(2)}]}{dt} = \frac{2\alpha\beta kK^2[X]^2}{1 + 2K_x[X] + \alpha K_x^2[X]^2}[P_t] - \frac{\alpha\beta kK_x[X]}{1 + \alpha K_x[X]}[PZ_t^{(2)}]$$

$$\frac{d[PZ_t^{(2)}Z^{(2)}]}{dt} = \frac{\alpha\beta kK_x[X]}{1 + \alpha K_x[X]}[PZ_t^{(2)}]$$

With sufficient excess of X over P we obtain a system of linear uniform differential equations of first order with constant coefficients which may be easily integrated. In a general case, only a numerical solution is possible.

Using these equations, the expression for the initial reaction rate v_o may be easily obtained as

$$v_o = \left(\frac{d[PZ_t^{(1)}]}{dt}\right)_{t=0} + \left(\frac{d[PZ_t^{(2)}]}{dt}\right)_{t=0} = \frac{2kK_x P_t[X]_o(1 + \alpha\beta K_x[X]_o)}{1 + 2K_x[X]_o + \alpha K_x^2[X]_o^2} \tag{83}$$

Here $[X]_o$ is equal to the total initial concentration of X, x_o if a reagent is present in sufficient excess. Otherwise, it should be expressed via x_o and p_o using the mass conservation law and the quasiequilibrium condition applied to the very beginning of the process.

$$[P]_o + [PX]_o + [PX_2]_o = p_o$$

$$[X]_o + [PX]_o + 2[PX_2]_o = x_o$$

$$\frac{[PX]_o}{[X]_o[P]_o} = 2K_x \qquad \frac{[PX_2]_o}{[X]_o[PX]_o} = \alpha K_x/2 \tag{84}$$

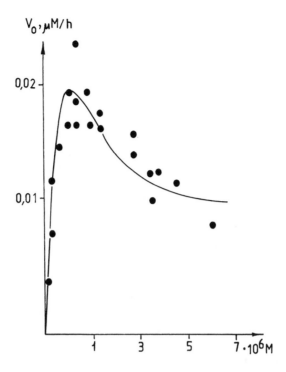

FIGURE 6. The dependence of the initial rate of alkylation of *E. coli* phenylalanine-tRNA-synthetase with *N*-bromoacetylphenylalanyl-tRNA on the initial concentration of the reagent. The solid line is calculated using Equation 83 with parameters $K_x = 7.15 \ \mu M^{-1}$, $k = 4.00 \times 10^{-2} \ hr^{-1}$, $\alpha = 0.65$, $\beta = 0.14$.

The dependence of v_o on $[X]_o$ may be of different appearance according to α and β values. Thus, at $\beta > 0.5$, $\alpha\beta < 2$, it is similar to hyperbolic, at $\beta > 0.5$, $\alpha\beta > 2$, it is sigmoid, at $\beta < 0.5$, $\alpha\beta < 2$, there exists a maximum, and at $\beta < 0.5$, $\alpha\beta > 2$, both the maximum and inflection point at the rising part of the curve exist. The equation was applied to the dependence of the alkylation rate of phenylalanyl-tRNA-synthetase with *N*-bromacetyl-Phe-tRNA^{Phe} and the parameters were estimated by minimizing a function

$$\chi^2 = \sum \left\{ \frac{v_e([X]_{oi}) - v_c([X]_{oi})}{v_e([X]_{oi})} \right\}^2$$

where v_e and v_c are the experimental and calculated (with the help of Equation 83) velocity values corresponding to the initial concentration of free reagent $[X]_{oi}$ in the i-th experiment, calculated by Equation 84 from the total initial concentrations of X and P, x_{oi} and p_{oi}, used in this experiment. The parameters were found to be $K_x = 7.15 \pm 1.9 \ \mu M^{-1}$, $k = (4.00 \pm 0.15) \times 10^{-2} \ hr^{-1}$, $\alpha = 0.65 \pm 0.17$, and $\beta = 0.14 \pm 0.03$. In accordance with condition $\beta < 0.5$, $\alpha\beta < 2$ there exists a maximum (Figure 6).

H. General Rules for Kinetic Treatment of Complicated Affinity Modification Processes

Concluding this part of the chapter we would like to summarize general rules for the derivation of kinetic equations suitable to the schemes of any level of complexity. We shall illustrate these rules by considering affinity modifcation by reagents forming reactive intermediates in the rate-limiting step.

1. After some scheme is proposed for the process it is necessary to determine the number of independent stages which in general should be described by independent differential equations. This number is equal to the number of reactions involved in the scheme minus the number of independent cycles. Thus, considering Scheme 54, we need five differential equations as we have seven reactions and two cycles in the reaction graph.

2. The second step of consideration is the estimation of the number of mass conservation equations which give algebraic expressions connecting the concentrations of the reaction participants. This number is the whole number of components in the scheme minus the number of independent differential equations. In Scheme 54 eight particles participate, namely P, X, L, PX, PL, PXL, PZ, and PZL. Therefore, we need three mass conservation equations which in this case are rather obvious as they describe the conservation of the sums of P and the products, X and the products, and L.

3. The rapid equilibrium stages should be determined. These stages are characterized by equilibrium constants. Each independent equilibrium is described by an algebraic equation which substitutes for one differential equation. All other stages should be characterized by rate constants. Statistical factors should be introduced if necessary (see Section G).

4. If two or more components are suggested to exist in quasiequilibrium it is not correct to describe the accumulation or consumption by separate kinetic equations. Quasiequilibrium suggests that instantaneous redistribution accompanies both formation and disappearance of either of participants of rapid equilibrium. Thus, for Scheme 54 it is not correct to use d[PZ]/dt or d[PZL]/dt as separate values. Only the sum of these derivatives may be considered and a respective kinetic equation should be presented as

$$d([PZ] + [PZL])/dt = k[PX] + k\beta[PXL]$$

(notations of the rate constants are taken from Scheme 78). At the same time expressions $d[PZ]/dt = k[PX]$ and $d[PZL]/dt = k\beta[PXL]$ are not correct. It is convenient to introduce for the intermediate calculations "supercomponents" representing all components connected by the rapid equilibrium process. Thus, describing Scheme 54 we may introduce supercomponent P_t representing P, PL, PX, and PLX, and supercomponent PZ_t representing PZ and PZL. Concentrations of participants of each supercomponent may be easily expressed via concentration of the latter using equilibrium conditions. For example,

$$[P_t] = [P] + [PL] + [PX] + [PLX]$$

$$\frac{[PL]}{[P][L]} = K_l \qquad \frac{[PX]}{[P][X]} = K_x \qquad \frac{[PLX]}{[PL][X]} = \alpha K_x$$

and, therefore,

$$[P] = [P_t]/(1 + K_l[L] + K_x[X] + \alpha K_x K_l[X][L])$$

$$[PX] = [P_t]K_x[X]/(1 + K_l[L] + K_x[X] + \alpha K_x K_l[X][L])$$

$$[PL] = [P_t]K_l[L]/(1 + K_l[L] + K_x[X] + \alpha K_x K_l[X][L])$$

$$[PLX] = [P_t]\alpha K_l K_x[L][X]/(1 + K_l[L] + K_x[X] + \alpha K_x K_l[X][L])$$

Similarly,

$$[PZ] = [PZ_t]/(1 + K_{zl}[L])$$

$$[PZL] = [PZ_t]K_{zl}[L]/(1 + K_{zl}[L])$$

where K_{zl} is the association constant of the process

$$PZ + L \rightleftarrows PZL$$

This permits the single differential equation describing the kinetic behavior of the system

$$\frac{d[PZ_t]}{dt} = k[PX] + k\beta[PXL] = \frac{k[P_t](K_x[X] + \alpha\beta\,K_xK_l[X][L])}{1 + K_l[L] + K_x[X] + \alpha K_xK_l[X][L]}$$

Taking into account that $[P_t] = p_o - [PZ_t]$ we obtain the expression identical to Equation 79.

5. Reactive intermediates assumed to be present in steady-state concentrations should be defined. Respective differential equations are transformed to algebraic ones. When using a steady-state approximation, one should keep in mind that steady-state concentrations of intermediates are very low and may be neglected in the mass conservation equations. The approximation simplifies the whole set of equations describing the system. At the same time, steady-state conditions result in the decrease in the number of independent kinetic parameters participating in the quantitative description of the system. This general statement is well illustrated by the steady-state version of the Kitz and Wilson equation where neither k_1 nor k_{-1} participate as separate parameters but only $K_x = k_1/(k_{-1} + k_2)$ as indicated in Equation 12. The latter statement is obviously true for a quasiequilibrium approximation where equilibrium constants are changed for the rate constants of direct and reversed reaction.

6. The system of equations obtained thus consists of the mass conservation equations, algebraic equations describing quasiequilibrium and steady-state conditions, and the remaining differential equations. The number of these equations is equal to the number of components or supercomponents participating in the scheme. As not all of them are of interest to the investigator, some of concentrations may be excluded thus lowering the number of variables and equations being subjected to further mathematical treatment.

7. The following treatment depends on the problem to be solved. The simplest is the solution of a direct problem which consists in the elucidation of the time course of affinity modification proceeding, according to the scheme, using the same data or assumptions concerning the numerical values of kinetic parameters. For a complicated scheme, numerical integration of differential equations is usually required. The reversed problem is the estimation of kinetic parameters providing for the best fit of experimental data to the scheme under consideration. Numerical minimization procedures are used in this case. At last, the choice between several hypothetic schemes often must be made. In this case some quantitative values characterizing the quality of the description by each of the considered schemes must be compared. Usually the used value is the SD of the experimental data from the theoretical curve calculated with the use of optimized parameters. The scheme providing the least SD is considered as the most probable.

To illustrate the whole set of derivation procedures we present the treatment of Scheme 53 describing affinity modification with the reagent forming the reactive intermediate I in the rate-limiting step

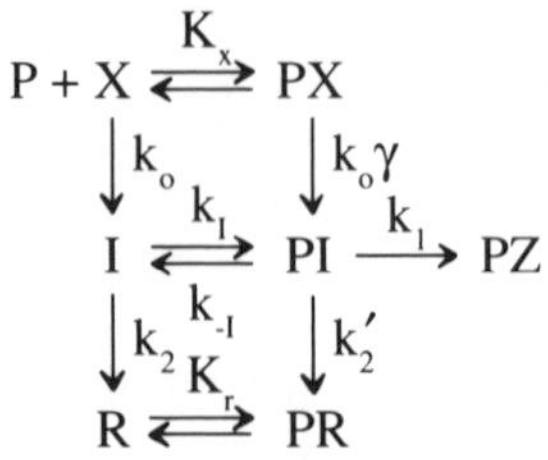

$$
\begin{array}{ccc}
& K_x & \\
P + X & \rightleftharpoons & PX \\
\downarrow k_o \quad k_1 & & \downarrow k_o\gamma \quad k_1 \\
I & \rightleftharpoons & PI \xrightarrow{\;k_1\;} PZ \\
& k_{-1} & \\
\downarrow k_2 \quad K_r & & \downarrow k_2' \\
R & \rightleftharpoons & PR
\end{array}
\qquad (85)
$$

This scheme was considered by one of the authors with collaborators in connection with the analysis of complementary addressed alkylation of nucleic acids with oligonucleotide derivatives bearing an aromatic 2-chloroethylamino group.[373,374] As told in Chapter 2, Section II.A, alkylation proceeds via the intermediate formation of a respective ethyleneimmonium cation. In this case P is the nucleic acid, X is the reactive derivative of an oligonucleotide complementary to some region of nucleic acid, I is the intermediate cation, and R is the product of hydrolysis of the reactive group, i.e., the 2-hydroxyethylamino derivative of the oligonucleotide. It is assumed that the rate constant of the rate-limiting step in the complex may differ from that in solution by a factor of γ. In the above-mentioned papers, k_2' was neglected and an additional unreactive competitor was introduced in the scheme and, therefore, the final expressions are slightly different.

It is seen that the scheme contains eight reactions and two independent cycles, and consequently six kinetic equations are necessary to describe the process. The number of components is also eight and, according to the second rule, the number of mass conservation equations is two. We shall suggest that P, X, and PX as well as P, R, and PR are present in quasiequilibrium concentrations and that the steady-state approximation may be applied to reactive intermediates I and PI. These assumptions result in the substitution of four algebraic equations for differential ones and only two differential equations remain in the final system of equations describing the kinetic behavior of the reaction components.

Prior to the presentation of the complete set of equations we shall introduce supercomponents X_t represented by X and PX and R_t represented by R and PR. The concentrations will be designated as x and r, respectively.

$$
x = [X] + [PX] \qquad r = [R] + [PR] \qquad (86)
$$

As the quasiequilibrium conditions are represented by the expressions

$$
[PX]/[P][X] = K_x
$$

$$
[PR]/[R][P] = K_r
$$

concentrations of [X] and [PX] may be written as

$$
[X] = x/(1 + K_x[P]) \qquad [PX] = K_x x[P]/(1 + K_x[P]) \qquad (87)
$$

and concentration of [PR] as

$$
[PR] = K_r r[P]/(1 + K_r[P]) \qquad (88)
$$

Six equations describing the kinetic behavior of concentrations of six components and supercomponents [P], [I], [PI], [PZ], x, and r may now be written as follows:

1. Mass conservation equations

$$[P] + [PX] + [PR] + [PZ] = p_o$$

$$[X] + [PX] + [R] + [PR] + [PZ] = x_o$$

where p_o and x_o are as usual total initial concentrations of the biopolymer and reagent. Small steady-state concentrations of reactive intermediates [I] and [PI] are neglected in the mass conservation equations. Taking into account Equations 86 to 88, concentrations of supercomponents x and r are easily substituted for [X], [PX], [R], and [PR] and, therefore,

$$[P] + K_x x[P]/(1 + K_x[P]) + K_r r[P]/(1 + K_r[P]) + [PZ] = p_o \qquad (89)$$

$$x + r + [PZ] = x_o \qquad (90)$$

2. Steady-state conditions

$$d[PI]/dt = k_o \gamma[PX] + k_I[P][I] - (k_{-I} + k_1 + k_2')[PI] = 0$$

$$d[I]/dt = k_o[X] + k_{-I}[PI] - (k_2 + k_I[P])[I] = 0$$

which may be written taking Equation 87 into account as

$$k_I[P][I] - (k_{-I} + k_1 + k_2')[PI] = k_o \gamma K_x x[P]/(1 + K_x[P]) \qquad (91)$$

$$-(k_2 + k_I[P])[I] + k_{-I}[PI] = k_o x/(1 + K_x[P]) \qquad (92)$$

3. Kinetic equations for the accumulation of PZ and the consumption of supercomponent X_t

$$d[PZ]/dt = k_1[PI] \qquad (93)$$

$$-dx/dt = k_o[X] + k_o \gamma[PX] \qquad (94)$$

The substitution of $x_o - x - [PZ]$ for r in Equation 89, the expression of [PI] in Equation 93 as a function of x and [P] with the help of steady-state conditions (Equations 91 and 92), and the substitution of the expressions in Equation 87 for [X] and [PX] in Equation 94 result after some transformations to the system of two differential and one algebraic equations for three variables, [PZ], x, and [P]

$$\frac{d[PZ]}{dt} = \frac{k_o x[P]}{1 + K_x[P]} \cdot \frac{(k_1/k_2) + \gamma K_x\{1 + (k_1/k_2)[P]\}}{(k_{-I}/k_1) + [(k_1 + k_2')/k_1]\{1 + (k_1/k_2)[P]\}} \qquad (95)$$

$$-\frac{dx}{dt} = \frac{k_o x}{1 + K_x[P]} \cdot (1 + \gamma K_x[P]) \qquad (96)$$

$$[P] + \frac{K_x x[P]}{1 + K_x[P]} + \frac{K_r[P](x_o - x - [PZ])}{1 + K_r[P]} + [PZ] = p_o \qquad (97)$$

Variable [P] is an auxiliary one which is expressed as an implicit function of the main variables x and [PZ] via Equation 97. It is seen that the independent parameters of the system are K_x, K_r, x_o, p_o, k_I/k_2, $(k_1 + k_2')/k_1$, and k_{-I}/k_1.) The three latter are combinations of five rate constants: k_1, k_2, k_2', k_I, and k_{-I}. Steady-state approximation does not permit the estimation of separate values of these parameters.

In general, the system of Equations 95 to 97 may be solved only by numerical integration. Easily integratable forms may be obtained for two-limit cases when one of the components is present in great excess over another under some additional assumptions. Most essential for the simplification of the solution is the assumption that the reactive intermediate I predominantly converts to R. This means that

$$k_I[P] \ll k_2 \tag{98}$$

and consequently Equation 95 is reduced to

$$d[PZ]/dt = k_o \gamma_{app} x[P]/(1 + K_x[P]) \tag{99}$$

where

$$\gamma_{app} = [(k_I/k_2) + \gamma K_x]/[(k_{-I}/k_1) + (k_1 + k_2')/k_1] \tag{100}$$

If $p_o \gg x_o$, p_o may be substituted for [P] in Equations 95 and 96. Therefore,

$$x = x_o \exp[-k_o(1 + \gamma K_x p_o)t/(1 + K_x p_o)]$$

and Equation 99 may be easily integrated giving

$$[PZ] = [PZ]_\infty \{1 - \exp[-k_o(1 + \gamma K_x p_o)t/(1 + K_x p_o)]\} \tag{101}$$

where

$$[PZ]_\infty = \frac{\gamma_{app} x_o p_o}{1 + \gamma K_x p_o} \tag{102}$$

If $x_o \gg p_o$, the following inequality holds

$$K_x[P] < K_x[P]_o = K_x p_o/(1 + K_x x_o) \ll K_x x_o/(1 + K_x x_o) < 1$$

and $K_x[P]$ may be neglected as compared with unity.

Assuming additionally that γ is not too great and $\gamma K_x[P] \ll 1$, that K_r does not exceed K_x significantly and, therefore, $K_r[P]$ may be omitted in the denominator of the third term in Equation 97, as well as taking into account that [PZ] may be neglected as compared with $x_o - x$ in Equation 97, we come to the system of equations

$$d[PZ]/dt = k_o \gamma_{app} x[P]$$

$$-dx/dt = k_o x$$

$$[P][1 + K_x x + K_r(x_o - x)] = p_o - [PZ]$$

which results in the final kinetic equation for accumulation of [PZ]

$$\frac{d[PZ]}{dt} = k_o \gamma_{app} x_o \exp(-k_o t) \frac{p_o - [PZ]}{1 + K_r x_o + (K_x - K_r)x_o \exp(-k_o t)} \qquad (103)$$

The appearance of this equation is identical with Equation 58 with the single difference that $k_o \gamma_{app}$ and k_o in Equation 103 substitute for $k_1 K_x$ and k_2 in Equation 58.

Therefore, expressions in Equation 59 if $K_x \neq K_r$ hold for this case after respective changes in the designation of parameters.

II. QUANTITATIVE CRITERIA OF AFFINITY MODIFICATION

A. General Direct Criteria

If chemical modification of a biopolymer is performed with some reagent bearing a structural similarity to a native ligand recognized by this biopolymer, it is reasonable to expect that the modification under consideration is an affinity process proceeding within the specific complex formed by the partners. However, as mentioned, affinity processes may proceed in some cases with reagents lacking a definite structural similarity to the native ligand. A vice versa existence in a chemical compound of both the reactive group and moiety suitable for recognition does not warrant that the reaction with the biopolymer would occur at all or that this reaction would proceed as an affinity process. Therefore, it is essential to have some criteria permitting the discrimination between the intracomplex and direct chemical conversions of the biopolymer. These criteria were formulated in several papers.[344,375-377] According to the latter, four main criteria should be considered.

1. Modification proceeds in the vicinity of the active center or in the active center itself thus leading to the loss of activity inherent to the biopolymer.
2. According to Equations 6 and 17, the pseudo first-order rate constant k_{app} or the initial rate of biopolymer modification v_o are hyperbolic functions of the initial concentration of reagent x_o and reach some limit value at a sufficiently high concentration of the latter.
3. The affinity of the reagent to the biopolymer estimated as the association constant K_x from the dependence of k_{app} or v_o on x_o determined according to Equation 6 or 17 should be the same as that determined from the experiments on the binding of a natural ligand with the reagent as a competitor.
4. The natural ligand or the unreactive analogue Y protects the biopolymer from modification, i.e., inhibits modification as follows from Equations 10 and 18. Using the dependence of v_o or k_{app} on x_o in the presence of a definite concentration of Y, the association constant K_y may be calculated. It should be within the limits of accuracy of experiments, the same as this value found from any direct measurement of the Y affinity towards a biopolymer, e.g., from equilibrium dialysis experiments or from enzymatic activity measurements if P is an enzyme and Y is a substrate.

Below we shall consider briefly all these criteria.

The coincidence of the extent of modification and of the residual activity usually gives evidence that the reaction proceeds within the specific complex. However, the same result may be observed if the biopolymer contains one unique group in the active center e.g., single exposed cysteine residue, serine residue with enhanced reactivity as in serine proteases, etc. At the same time, the lack of such a coincidence does not prove that the process under investigation is not an affinity modification. The inactivation of multisubunit proteins may occur at the level of modification lower than one modified residue per subunit due to a strong negative cooperativity. On the contrary, it may happen that affinity modification does not result in the complete inactivation of a biopolymer. Respective examples will be given in Section C. In this case, the extent of modification may exceed that of inactivation in spite of the intracomplex nature of the process. Thus, it is seen that the first criterion should be used with some caution, and fulfillment does not give final proof of the mechanism of chemical modification.

The second criterion indicates definitely that we deal with some complicated process, and complex formation is involved in this process. In most cases, the hyperbolic dependence of either v_o or k_{app} on x_o really proves that affinity modification occurs. However, some special cases may be considered as an alternative interpretation. Although low-probable, they cannot be completely excluded from consideration. As an example we present the process proceeding by the scheme

$$P + X \rightarrow PZ(k)$$

$$P + X \rightleftarrows PX(K_x)$$

called pseudoaffinity labeling. According to this scheme, the reaction proceeds as a direct attack of the reagent at the biopolymer but there exists a parallel formation of the unreactive complex. Quasiequilibrium treatment of this scheme results in the expression

$$d[PZ]/dt = k_{app}(p_o - [PZ])$$

with

$$k_{app} = \frac{kx_o}{1 + K_x x_o}$$

The single difference as compared with Equation 6 is that the limit k_{app} value in this case is kK_x instead of k_2 for Equation 6.

For example, *p*-carboxyphenyl glyoxal inactivates pig liver aldehyde reductase and Equation 6 is fulfilled for the dependence of k_{app} of the reaction on the initial concentration of the reagent.[378] It could be suggested that the reaction proceeds via intermediate formation of the complex of the reagent with the anion binding site of the enzyme. However, the inactivation rate is of the same order of magnitude as with phenyl glyoxal and 2,3-butanedione lacking the anionic group. It is reasonable to suggest that in this case we deal with the direct modification complicated by a parallel formation of unreactive complex of *p*-carboxyphenyl glyoxal with the enzyme.

The protective action of a biopolymer by a native ligand or the unreactive analogue against modification also cannot be considered as final proof of the affinity nature of the process. The formation of a specific complex between such a ligand and biopolymer may protect against direct attack at the same residue participating in recognition. It was also shown that, for the process described by Scheme 85, theoretically there exists a range of kinetic parameters which favors the process

$$X \longrightarrow I \longrightarrow PI \longrightarrow PZ$$

and in this case the kinetics of PZ accumulation may be low-sensitive to the presence of some unreactive competitor L, forming the complex PL.[374]

Thus, it is seen that neither of the commonly used criteria is sufficient for the elucidation of a mechanism of modification and some additional criteria may be useful for this aim.

A specific criterion was proposed by Groman et al.[376] which was called by the authors as catalytic competence. The criterion is defined as the ability of the affinity-modified enzyme to catalyze a normal reaction on the covalently bound ligand. Several examples of the fulfillment of this criterion are described.

The above-mentioned authors have studied the modification of estradiol-17β-dehydrogenase (EC 1.1.1.62) from human placenta with 3-bromoacetoxy estron. The enzyme catalyzes the reduction of estron to 17β-estradiol.

$$\text{(estron)} + \text{NADH} \rightleftharpoons \text{(estradiol, OH)} + \text{NAD}^+ \qquad (104)$$

The reagent rapidly inactivates the enzyme. However, if NADH is added to the covalent adduct formed in the course of affinity modification, the estron moiety is reduced to estradiol one. The reaction proceeds with the same stereospecificity as with the unbound substrate; deuterium is transferred to covalently bound estradiol only from the 4R isomer of [4-³H] NADH. A similar result was obtained with estradiol-17β-dehydrogenase modified with 3-(4′-azido-2′-nitrophenyl-β-alanyl)-estron.[379]

NAD covalently bound to yeast alcohol dehydrogenase via N^6-[N-(6-aminohexyl) carbamoylmethyl] spacer to Asp-173 of the enzyme retains the coenzyme properties. Thus, it efficiently catalyzes the oxidation of n-butanol to n-butanal with p-nitrosodimethyl-aniline according to the scheme.[380]

$$ON\!-\!\langle\rangle\!-\!N(CH_3)_2 + E\!-\!NADH \xrightarrow{(H^+)} HN\!=\!\langle\rangle\!=\!\overset{+}{N}(CH_3)_2 + E\!-\!NAD^+ \qquad (105)$$

$$E\!-\!NAD^+ + CH_3CH_2CH_2CH_2OH \longrightarrow E\!-\!NADH + CH_3CH_2CH_2\overset{O}{\underset{H}{C}} + H^+$$

Cytochrome c oxidase from rat liver mitochondria efficiently catalyzes the oxidation of N,N,N',N'-tetramethyl-p-phenylenediamine/ascorbate with O_2 in the presence of cytochrome c. A photoreactive derivative of cytochrome c obtained by the treatment with 2,4-dinitro-5-fluorophenyl azide binds covalently to cytochrome c oxidase under irradiation of the mixture of this derivative with cytochrome c-depleted mitochondria. The adduct formed catalyzes the oxidation of the same electron donor without the addition of cytochrome c. This means that both components retain activity in the covalently fixed state.[381]

Some other examples of catalytic competence will be presented in the next chapter when affinity modification of RNA-polymerase and ribosomes will be considered.

However, it is obvious that the criterion of catalytic competence is not a universal one as it suggests that affinity modification proceeds at some residue nonessential for catalysis. It also suggests that the covalently bound substrate is capable of achieving the orientation required for enzymatic catalysis. Therefore, the fulfillment of this highly informative criterion is rather an exception than a rule.

B. Comparative Criteria of Affinity Modification

An additional criterion of the affinity mechanism of modification is the high reaction rate of the process under consideration as compared with the rate of a similar process definitely proceeding without a specific complex formation.

If there exists a set of enzymes with a similar structure of catalytic center but differing in specificity, reaction rates of a reagent with different enzymes may be compared. The most studied group of such enzymes is presented by serine proteases. Thus, *N*-tosylphen-ylalanylchloromethane (XX, X = $C_6H_5CH_2$ and R = *p*-$CH_3C_6H_4SO_2$) alkylates chymo-trypsin which is specific to hydrophobic aminoacyl residues and does not inactivate trypsin specific for a strongly basic aminoacyl moiety. Vice versa, *N*-tosyllysylchloromethane readily inactivates trypsin and is rather inert towards chymotrypsin.[97] Consequently, it may be concluded that the reactions of the phenylalanine analogue with chymotrypsin and of the lysine analogue with trypsin are affinity processes. However, this criterion has rather re-stricted applicability since it suggests the existence of a set of enzymes with an elucidated mechanism and known structural features of the catalytic center.

The comparison of the rate of modification of a biopolymer under investigation by a reagent studied with the rate of modification by a reference compound lacking specificity towards this biopolymer may find a significantly wider application. Two problems should be solved for such a comparison.

First, the reference compound should be chosen which has the same reactivity towards low molecular weight targets or nonspecific biopolymers as the reagent under investigation. For example, in the studies of complementary-addressed alkylation of nucleic acids with the reactive oligonucleotide derivatives CXVII in Chapter 2 obtained by Reaction 78, the comparison with the starting *p*-(*N*-2-chloroethyl-*N*-methylamino)benzaldehyde is incorrect for at least two reasons. First, due to the strong electron acceptor character of carbonyl group formation of the intermediate, the ethyleneimmonium cation is hindered as compared with acetals CXVII in Chapter 2. The second reason is that the ethyleneimmonium cation formed in the rate-determining step has enhanced reactivity to nucleic acids due to a strong electrostatic interaction of a positively charged intermediate with the polyanion polynucleo-tide chain. For the second reason, a nucleoside derivative of the type of acetal (CXVII in Chapter 2) also cannot be used as a reference compound. The simplest derivative which may be used is a mononucleotide derivative, e.g., the derivative IX in Chapter 1 of uridine-5′-methylphosphate which produces an electroneutral (zwitterionic) intermediate. Still more correct, there should be a comparison with some nonaddressed derivative containing several negatively charged fragments since the presence of additional negative charges influences to some extent the efficiency of alkylation.[303]

The second problem is the choice of appropriate quantitative characteristics for such a comparison. The simplest case is the comparison of reaction rates of processes proceeding via a one-step mechanism, e.g., alkylation by S_N2 mechanism. Alkylation with haloacetic acid derivatives may represent this type of reaction. The reaction rate of addressed modi-fication

$$v_a = k[PX] = kK_x[P][X]$$

and if as usual, the initial concentration of X exceeds significantly that of P ($x_o \gg p_o$) in the beginning of the process

$$v_a = \frac{kK_xp_ox_o}{1 + K_xx_o}$$

Reaction with reference compound X′ proceeds at the initial rate

$$v_n = k'p_ox_o'$$

Taking the ratio v_a/v_n as a criterion of affinity mechanism we come to expression

$$v_a/v_n = (kK_x/k')[x_o/x_o'(1 + K_xx_o)] \tag{106}$$

This ratio may be referred to as the efficiency of affinity modification. As it is seen from Equation 106, it depends on the concentration of affinity reagent used. This dependence disappears at low X concentrations

$$v_a/v_n = (kK_x/k')(x_o/x_o') \tag{107}$$

As an example we present the reaction of the inactivation of catechol-O-methyltransferase with N-iodoacetyl-3,5-dimethoxy-4-hydroxyphenylethylamide.

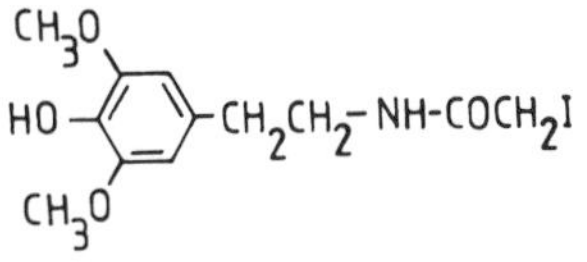

(CXL)

Iodoacetamide was chosen as a reference compound. At identical 0.415 mM concentrations of CXL and ICH_2CONH_2, the ratio of reaction rates is 9.1.[382] Since the K_x for CXL was found to be 0.9 mM^{-1} at the same conditions, the maximal efficiency value achieved at low concentrations is

$$9.1 \times (1 + K_x x_o) = 9.1 \times (1 + 0.9 \times 0.415) = 12.5$$

If reaction is described by some multistep scheme more complicated treatment is necessary. We shall illustrate the approaches used by two examples of complementary-addressed alkylation of synthetic deoxyribooligonucleotide and *Escherichia coli* tRNA[Phe]. Both cases needed additional proof of the intracomplex mechanism of alkylation. Aromatic 2-chloroethylamines were used as alkylating groups and therefore addressed modification proceeded according to Scheme 85. The reference nonaddressed reaction proceeds as follows

$$X' \rightarrow I'(k_o')$$

$$I' + P \rightarrow PZ'(k_1')$$

$$I' \rightarrow R'(k_2') \tag{108}$$

where X' is the reference compound, I' is the reactive intermediate, and R' and PZ' are the products of hydrolysis and of the reaction with the biopolymer.

It was shown that in steady-state approximation the initial rate of the reaction proceeding in accordance with Scheme 108 is described as

$$(v_n)_o = \frac{k_o' p_o x_o' r}{r p_o + 1} \tag{109}$$

where $r = k_1'/k_2'$ is a competition factor.[383] This value may be easily determined from the kinetic measurements of the nonaddressed reaction.

The reaction of the derivative pdApdApdCrA > CHRCl (here and below RCl is a p-N-2-chloroethyl-N-methylaminophenyl fragment) with the oligonucleotide d(pCpCpCp CpGpTpTpTpGpGpC) was studied using the excess of template.[384] As a reference compound pdTpdTpdTprU > CHRCl was taken. A competition factor for this nonaddressed reagent lacking any complementarity towards the template was found to be 10 M^{-1} per one guanine residue. According to Equation 99 in these conditions the reaction rate of the addressed reaction is

$$(v_a)_o = k_o \gamma_{app} x_o p_o/(1 + K_x p_o)$$

Using Equation 109 we easily obtain for the efficiency of reaction

$$(v_a)_o/(v_n)_o = \frac{k_o}{k_o'} \cdot \frac{\gamma_{app}}{1 + K_x p_o} \cdot \frac{rp_o + 1}{r} \cdot \frac{x_o}{x_o'} \tag{110}$$

This expression may be significantly simplified if we take into account that the rate constant of the ethyleneimmonium formation of CHRCl derivatives is nearly the same for nucleosides, nucleotides, and oligonucleotides and does not change in the complex.

This means that $k_o = k_o'$ and $\gamma = 1$. Consequently, taking into account Equation 102 we easily come to expression

$$(v_a)_o/(v_n)_o = ([PZ]_\infty/x_o) \times [(rp_o + 1)/rp_o] \times (x_o/x_o') \tag{111}$$

where $[PZ]_\infty/x_o$ is the extent of consumption of the reagent in the modification process at the end of the reaction. Using this expression and finding $[PZ]_\infty$ it was shown that at $p_o = 1 \times 10^{-3}$ M this ratio is 7.1 and, therefore, in spite of a very short addressing part, the reaction proceeds mainly within the complex.

As another example we shall present the alkylation of tRNA^{Phe} from *E. coli* with the reagent ClRCH$_2$NH-pd(ApTpTpTpTpCpA) which has an addressing part complementary to the anticodon loop of this tRNA.[385] These experiments were carried out with the excess of alkylating reagent. Under the same assumptions as in the previous case using kinetic Equation 103 we may express the initial alkylation rate as follows:

$$(v_a)_o = k_o \gamma_{app} x_o p_o / (1 + K_x x_o)$$

As mentioned, differential Equation 103 differs from Equation 58 only by the meaning of parameters. We may expect in our case that the reagent and the product of hydrolysis at the C–Cl bond have the same affinity to tRNA, i.e., $K_r = K_x$. Therefore, solution in the form of Equation 60 may be used with a substitution of $k_o \gamma_{app}$ for $k_1 K_x$ and k_o for k_2

$$PZ/p_o = 1 - \exp[-\gamma_{app} x_o / (1 + K_x x_o)]$$

Consequently,

$$(v_a)_o = -k_o p_o \ln(1 - [PZ]_\infty/p_o)$$

If p_o is sufficiently small and $p_o r \ll 1$ as usually is the case

$$(v_a)_o/(v_n)_o = -\ln(1 - [PZ]_\infty/p_o)/rx_o' \tag{112}$$

In the described reaction, G28/G29 in the anticodon stem as well as G10 and G24 in the D stem of tRNA^{Phe} were modified. Modification extents $[PZ]_\infty/p_o$ were found to be 0.108 for G10, 0.247 for G24, and 0.09 for G28/G29 at an initial concentration of reagent of 9×10^{-5} M. Competition factors were taken from the paper of Vlassov and Skobeltsyna[386] who have determined these values separately for each guanine residue alkylated with MepUCHRCl: 44.2 for G10, 40.3 for G24, and 2.4 for G28/G29. Using these values, the efficiencies of addressed modification were calculated

G10 $\quad (v_a)_o/(v_n)_o = -\ln(1 - 0.108)/(9.10^{-5} \times 44.2) = 29$

G24 $\quad (v_a)_o/(v_n)_o = -\ln(1 - 0.247)/(9.10^{-5} \times 40.3) = 78$

G28/G29 $\quad (v_a)_o/(v_n)_o = -\ln(1 - 0.09)/(9.10^{-5} \times 2.4) = 440$

It is seen that for all reactions addressed the process strongly predominates.

C. Dynamic Aspects of Affinity Modification

Recognition of a specific ligand L by a biopolymer P is considered as proceeding due to mutual interactions of the partners at a number of definite points. Both participants are present in the specific complex in conformations most favorable for these interactions. However, this complex is not absolutely rigid. Some conformational degrees of freedom still remain as well as the possibility of the formation of complexes with an incomplete number of contacts. This means that what we consider to be a definite particle PL is a dynamic system represented by a number of different states PL. Each of these states may be characterized by having a partial association constant

$$K_{li} = [PL_i]/[P][L]$$

The full association constant which is obtained from the direct binding measurements or some competition experiments is a sum of these partial constants.

$$K_l = [PL]/[P][L] = \sum_{i=1}^{N} [PL_i]/[P][L] = \sum_{i=1}^{N} K_{li}$$

This is true irrespective of whether L is substrate, effector, competitive inhibitor, or affinity reagent.

Any functional property of a specific complex may be different for different states. For example, if we deal with the complex ES of enzyme E with substrate S, the reaction rate of an enzymatic process in quasiequilibrium approximation should be presented as

$$v = k_{cat}[ES] = \sum_{i=1}^{N} k_{cat}^{(i)}[ES_i] = \left(\sum k_{cat}^{(i)} K_{si}/K_s\right)[ES]$$

A similar expression can be written for any biological response to the specific complex formation. For affinity modification of biopolymer P with reagent X

$$v = k_x[PX] = \left(\sum_{i=1}^{N} k_{xi} K_{xi}/K_x\right)[PX]$$

where K_{xi}/K_x is the probability of the i-th state and k_{xi} is the probability of the reaction within the complex PX_i per unit time.

It is well known that the reaction probability or rate constant depends strongly on the mutual orientation of the reacting partners.[387] Therefore, the values of k_{xi} may differ by several orders of magnitude. The reaction should proceed within the states with the most favorable mutual orientation of the reagent and one of the potentially reactive biopolymer residues. These states are not necessarily identical, either with the main functional states of a similar complex with the natural ligand (e.g., the state of the enzyme-substrate complex advantageous for catalysis) or with the most populated state. Moreover, different states may in some cases favor a reaction with different fragments of a biopolymer. Several conclusions follow from this consideration.

1. Modification may proceed at a number of nonadjacent points of a biopolymer including groups rather far from the active site. The low probability of the occasional approach of some of these groups to the reactive moiety of the reagent may be compensated by favorable mutual orientation of the reacting fragments.

2.   If the intracomplex reaction took place in some low-probable state, the addressing part of the reagent may be found covalently attached in an arrangement different from that favorable for the normal function of the complex.

3.   As a limit case the reagent covalently bound in a nonfunctional state may be removed from the active center and consequently, biopolymer may retain activity at least partially.

At the present state of our knowledge, the dynamic aspects of affinity modification were not attacked systematically. However, some experimental evidence in favor of the above statements may be presented. Thus, retention of functional activity is described in a number of cases. For example, rabbit muscle creatine kinase catalyzing a phosphate transfer from creatine phosphate to ADP is inactivated by the ATP and ADP analogues ethenoadenosinedi- and triphosphates. However, a modified enzyme, although enzymatically inactive, retains the ability to form specific complexes with ADP and ATP with nearly the same affinity as the native enzyme. This means that covalent binding of εATP or εADP is followed by the removal of the bound residues from the nucleotide binding site.[388] A similar explanation was given for the results of affinity modification of *E. coli* phenylalanyl-tRNA-synthetase (PheRS) with adenosine-5'-trimetaphosphate (ATMP). ATP is known to enhance strongly the affinity of amino acids and analogues to aminoacyl-tRNA-synthetases. In the presence of 1 mM ATP the affinity of 3-phenyl-2-amino-propanol-1 (phenylalaninol) is characterized by the association constant $2 \times 10^5\ M^{-1}$. However, the enzyme treated with ATMP and therefore containing an ATP fragment, covalently bound to PheRS via γ-phosphate has an association constant with the same analogue, $6.7 \times 10^3\ M^{-1}$. This value is only seven times higher than that in the absence of ATP. The binding of phenylalanine to PheRS in the absence of ATP and to modified PheRS is nearly the same, the association constant being $3.3 \times 10^4\ M^{-1}$. This means that the influence of ATP strongly differs in free and covalently attached states. The authors of a respective paper explain these results as a consequence of the removal of an ATP fragment from an ATP binding area.[389]

Still more striking results were obtained in the experiments with an alkylating derivative of pdTpdA, namely *p*-(*N*-2-chloroethyl-*N*-methylamino)benzyl(N → P)amide, studied as an affinity reagent for pancreatic RNAase. It was already mentioned in Chapter 3, Section IV.B that modification results in the formation of four different forms of the monomodified enzyme. In spite of a rather high degree of modification, no double-modified species were found. Therefore, it is reasonable to conclude that all these forms do not result from several complexes with different sites occupied by a reagent but rather originate in parallel from some single dynamic structure. All these forms were found to retain 23 to 60% of activity inherent to the native enzyme as revealed by measuring the rates of hydrolysis of cytidine 2',3'-cyclophosphate catalyzed by the enzyme.[115]

The examples presented do not prove the reliability of the above consideration. However, these data indicate that the dynamic aspects should not be neglected in the interpretation of the data obtained in affinity-labeling experiments. At the same time it is seen that the results of affinity modification reflect in some way the dynamic behavior of the macromolecular system and may serve as a tool for the investigation of the dynamics of biopolymers.

Chapter 5

AFFINITY MODIFICATION IN BIOCHEMISTRY, BIOLOGY, AND APPLIED SCIENCES

I. INTRODUCTION

In the three preceding chapters we have presented systematically a general methodology of affinity modification — preparation of reagents, performance of the process, analysis of time course and results, and quantitative treatment of kinetics of the process. It is impossible within the framework of this book to describe in a similar systematic way all the systems subjected to affinity modification due to the huge number. Moreover, such a description would be unreasonable since every system studied by affinity modification has been investigated as a rule by other methods and only a combination of them has resulted in some definite biochemical or biological conclusions. What we shall try to do in this chapter is to present the most promising fields of applications of the method, to describe the methodology used, and to illustrate the main approaches by respective examples. These fields may be tentatively divided into three groups: the study of the structure and functions of biopolymers and complexes of biopolymers (biochemical applications), the study of the distribution of biopolymers of definite specificity within the cells, between tissues and organs of living organisms (biological applications), and the attempts to affect some specific biopolymers in the living systems in order to either suppress or stimulate definite biochemical processes for some medical purpose (pharmacological applications).

II. BIOCHEMICAL APPLICATIONS OF AFFINITY MODIFICATION

A. Structural Studies of the Active Centers of Enzymes

Affinity modification was born as a method of the active site-directed irreversible inhibition of enzymes. Although during the last two decades a great number of other functional proteins were studied by this method, enzymology still remains the most advanced field of application of affinity modification. Therefore, the general methodology of the method as directed to the studies of proteins will be illustrated by enzymological problems.

The treatment of enzymes with the reactive substrate analogues usually leads to a covalent attachment of the reagent to the enzyme in the region of the active center. An elucidation of the modification point gives some information concerning the active center. However, this information is usually rather poor. In a number of cases modification of a cofactor takes place. A significant number of such cases is known. We shall present a few typical examples.

Thiaminpyrophosphate participates as a cofactor of the enzymes catalyzing decarboxylation of α-keto acids. (E)-4-(4-chlorophenyl)-2-oxo-3-butenoic acid was found to be a suicide inhibitor of the brewer's yeast pyruvate decarboxylase, catalyzing the reaction

$$CH_3COCOO^- \longrightarrow CH_3CHO + CO_2 \tag{1}$$

It was shown that the transformation of thiaminpyrophosphate (CXLI) to the derivative CXLII is the reason of inactivation.[390]

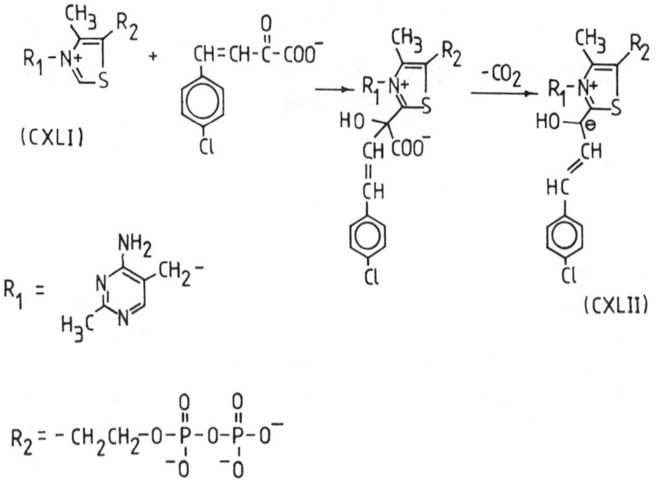

(CXLI)

$R_1 =$

$R_2 = -CH_2CH_2-O-\overset{O}{\underset{O^-}{P}}-O-\overset{O}{\underset{O^-}{P}}-O^-$

(CXLII)

Mammalian lactoperoxidase is a hemoprotein containing heme (LVII in Chapter 2) as a cofactor. Bovine enzyme is inactivated by SH^-, 2-mercaptobenzimidazole, and 1-methyl-2-mercaptoimidazole via suicide mechanism. It is shown that the cause of inactivation is a reaction of heme with intermediates formed by S-oxygenation.[391]

General acyl-CoA dehydrogenase is a flavoprotein catalyzing the essential step of mitochondrial oxidative degradation of fatty acids

$$R-CH_2-CH_2-CO-SCoA + EFl_{ox} \longrightarrow EFl_{red} + R-CH=CH-CO-SCoA \tag{2}$$

(EFl is the enzyme containing an oxidized or reduced form of flavinadeninedinucleotide). The allenic substrate analogue 3,4-pentadienoyl-CoA readily inactivates the pig kidney enzyme forming the covalent flavin adduct CXLIII.[392]

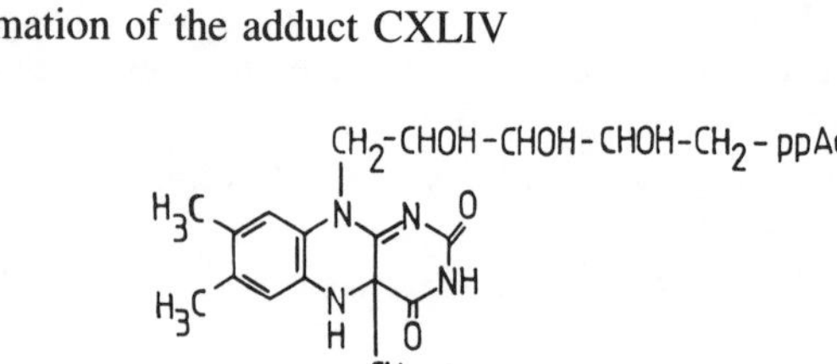

(CXLIII)

Similarly, formation of the adduct CXLIV

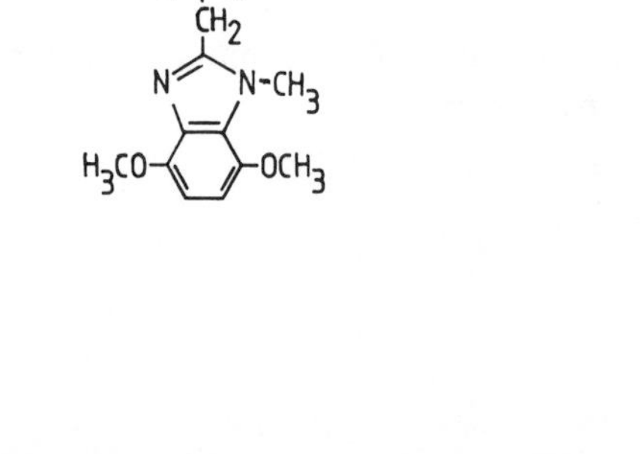

(CXLIV)

is suggested to be the reason of inhibition of buttermilk xanthine oxidase by 1-methyl-2-(bromoethyl)-4,7-dimethoxybenzimidazole.[393]

Uridinediphosphategalactose 4-epimerase catalyzing the mutual conversion of UDP-galactose and UDP-glucose contains tightly bound NAD as a cofactor. The reactive substrate analogue p-(bromoacetamido)phenyluridyl pyrophosphate

$$BrCH_2CO-NH-\langle \bigcirc \rangle-O-ppUrd$$

irreversibly inactivates the enzyme from *Escherichia coli*. The adenine ring of the AMP moiety of NAD was found to be alkylated.[394]

The cases are described when strong inhibition due to coordination of the ligand to the metal ion in the active center takes place. Thus, 2-benzyl-3-mercaptopropanoic acid (SQ 14,603) is a strong inhibitor of carboxypeptidase A, specifically cleaving C-terminal aromatic amino acid residues (see Chapter 1, Section I.A). The dissociation constant of the enzyme-inhibitor complex is $K_i = 1.1 \times 10^{-8} M$, whereas for carboxypeptidase B specific to C-terminal positively charged amino acid residues $K_i = 1.6 \times 10^{-4} M$. At the same time 2-mercaptomethyl-5-guanidinopentanoic acid (SQ 24,798) has a K_i of $4 \times 10^{-10} M$ vs. carboxypeptidase B and a K_i of $1.2 \times 10^{-5} M$ vs. carboxypeptidase A. In this case the inhibitory action is due to coordination to a catalytically essential Zn^{2+} ion.[395]

In these cases the value of information presented by the affinity modification experiment is very low since it is usually well known without any modification experiments that respective cofactors are obligatory participants of the active center.

In most cases modification results in the covalent attachment of the reagent to one definite or to several amino acid residues. However, this result does not mean that we have found the residue in the active center. This may be a group residing in the vicinity of this center, preferentially modified due to enhanced reactivity as well as to the dynamic nature of the enzyme-reagent complex. To our knowledge today there is no case when affinity modification has played a decisive role in the elucidation of the structure of the enzyme active center. Informativity of affinity modification as far as this problem is concerned is out of competition with the data obtained by X-ray analysis of crystalline enzymes and enzyme-substrate complexes. This does not mean that the approach is without any perspective as a tool of the structural investigations of the active center. In our opinion this may be done by a more systematic investigation, especially by the use of a number of specially designed reagents including those capable of specific cross-linking of amino acid residues of the active center belonging to different parts of the polypeptide chain of the enzyme or even to different polypeptides in the case of multisubunit enzymes.

Much more promising affinity modification seems to be a tool of elucidation of the distribution of the active sites for substrates and effectors between different subunits of multisubunit enzymes. If a reactive analogue of a substrate or effector modifies one definite subunit, it is reasonable to conclude that the active center recognizing this ligand resides at this subunit. If two or more subunits are labeled with the analogue, it may be suggested that the active site for the respective ligand resides in the region of the intersubunit contact.

A typical example is the elucidation of the distribution of active centers between subunits of *E. coli* phenylalanyl-tRNA-synthetase. This enzyme contains two identical sets of two subunits called α and β ($\alpha_2\beta_2$-enzyme). The reaction catalyzed by aminoacyl(AA)-tRNA-synthetases is presented by the Equations 10 and 11 in Chapter 1, Section II.A. It is seen that three substrates participate in the whole process: amino acid, ATP, and transfer (t)RNA. A reaction of the enzyme with the reactive ATP analogue adenosine-5′-trimetaphosphate (ATMP) resulted in modification of both subunits. UV irradiation of the enzyme in the presence of the amino acid analog phenylalanyl-p-azidoanilide led to a labeling of the α subunit. To study the distribution of the tRNA-binding area three reactive tRNA derivatives were used: tRNA^{Phe} containing arylazido residues scattered statistically over guanine residues obtained with the help of reactions (Chapter 2, Equation 83) with reactive groups in the internal part of the molecule; two derivatives containing a reactive group attached to the α-NH$_2$ group of the phenylalanyl moiety of the Phe-tRNA^{Phe}, namely bromoacetyl residue (CXXXIb) and chlorambucilyl residue (ClCH$_2$CH$_2$)$_2$NC$_6$H$_4$(CH$_2$)$_3$CO (CXXXIa). With the first two reagents, labeling of the β-subunit was achieved. With the latter reagent, a reaction with the α subunit strongly predominated. According to these results it was concluded that the α subunit contains an amino acid binding center, the β subunit contains all the tRNA-binding area with the 3′ end residing not far from the subunits interface, and the ATP-binding center resides in the area of the subunit contacts.[396,397]

Another enzyme extensively studied by a similar approach is mitochondrial F$_1$ ATPase (see Chapter 1, Section I.F). The mitochondrial enzyme contains five different subunits with stoichiometry α$_3$β$_3$γδε. Six nucleotide binding sites reside at the enzyme. Three of them tightly bind ADP or ATP and exert most probably regulatory function. Three others are thought to be catalytic sites. In most cases labeling of the enzyme with ADP and ATP analogues results in the modification of both subunits. Thus, photoirradiation of the pig heart enzyme in the presence of 3′-O-[3-(4-azido-2-nitrophenyl)propionyl] ADP (3′-O-NAP-β-alanyl-ADP) results in the complete inactivation after the level of modification achieves 3 mol/mol F$_1$ ATPase. Distribution of the label between α and β subunits is 3:7.[398] Similar distribution (23:77) was found when 8-azido-ATP and 8-azido-ADP were used as photoaffinity labels for bovine heart enzyme within submitochondrial subparticles.[399] More essential labeling of the α subunit (45% of the total label) was found after UV irradiation of bovine heart enzyme in the presence of 3′-O-[5-azidonaphthoyl]-ADP.[400] Distribution of the label depends on the concentration of the reagent. Thus, it was demonstrated that at concentrations of 8-azido-ADP or 8-azido-ATP of 300 to 1000 μM labeling of α and β subunits proceeds at the ratio 1:3. At lower concentrations (20 μM) only the β subunit is modified.[401] It was concluded that one catalytic site is located on the β subunit and the other is located on the α/β interface.

To prove intersubunit location of the ligand binding site, it may be useful to design affinity reagents with two reactive groups bound to the same address. This approach was applied to F$_1$ ATPases from bovine heart mitochondria[263] as well as for similar bacterial enzymes.[402,403] 3′-O-(NAP-β-alanyl)-8-N$_3$ATP (CXXII) was used for this goal bearing photoreactive groups both at the ribose and adenine fragments. UV irradiation of all these enzymes in the presence of the reagent resulted in the cross-linking of α and β subunits, thus indicating that the ATP-binding site is situated at or near the contact area of α and β subunits.

A number of cases is known when several different reactions are catalyzed by the same enzyme preparation. It is essential to elucidate whether we deal with one active center bearing two activities or with two different or only partially overlapping regions of this preparation. Affinity modification permits discrimination of these possibilities. If both activities are inherent to one active center, these activities should be eliminated in parallel in the course of affinity modification. Below we present some typical examples.

Cortison reductase (EC 1.1.1.53) from *Streptomyces hydrogenans* catalyzes the reduction of the 20-oxogroup of cortison (CXLV) with NADH

(CXLV)   (CXLVI)   (CXLVII)

The enzyme behaves also as 3α-hydroxysteroid reductase. Both activities disappear in parallel when the enzyme is treated with a suicide analogue of CXLV, 17β-[(1S)-1-hydroxy-2-propynyl]-androst-4-en-3-on (CXLVI).[404] Oxidation of the 20-hydroxy group with NAD$^+$ converts oxopropynyl residue to an alkylating moiety which inactivates both activities. The same is the case with other analogues, e.g., 17-bromoacetoxy derivative of progesterone (CXLVII).[405]

17β-Estradiol dehydrogenase (EC 1.1.1.62) from human placenta also has 20-hydroxysteroid dehydrogenase activity. Both activities disappear in parallel when the enzyme is treated with steroid analogues CXLVI (1R isomer)[406] and 16α-bromoacetoxy-progesterone[407] and with NAD analogues 5′-[p-(fluorosulfonyl)-benzoyl adenosine.[408]

Mammalian and avian liver as well as yeast contain the enzyme preparation catalyzing three different activities dealing with tetrahydrofolate derivatives,

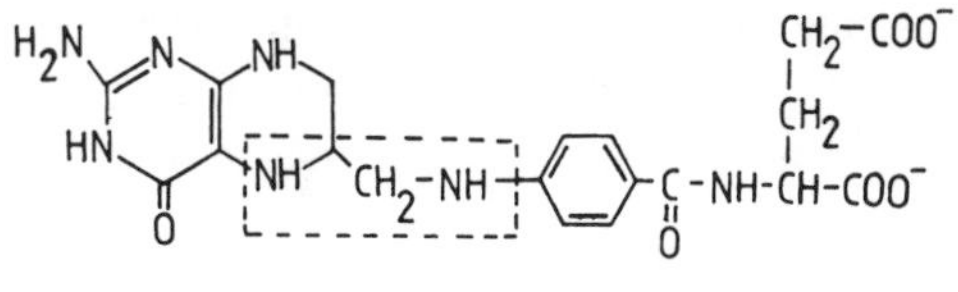

tetrahydrofolate (THF)

namely formyl-THF-synthetase (EC 6.3.4.3), methenyl-THF cyclohydrolase (EC 3.5.4.9), and methylene-THF dehydrogenase (EC 1.5.1.5).

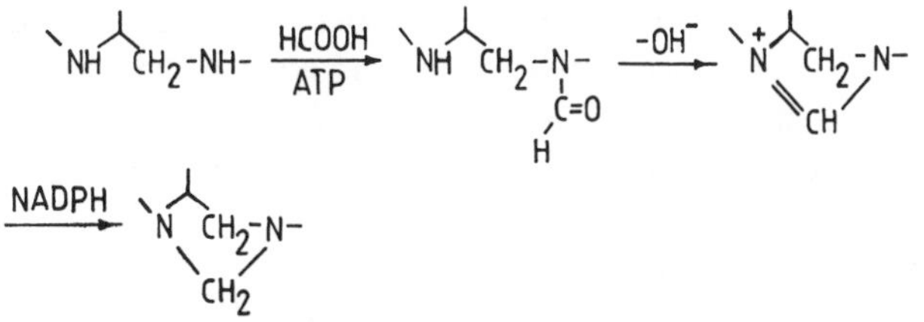

(3)

The two latter activities were destroyed in parallel by treatment of the porcine liver enzyme with the reactive folic acid derivative XXVII in Chapter 2. This means that at least these two activities operate with the same THF-binding area.[105]

tRNA-nucleotidyl transferase (EC 2.7.7.21, 2.7.7.25) catalyzes the reversible pyrophosphorolysis of three universal 3′-terminal nucleotides of all tRNAs by consecutive reactions.

$$tRNA - pCpCpA + pp_i \rightleftharpoons tRNA - pCpC + ATP$$
$$tRNA - pCpC + pp_i \rightleftharpoons tRNA - pC + CTP$$
$$tRNA - pC + pp_i \rightleftharpoons tRNA_{(-CCA)} + CTP$$

Both ATP γ-p-[N-(2-chloroethyl-2-methylamino)benzylamide] bearing alkylating fragment attached to γ-phosphate and ATMP inactivate the *E. coli* enzyme with parallel disappearance of both activities. It may be concluded that at least some essential groups of the catalytic center of the enzyme are common for both activities.[409]

As an alternative example we present the inactivation of the mammalian glycogen debranching enzyme. This enzyme is known to catalyze two types of reaction: the transfer of the fragments of side chains of branched glycogen to the terminal residue of the main chain (transferase activity EC 2.4.1.25) as well as the hydrolysis of residual branched glucose residue (hydrolase activity EC 3.2.1.33). 1,5-Dimethylarsino-1-thio-β-D-glucopyranoside inactivates both activities. However, the reversible inhibitor $(HOC_2H_4)_2NC(CH_2OH)_3$ (bis-tris) selectively protects glucosidase against inactivation, thus demonstrating that these activities are inherent to different active centers.[410]

B. Cooperative Interactions in Enzymes as Revealed by Affinity Modification

In Chapter 4, Section I.G we presented several examples demonstrating that the presence of a ligand in one active center of the enzyme may influence the affinity modification proceeding in the other active center. This may be the influence of one substrate of bi- or trisubstrate enzymes on the reaction with the analogue of the other substrate. This may be also the influence of the effector on the affinity modification of the catalytic center and vice versa. Many such cases are already known and we shall describe only a few typical examples.

The inactivation of a multifunctional enzyme catalyzing the reaction set (Equation 3) with reactive folate derivative is strongly enhanced in the presence of NADP, the substrate of the last reaction.[105]

Undecaprenylpyrophosphate synthase, a bacterial enzyme catalyzing the oligomerization of isopentenyl pyrophosphate, is photoinactivated by (E,E)-(2-diazo-3-trifluoropropionyloxy)geranyl pyrophosphate, the analogue of the growing chain.

$$CF_3-C(N_2)-COO-CH_2-CH(CH_3)=CH-CH_2-CH_2-C(CH_3)=CH-CH_2-OP_2O_7^{\ominus\ominus\ominus}$$

However, the reaction is significantly favored by the presence of magnesium salt of a monomer.[411]

The enzyme ribonucleotide reductase (EC 1.17.4.1) catalyzes the reduction of the ribose moiety of ribonucleoside diphosphates to deoxyribose one, thus supplying DNA synthesis with monomers. Deoxynucleoside triphosphates are known to be allosteric effectors of the enzyme. According to this, dTTP is directly cross-linked to the *E. coli* enzyme under irradiation at 254 nm. The presence of the substrate GDP approximately threefold stimulates the cross-linking.[412]

N^α-Tosyl-L-lysylmethylene chloride, a reactive analogue of the fragment of protein substrate, inactivates bovine lung cAMP-dependent protein kinase. The presence of cAMP tenfold stimulates affinity modification of this system.[413]

Negative cooperativity often operates between identical subunits of multisubunit enzymes. The limit case of this phenomenon is known as a "half-of-the-sites-reactivity".[414,415] Affinity modification permits negative cooperativity at least in two experimental versions to be revealed. First, modification of half of the identical subunits may strongly or even completely suppress enzymatic activity. Second, affinity modification may be significantly slowed or even ceased after reaction with half of the subunits took place. Again numerous examples are known and we shall present only several examples.

F_1 ATPase from *Micrococcus luteus* is the enzyme with the subunit structure $\alpha_2\beta_2\gamma\delta\epsilon$. Photoaffinity modification of one β subunit with 8-azido-ATP completely inactivates the enzyme.[416] Inactivation of $NADP^+$-specific isocitrate hydrogenase, the enzyme involved in the tricarbonic acids cycle and catalyzing the reaction shown in Chapter 3, Equation 3, performed using the NADP analogue 2-[(4-bromo-2,3-dioxobutyl)thio]-1,N^6-ethenoadenosine 2',5'-biphosphate proceeds as a biphasic process. The rapid phase corresponds to the incorporation of one reagent residue per dimeric enzyme. This is accompanied with a tenfold decrease of the enzymatic activity. Thus, modification of one subunit results in a significant loss of both activity of the second subunit and the reactivity towards the modifying reagent.[304]

The study of modification of malic enzyme (EC 1.1.1.40) from pigeon liver catalyzing the reaction

$$^-OOC-CH_2-CHOH-COO^- + NADP^+ \longrightarrow CH_3COCOO^- + CO_2 + NADPH$$

with NADP oxidized with periodate has demonstrated that the process stops after the incorporation of two molecules of the reagent into tetrameric enzyme.[417]

Alkylation of beef pancreas tryptophanyl-tRNA-synthetase with chlorambucilyl-Trp-tRNA^{Trp} (CXXXIa in Chapter 2) results in the modification of one subunit of dimeric enzyme. This modified enzyme is completely inactive in the aminoacylation of tRNA^{Trp}. However, it retains 50% of the activity in the tRNA-independent reaction of ATP-[^{32}P]pyrophosphate exchange in the presence of tryptophane (see Chapter 1, Section II.A).[418]

The latter example demonstrates that affinity modification permits preparation of a specific enzyme derivatives lacking some activities with the retention of others. Such modified enzymes may serve as simplified models in kinetic investigations. Thus, the kinetic parameters of the above-described derivative were compared with those for tryptophane-dependent ATP-[^{32}P]pyrophosphate exchange catalyzed by a nonmodified enzyme in the absence of tRNA. It was shown previously that the enzyme binds substrates in the order: ATP, tryptophane. However, the unproductive complex of an enzyme with tryptophane lacking ATP is formed according to kinetic data. The presence of covalently bound chlorambucilyl-Trp-tRNA^{Trp} at one subunit significantly suppresses the affinity of enzyme to tryptophane, the dissociation constant rising from 1.2 to 13×10^{-7} M. The same is the case for the affinity of the enzyme towards the tryptophane analogue tryptamine (decarboxylated tryptophane) — modification of one subunit with the tRNA derivative changes the dissociation constant from 2.1×10^{-6} to 5.0×10^{-4} M. This means that the presence of tRNA at one subunit enhances the specificity of the processes proceeding at the second subunit.[419]

Modification of allosteric sites switches off the influence of the external ligand on the enzymatic reaction. It may be especially useful when similar ligands behave both as substrate and effector thus complicating enzyme kinetics. As an example we may consider additional nucleotide binding centers found at *E. coli* phenylalanyl-tRNA-synthetase and beef pancreas tryptophanyl-tRNA-synthetase.[313,420] It was found that small concentrations of different nucleotides enhance the aminoacylation rate. As an example, in Figure 1 the dependence of the rate of Phe-tRNA^{Phe} formation on the CMP concentration is presented. It is seen that up to 5×10^{-6} M CMP accelerates aminoacylation. UV irradiation of the enzyme in the presence of GDP γ-*p*-azidoanilide results in the disappearance of the sensitivity of the enzyme to CMP at least at low concentrations.

Sheep heart phosphofructokinase (see Chapter 1, Section I.D) when treated with 5′-*p*-fluorosulfonyl adenosine is modified at allosteric sites. The activity of the modified enzyme is not influenced, either by allosteric inhibitor ATP or by allosteric activators AMP and ADP.[421] The model was used for the determination of the free energy coupling between the ATP and affinity label attached to the regulatory site of the enzyme.[422]

Another enzyme subjected to multiple regulation with nucleotides is glutamate dehydrogenase, catalyzing reaction

$$^-OOC-CH(NH_3^+)-CH_2-CH_2-COO^- + NAD^+ + H_2O \rightleftharpoons$$
$$\rightleftharpoons\ ^-OOC-CO-CH_2-CH_2-COO^- + NH_4^+ + NADH$$

GTP is known to be one allosteric inhibitor of the enzyme. *p*-Fluoro-sulfonyl-benzoylguanosine[423] and *p*-fluorosulfonylbenzoyl-1,N^6-ethenoadenosine[356] were shown to modify bovine liver enzyme without affecting enzymatic activity but with the loss of sensitivity of GTP inhibition. NAD-dependent isocitrate dehydrogenase is activated by ADP. Similar to the case of phosphofructokinase (see Chapter 1, Section I.D), ADP warns that there is a deficit of ATP in the cell and the supply of the electron transport chain with NADH should be intensified. The ADP analog 6-[(4-bromo-2,3-dioxobutyl)thio]-6-deaminoadenosine 5′-monophosphate reacts irreversibly with the pig heart enzyme, and the process is accompanied with the loss of sensitivity to ADP activation.[424]

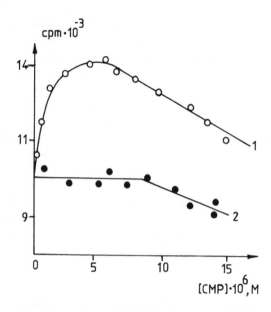

FIGURE 1. The dependence of the initial rate of enzymatic aminoacylation of *E. coli* tRNA^{Phe} on the concentration of cytidine-5'-monophosphate. (1) Native enzyme and (2) enzyme irradiated in the presence of GDP γ-*p*-azidoanilide (10^{-4} *M*). (From Nevinsky, G. A., Lavrik, O. I., Favorova, O. O., and Kisselev, L. L., *Bioorg. Khim. (USSR)*, 5, 352, 1979.)

Glycogen is not only a substrate but also an activator of rabbit muscle glycogen phosphorylase. Activation proceeds as a result of binding to a special glycogen storage site. Maltooligosaccharide activated with BrCN covalently binds to the enzyme without a loss of activity. This means that attachment takes place outside the substrate binding site. At the same time the enzyme loses the sensitivity to activation by glycogen, thus indicating that reaction proceeds in the glycogen storage site.[425]

This is not the complete list of the functionally modified enzymes prepared by affinity modification. But this list is sufficient to illustrate the possibilities of the approach.

Enzymology is the most advanced field of application of affinity modification. Many approaches elaborated here may be easily transferred to other functional proteins. To facilitate the acquaintance with main investigations performed in this field we have supplied this book with a special table (see the Appendix, Table 2) where many of enzymes are surveyed which were subjected to the study by affinity modification.

C. Affinity Modification of the Nucleic Acid Polymerases

Two main nucleic acid polymerases, DNA-dependent RNA polymerase and DNA-dependent DNA polymerase, catalyze the synthesis of new polynucleotide chains directed by template DNA. Most catalytic events carried out by these enzymes represent an elongation step, proceeding in accordance with the reaction shown in Chapter 1, Section I.D, Equation 1. Ribonucleoside-5'-triphosphates (NTPs) and deoxyribonucleoside-5'-triphosphates (dNTPs) are respective substrates of these enzymes. Using abbreviated symbols of phosphate and nucleoside moieties, the elongation step may be written as

$$\sim\!\!\!\sim pN_i + pppN_{i+1} \rightarrow \sim\!\!\!\sim pN_ipN_{i+1} + pp \tag{4}$$

where N_i and N_{i+1} are the nucleoside fragments with 5'-phosphate at the left side and 3'-phosphate at the right side.

In spite of the similarity of the reactions catalyzed by these two types of enzymes there exists a strong difference between these polymerases.

The first one concerns initiation of the process. RNA polymerases can initiate polymerization which starts from the reaction between two NTPs at definite points of the DNA template according to the reaction shown in Chapter 1, Section I.E, Equation 2. In abbreviated symbols the reaction may be written as

$$pppN_1 + pppN_2 \rightarrow pppN_1pN_2 + pp \qquad (5)$$

Therefore, RNA polymerases are able to recognize definite regions of double-stranded DNAs, usually called promoters. These regions strongly differ for RNA polymerases of different origin. Promoters used by some definite polymerase bear common features. Thus, all promoters recognized by the most-studied *E. coli* RNA polymerase have rather similar sequences in the region preceding the transcribed part of the template (Pribnow box) and in the region residing nearly 35 nucleotides prior to the initiation point as shown in Chapter 1, Figure 16. At the same time, DNA polymerase is unable to initiate formation of new chains and needs in a preexisting oligonucleotide (primer) complementary to some region of the template DNA. In experimental conditions this may be achieved by the use of synthetic oligonucleotides. In living systems replication starts by special RNA polymerases called primases which initiate the synthesis of short RNA primers further elongated by DNA polymerase. Primase recognizes definite points at the template DNA which thus are the points of origin of replication. At the final steps of replication short RNA pieces are eliminated by special enzymes.

The second difference concerns the forms of the template-entering polymerase molecule and of the product leaving the enzyme. RNA polymerase interacts with double-stranded DNA which is locally unwound in the region of the active center. The immediate product of the reaction is most probably a short duplex between single-stranded template DNA (transcribed strand) and nascent RNA. However, after several steps of RNA synthesis, the latter is displaced by the second (nontranscribed) DNA chain and, therefore, RNA leaves the enzyme in the single-stranded form. For some period of the process, the nascent chain still remains in contact with the enzyme. The respective region of the enzyme is often referred to as the exit corridor. DNA polymerase deals with single-stranded DNA obtained by preceding unwinding with the help of special auxiliary proteins. Contrary to transcription, when only temporary separation of the nucleic acid strands occurs, in replication DNA strands separate forever, each of them directing the synthesis of the daughter molecule which leaves the enzyme as duplex formed with a maternal strand.

Most detailed investigations of RNA polymerases by affinity modification were performed with *E. coli* DNA polymerase. This enzyme has the subunit structure $\alpha_2\beta\beta'\sigma$, the latter being an easily separable subunit essential for correct initiation.

To elucidate subunits interacting with the template, a number of photocross-linking studies was carried out. This was done with a self-complementary polymer with alternating dAMP and dTMP residues, poly d(A-T),[426] *E. coli* DNA;[427] with oligo- and polynucleotide containing some special photoreactive analogues of natural mononucleotides: oligo-4-thiothymidilate,[428] oligothymidilates with 5-bromodeoxyuridine dbr⁵U in definite positions, namely (pdT)₅pdbr⁵U(pdT)₄ and (pdT)₉pdbr⁵U;[429] and copolymers poly d(A,8-N₃A)[283] and poly d(A-5-IU).[430] In all cases either β and β′ subunits or ββ′ and σ subunits were modified. In no one case was the direct contact of DNA with an α subunit observed. Recently, bromocyan fragments of β and β′ subunits photocross-linked with (pdT)₅pdbr⁵U(pdT)₄ were isolated and the points of cross-linking were tentatively identified as Tyr-726 of the β subunit and Phe-988 of the β′ subunit of *E. coli* RNA polymerase.[431]

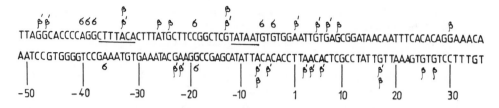

FIGURE 2.   Points of contact of lac UV5 promoter region with subunits of *E. coli* RNA polymerase as revealed by chemical cross-linking. Pribnow box and "−35" region of nontranscribed chain are underlined. (From Chenchik, A. S., Beabealashvilli, R. S., and Mirzabekov, A. D., *FEBS Lett.*, 128, 46,1981.)

Studies performed with promoter-containing systems could elucidate the regions of the promoter which are in direct contact with RNA polymerase. It was shown that UV irradiation of the complex of *E. coli* RNA polymerase with DNA of bacteriophage T7 taken in stoichiometry 3:1, the conditions favoring formation of the complex with promoter regions, results in cross-linking of β and σ subunits with DNA.[432,433] The use of DNA fragments containing promoter regions with 5-bromodeoxyuridine substituted for thymidine presented the possibility of revealing the promoter regions which are in contact with the enzyme.[434,435] Chemical cross-linking of the fragment containing lac UV5 promoter (mutant promoter of lactose operon) with *E. coli* RNA polymerase was performed using a partially apurinic DNA fragment by Scheme 84 described in Chapter 2. RNA polymerase subunits containing covalently attached ^{32}P-labeled promoter fragments were separated by electrophoresis in polyacrylamide gel containing sodium dodecyl sulfate (SDS). To isolate oligonucleotides bound to each subunit the conjugates were digested by pronase. The subsequent identification of oligonucleotides permitted the points of attachment of both promoter strands to each of the subunits to be found. The resulting distribution of the points of contacts are given in Figure 2.[436]

An intriguing question arises as to how the enzyme finds the promoter regions. As the search of the promoter proceeds within a few minutes, only fast kinetic measurements are suitable for this purpose. The respective study was performed using stopped-flow apparatus for rapid mixing of phage T7 DNA and *E. coli* RNA polymerase. After various times after mixing the solution was irradiated with a 10-μs pulse of UV light. Investigation of the cross-linking pattern and kinetics of cross-linking has provided an insight into the mechanism of RNA polymerase-DNA binding and the mechanism by which the enzyme searches for the promoter. It was found that the initial binding of the enzyme to the DNA is a fast diffusion-controlled process. Initial contacts of RNA polymerase subunits β,β′ and σ identified by cross-linking are proportional to the size of subunits; the largest β and β′ subunits were cross-linked to DNA predominantly. It was concluded that the complex formation occurs through a random collision between the complex components since, in this case, the degree of DNA contact with the subunits should be determined by surface area and arrangement in the enzyme. Fast mixing-cross-linking experiments also allowed insight into the mechanism of the promoter search. After the initial random binding, the enzyme must find the promoter sites either by searching within the same DNA molecule or among different DNA molecules. To investigate the mechanism of the search, an initial random complex of radiolabeled DNA and the enzyme was mixed with an excess of unlabeled DNA and cross-linked at various times after the mixing, to study the process of transfer of the protein from the preformed random complex to the added DNA. It was found that this transfer process is slower than the formation of the specific RNA polymerase-T7 promoter complex. Therefore, it was concluded that the enzyme is undergoing predominantly intramolecular transfer in the course of the promoter search. This suggests that the promoter search is not a purely random collision process as is the initial binding. Kinetic experiments have also demonstrated that complexes with nonspecific binding sites are intermediates in the search of the promoter sites and that this process is a one-dimensional diffusion. The distribution of the enzyme along the DNA was studied by restriction endonucleases digestion and analysis of the fragments bearing the cross-linked enzyme. Analysis of distribution of RNA polymerase on the DNA chain at various times after the initial binding showed that the initial binding occurs randomly along the DNA molecule and that, following the initial binding, the enzyme molecules were transferred from nonpromoter- to promoter-containing fragments of T7 DNA.[301,437]

To study the exit corridor, two main approaches were proposed. The first one is based upon the observation of Grachev and Zaychikov that *E. coli* RNA polymerase may use NTPs bearing rather bulky groups at γ-phosphate both as the substrate and initiator. This means that the enzyme catalyzes the modified version of Equation 5

$$XpppN_1 + XpppN_2 \rightarrow XpppN_1pN_2 + Xpp$$

where X is some substituent.[438] A similar reaction proceeds at the following elongation steps retaining one X residue at the 5′ end of the growing reaction chain. This group moves along the exit corridor till the moment when the 5′ end of RNA transcript leaves the enzyme. Using γ-*p*-azidoanilidates of ATP and GTP, nascent RNA with the photoreactive azido group was obtained thus permitting the study of the exit corridor.[439] Further investigations performed with the promoter-containing fragment of phage T7 DNA have demonstrated that β and σ subunits of the enzyme are labeled. It was also found that transcripts of the length of 2, 6, or 10 nucleotide residues were mainly photocross-linked with the β subunit, and those of the length of 2 and 6 were cross-linked with the σ-subunit.[440]

A similar method based upon the use of the diribonucleotide primer 5′-(4-azidophenacylthio)-pApU to initiate the RNA chain growth was applied to an *E. coli* RNA polymerase/T7 DNA transcription complex. Irradiation of the system containing the transcripts of different length has shown that with nascent RNA up to 12 nucleotides in length DNA is labeled predominantly. Afterwards, the 5′ end diverged from DNA and only the β and β′ subunits were labeled thereafter by reactive RNA chains up to 94 nucleotides in length.[441]

The first part of the nascent RNA chain just preceding the point of growth at the 3′ end is believed to remain in the form of a double strand formed with a coding chain of DNA. An artificial model of this system may be prepared using complementary oligonucleotides covering completely or partially the melted fragment of the promoter region of DNA. It was shown that such oligonucleotides serve as efficient primers in the RNA polymerase reaction, thus indicating that they form duplexes with the coding chain.[442] The study of the exit corridor based upon this observation was performed using oligonucleotides complementary to the A2 promoter region of T7 phage DNA. The product of initiation at this primer is pppGpC. With the reactive analogue bearing *N*-methylimidazole residue (MeIm) at 5′-phosphate, MeImpdGpC, it was found that histidine and lysine residues of the β subunit are modified. Using a number of such derivatives with additional nucleotide residues complementary to preceding region of coding promoter strands from tri- to heptanucleotide (from MeImpdCpdGpC to MeImpdApdApdApdTpdCpdGpC), it was shown that tri- and tetranucleotide derivatives modify only the same residues of the β subunit; the derivative of pentanucleotide additionally modifies the lysine residue of β′ subunit and starting from hexanucleotide essential labeling of histidine residue of the σ subunit is observed. At present, this is the most detailed information concerning the exit corridor.[443]

An interesting observation was done by Mustaev et al.[323,444] It was found that ATMP being added to an *E. coli* RNA polymerase/T7 DNA fragment complex modifies the enzyme in a catalytically competent manner, thus permitting the following phosphorylation by a [^{32}P] UMP fragment upon the addition of [α-^{32}P] UTP. Therefore, with the cold reagent only competently bound ATP fragments become radioactive and the location at subunits and definite amino acid residues may be identified. This opened the possibility of highly selective radioactive labeling of RNA polymerase. Further, a similar approach with the label transfer to a competently bound affinity reagent was applied to a number of more stable reactive analogues of initiator nucleotides, including nucleotide 5'-imidazolides.[324] This permitted to elucidate the point of labeling of a β-subunit of *E. coli* RNA polymerase by the method described in Chapter 3, Section IV.B. The same approach of highly specific affinity modification was recently used for the labeling of wheat germ RNA polymerase II (in eukaryotes different enzymes are responsible for synthesis of large ribosomal (r) and messenger (m) RNAs, RNA polymerase I and II, respectively). A subunit of molecular mass 140,000 was found to be modified.[445]

The result strongly differs from those obtained for other eukaryotic RNA polymerase II. Photoaffinity modification of enzymes from calf thymus[446] and term human placenta[447] with 8-azido-ATP resulted in the labeling of one small subunit with a molecular mass 37,000. However, it should be noted that these enzymes were labeled in the absence of a template. It cannot be excluded that labeling of a template or product binding site occured in this case similar to results obtained for DNA polymerases (see below).

Only a few papers dealing with affinity modification of DNA polymerases have appeared. Enzyme sources and reagents used in the described investigations are surveyed in Table 2 in the Appendix. More systematic studies were started in last few years. To study the mechanism of interaction of a template with DNA polymerase, affinity modification of the enzyme was performed with the derivative of an oligonucleotide $(pdT)_2pdC(pdT)_7$ obtained by the treatment of the latter with $Pt(H_2O)(OH^-)(NH_3)_2^+$. This derivative has single cytosine residue with a moderately reactive platinating group.

R is the deoxyribose fragment of oligonucleotide.

It was shown to inactivate DNA polymerase α from human placenta with $K_x = 2 \times 10^6$ M^{-1} at 30°C. Oligothymidylates of different length were used as competitors and the affinities to the enzyme were determined. It was found that oligonucleotide $(dTp)_9dT$ and a similar oligonucleotide with 3'-terminal 2', 3'-dideoxythymidine residue lacking a 3'-hydroxy group compete with the reagent with equal efficiency, K_y being 6.6×10^6 M^{-1}. The identity of K_y values for both oligothymidylates indicates that the 3'-hydroxy group is unessential in the above interaction. Consequently, this cannot be in competition with the primer site since the 3'-hydroxy group of the primer should participate in the recognition of the enzyme. A number of oligothymidylates were studied as competitors of the platinating reagent and K_y values were found to change from 1.9×10^5 M^{-1} for $(pdT)_3$ to 1.09×10^7 M^{-1} for $(dTp)_{14}dT$. These values permitted an estimation of the free energy $\Delta G°$ of interaction of one monomeric pdT unit with the enzyme which was found to be 1.35 kJ/mol. This value is too small to be related to electrostatic interactions or hydrogen binding. The predominating role of hydrophobic interactions seems more probable. The dependence of $\Delta G°$ on the number of thymidylate residue is presented in Figure 3. Extrapolation to monomeric pdT demonstrates that $\Delta G°$ for the first nucleotide unit is sufficiently greater than that for the next ones. This means that electrostatic contacts may play a significant role in the interaction of one or two nucleotide units of the template with the enzyme. This also means that the mononucleotide should have significant affinity to the template site and, therefore, affinity modification with dNTP analogues may proceed in this site. Therefore, correct investigations of the substrate

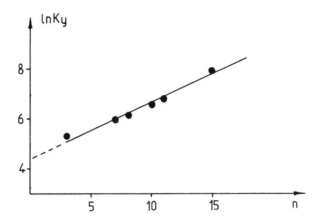

FIGURE 3. The dependence of association constant K_y of oligo-thymidylates $d(pT)_n$ with human placenta DNA polymerase on the length of oligomers as revealed from the competition experiments on affinity modification of the enzyme with platinating oligonucleotide derivative. (From Nevinsky, G. A., Podust, V. N., Levina, A. S., Khalabuda, O. V., and Lavrik, O. I., *Bioorg. Khim. (USSR)*, 12, 357, 1986.)

site of DNA polymerases should be performed in the presence of the complex template-primer.[448]

This conclusion was supported by further investigations of affinity modification of *E. coli* DNA polymerase I with γ-imidazolides of dNTPs. It was demonstrated that the presence of the template-primer complex is absolutely necessary for the inactivation of the enzyme, and inactivation proceeds efficiently only with derivatives of dNTPs complementary to the template, e.g., dCTP imidazolide in the presence of complex of poly(dG) · poly(dC).[449]

Data presented in this section clearly demonstrate that affinity modification becomes an efficient tool of investigation of nucleic acid polymerases, and systematic use of this tool may help significantly in the elucidation of the mechanism of template catalysis.

D. Affinity Modification as a Tool of the Study of Functional Topography of Ribosomes

Ribosomes are multicomponent nucleoproteins which govern the translation of the genetic message delivered by mRNA as a sequence of codons into the respective amino acid sequence of the protein to be synthesized. The most essential features of the structure of ribosomes were described in Chapter 1, Section I.C. The main chemical process carried out by ribosomes is the formation of new peptide bonds which proceeds in accordance with the following equation:

$$XCHR_iCO-\ldots NHCHR_iCO\text{-}tRNA^{(i)} + NH_3^+CHR_{i+1}CO\text{-}tRNA^{(i+1)}$$

$$\rightarrow XCHR_iCO-\ldots NHCHR_iCO\text{-}NHCHR_{i+1}CO\text{-}tRNA^{(i+1)} + tRNA^{(i)} \qquad (6)$$

X is the protonated amino group in eukaryotic systems and formylamino group HCONH in prokaryotic systems.

At each step AA-tRNA is selected in accordance with the codon to be translated. Reaction 6 is a transfer of the peptidyl fragment to the amino group of the newly selected AA-tRNA and is called a peptidyltransferase reaction. During this reaction, two RNA derivatives are present simultaneously at the ribosome. Consequently, at least two tRNA binding sites should exist. They will be further referred to as the P (peptidyl) and A (aminoacyl) sites in accordance with the fragment attached to tRNA at the moment of peptidyl transfer. Therefore, four

$$\text{Rs.mRNA.pept}_i\text{-tRNA}^{(i)}(\text{P}) \tag{a}$$

$$\downarrow \quad \begin{array}{l} + \text{ EF-1·GTP·aa}_{i+1}\text{-tRNA}^{(i+1)} \\ \quad (\text{selection}) \end{array}$$

$$\text{Rs.mRNA.pept}_i\text{-tRNA}(\text{P}).\text{EF-1·GTP·aa}_{i+1}\text{-tRNA}^{(i+1)} \tag{b}$$

$$\downarrow \quad - \text{ EF-1·GDP} + \text{P}_i$$

$$\text{Rs.mRNA.pept}_i\text{-tRNA}^{(i)}(\text{P}).\text{aa}_{i+1}\text{-tRNA}^{(i+1)}(\text{A}) \tag{c}$$

$$\downarrow \quad \text{peptidyl transfer}$$

$$\text{Rs.mRNA.tRNA}^{(i)}(\text{P}).\text{pept}_{i+1}\text{-tRNA}^{(i+1)}(\text{A}) \tag{d}$$

$$\begin{array}{c} \text{EF-2·GTP} \\ \downarrow \quad \text{EF-2·GDP} + \text{P}_i \end{array}$$

$$\text{Rs.mRNA.pept}_{i+1}\text{-tRNA}^{(i+1)}(\text{P}) + \text{tRNA}^{(i)} \tag{a}$$

FIGURE 4.   General scheme of elongation cycle. Rs, ribosomes; pept$_i$, pept$_{i+1}$, peptidyl fragments containing i; and i + 1, amino acid residues. The states occupied by tRNA derivatives are indicated in the brackets. States (b) and (c) may not be realized in the fully native system since the addition of ternary complex EF-1·GTP·AA-tRNA results directly in the formation of posttranslational state (d). Artificial conditions leading to states (b) and (c) are described in the text.

most essential partially overlapping areas should exist on ribosomes: mRNA binding area, P site, A site, and peptidyltransferase center.

In reality, functional sites of ribosomes are significantly more numerous. The complete set of our knowledge of the problem may be found in special manuals.[450] Here we must only mention the participation of special elongation factors in the natural translation process. AA-tRNAs are selected in the form of the ternary complex with the special elongation factor EF-1 (in prokaryotes usually referred to as EF-Tu) and GTP. After formation of the complex,

$$\text{Ribosome} \cdot \text{mRNA} \cdot \text{peptidyl-tRNA} \cdot \text{EF-1} \cdot \text{AA-tRNA} \cdot \text{GTP}$$

the latter component is hydrolyzed to GDP and inorganic phosphate, P_i, and simultaneously EF-1 is removed from the ribosome. This process is followed by peptidyl transfer (Equation 6). Now peptidyl-tRNA^{i+1} occupies the A site. To carry out the selection of the next AA-tRNA, the ribosome must translocate peptidyl-tRNA^{i+1} to the P site. This step proceeds with the help of the complex of elongation factor EF-2 (in prokaryotes referred to as EF-G) with GTP. Prior to the selection of the next AA-tRNA, GTP is hydrolyzed to GDP and P_i with subsequent dissociation of EF-2 off the ribosome, thus closing the cycle. The general scheme of these processes is presented in Figure 4. From the above consideration, it follows that the ribosomes also contain specific areas for the binding of elongation factors and participate in some way in the GTP hydrolysis. Additional factors participate in the initiation and termination of protein biosynthesis. Besides this, the action of many important antibiotics (tetracycline, streptomycin, chloramphenicol, etc.) is known to be based upon the specific interaction with prokaryotic ribosomes and consequently the latter contain some antibiotic binding sites. The arrangement of all above-mentioned sites is referred to as functional topography of ribosomes. It is evident that affinity modification of ribosomes may serve as a powerful tool in the study of the functional topography. In this section, we shall present the results of most systematic investigations in this field concerning the main functional centers of ribosomes.

These investigations started from the study of the peptidyl-transferase center of ribosomes with tRNA derivatives bearing the reactive group attached to the α-amino group of aminoacyl moiety. The first reagents proposed by three groups of the authors were chlorambucilyl-Phe-tRNA (CXXXIa),[273] bromoacetyl-Phe-tRNA (CXXXIb),[451] and p-nitrophenoxycarbonyl-Phe-tRNA (CXXXIc).[452] The reagents were used to modify $E. coli$ ribosomes within the ternary complex with poly(U) coding for polyphenylalanine. In all three cases, only the 50S subunit was modified, 23S rRNA being attached preferentially by the first two reagents. Interestingly, in the first case a catalytically competent attachment took place; chlorambucilyl-Phe residue covalently attached to 23S rRNA within $E. coli$ ribosomes could participate in a number of subsequent elongation cycles giving rise to chlorambucilyl-oligophenylalanyl-tRNA.[453]

The complete survey of the results obtained in the studies of the peptidyltransferase center may be found in special reviews.[454,455] Here we shall mention the most essential results. It is obvious now that to study functional topography of ribosomes it is strongly preferable to use an elongation factor-dependent system. The use of the ternary complex EF-Tu · GTP · (ϵ-BrCH$_2$CO)Lys-tRNA^{Lys} to put AA-tRNA in the A site of $E. coli$ ribosomes programmed with poly(A) resulted in alkylation of the single protein L27 (for designation of ribosomal proteins here and below see Chapter 1, Section I.C). After addition of EF-G and GTP to perform translocation, proteins L2, L13/L14/L15 (not separated), L24, and L32/L33 were found to be alkylated.[456] Some of these proteins were determined as alkylation points in the study of the exit corridor of the growing peptide chain similar to that discussed in the RNA polymerase studies described in the previous section. To perform this investigation Br-CH$_2$CO-(Gly)$_n$ was attached to an α-amino group of Phe-tRNA^{Phe} with n in the range of 1 to 16. The derivatives obtained were used for affinity modification of $E. coli$ ribosomes in the ternary complex with poly(U). An oligoglycine fragment served as a model of the growing polypeptide chain. The pattern of alkylated proteins for oligopeptides of various length permitted a qualitative evaluation of the distance of the proteins modified from the point of residence of the tRNA 3' end. Thus, L2 was modified preferentially by derivatives with short oligopeptide fragments, thus indicating that this protein is the closest one. L32/L33 was preferentially modified with the oligopeptide moieties of moderate length, and L24 was modified mainly with the longest ones.[457]

As mentioned, 23S rRNA was found to be modified by a number of reactive AA-tRNA derivatives with the reactive group residing in the region of the peptidyltransferase center. Using the approach described in Chapter 3, Section IV.B, Barta et al.[279] have identified the point of modification of 23S rRNA within the UV-irradiated ternary complex of $E. coli$ ribosomes and poly(U) with photoreactive tRNA derivative 3-(4'-benzoylphenyl)propionyl-Phe-tRNA as U2584 or U2585.

The derivatives bearing reactive groups in these regions were used to study the internal regions of tRNA interacting with ribosomes. The most systematic study was carried out by Karpova et al.[288] with photoreactive derivatives of $E. coli$ tRNA^{Phe} bearing arylazido groups scattered statistically over guanosine residues (see Chapter 2, Section III.D). Further, these derivatives will be referred to as tRNA$^{Phe}_{az}$. These derivatives may be enzymatically aminoacylated giving Phe-tRNA$^{Phe}_{az}$. Five complexes differing in the steps of elongation cycle and the residence of photoreactive tRNA moiety were obtained. All these complexes were subjected to irradiation and the patterns of modified proteins were studied. We shall briefly characterize these complexes.

The addition of ternary complex EF-Tu · GTP · Phe-tRNA$_{az}^{Phe}$ to ribosomes programmed with poly(U) and containing the peptidyl-tRNA analogue AcPhe-tRNA^{Phe} in the P site results in the selection of tRNA$_{az}^{Phe}$ and subsequent peptidyl transfer. Thus, the state (Figure 4d) is formed with AcPhe-Phe-tRNA$_{az}^{Phe}$ in the A site.[458] The state (a) was obtained performing polyphenylalanine synthesis in the poly(U)-directed system supplied both with EF-Tu and EF-G up to complete exhaustion of Phe-tRNA$_{az}^{Phe}$. This state contains (Phe)$_n$-tRNA$_{az}^{Phe}$ in the P site, the A site being vacant.[288] Following addition of the ternary complex with nonmodified tRNA^{Phe} EF-Tu·GTP·Phe-tRNA^{Phe} leads to selection of the next Phe-tRNA^{Phe} followed by peptidyl transfer, thus giving (Phe)$_{n+1}$-tRNA^{Phe} in the A site and leaving nonacylated tRNA$_{az}^{Phe}$ in the P site. This is again state (d) but with photoreactive tRNA in the P site.[288] States (b) and (c) may be also obtained in somewhat artificial conditions. Thus, using a ternary complex containing the nonhydrolyzable GTP analogue GPPCP with the methylene group substituted for the oxygen bridging, β- and γ-phosphate, it is possible to obtain state (b). To stop the process at the state (c) AcPhe-tRNA$_{ox-red}^{Phe}$ may be used, obtained by consecutive oxidation of tRNA^{Phe} with IO_4^- and reduction with $NaBH_4$. This tRNA$_{ox-red}^{Phe}$ retains the ability of enzymatic acylation forming Phe-tRNA$_{ox-red}^{Phe}$ but the latter cannot serve as a peptide donor in Equation 6.[458] The full set of the results obtained may be found in the original papers. In Figure 5 we present the patterns of modified proteins of 30S subunit for the above-described five complexes. It is seen that these patterns differ strongly, thus indicating the significant displacement of tRNA along the ribosome in all steps of elongation cycle.

Similar results were obtained with some tRNA derivatives containing photoactivated arylazido fragments attached to minor bases. A derivative of *E. coli* tRNA containing p-N_3-phenyl residue bound to s^4U-8 via $COCH_2$ or $COCH_2OCOCH_2$ spacers modified ribosomes being introduced into poly(U)-programmed ribosomes in the form of a ternary complex with EF-Tu and GTP (state d) but did not react with ribosomes either when added as a ternary complex with EF-Tu and GPPCP (state b) or after translocation into P site (state a).[459] Single protein S19 was found to be modified in the first case.[460]

Several species of *E. coli* tRNA$_1^{Val}$ containing 5-carboxymethoxyuridine in position 34 of the anticodon with attached 2-nitro-4-azidophenyl (NAP) groups were used for modification of ribosomes in the presence of copolymer poly(U,G) (GUU codes for valine). It was found that photomodification proceeded both with AcVal-tRNA^{Val} derivatives put in the absence of elongation factors in the P site and with Val-tRNA^{Val} derivatives introduced as the ternary complex with EF-Tu and GTP in the A site. However, 16S rRNA was labeled predominantly with derivatives put in the A site whereas only 30S proteins were subjected to modification with derivative in the P site.[461] Using derivatives containing NAP-$NH(CH_2)_5CONHCH_2$ $CH_2NHCOCH_2O$-U34 and NAP-$NH(CH_2)_2SS(CH_2)_2CONH(CH_2)_2NHCOCH_2O$-U34, it was demonstrated that C1400 is the point of labeling of 16S rRNA.[462]

Among other approaches direct UV cross-linking was used for the same purpose. The survey of results obtained under irradiation at 254 nm by Abdurashidova et al.[463] may be found in the review paper. Specific cross-linking of 5-carboxymethoxyuridine with C1400 of 16S rRNA was observed with unmodified AcVal-tRNA$_1^{Val}$ localized in the P site of *E. coli* ribosomes programmed with poly(U,G) under mild UV-irradiation ($\lambda > 310$ nm).[150]

The study of mRNA binding area of ribosomes was started nearly at the same time as the study of peptidyltransferase center. In the first experiments performed with an alkylating derivative of oligoadenylate (Ap)$_5$A > CHRCl (n = 5), RCl — p-(N-2-chloroethyl-N-methylamino) phenyl residue, it was found that modification of *E. coli* ribosomes in the presence of respective tRNA, namely tRNA^{Lys}, readily takes place. Moreover, it was demonstrated that modified ribosomes bind Lys-tRNA^{Lys} in the absence of exogenous poly(A). This means that modification proceeds in a functionally competent manner.[464] A similar result was obtained with an alkylating derivative of polyuridylate (Up)$_3$ (5-ICH_2CONH)U. Modified ribosomes could bind Phe-tRNA^{Phe} in the absence of poly(U).[465]

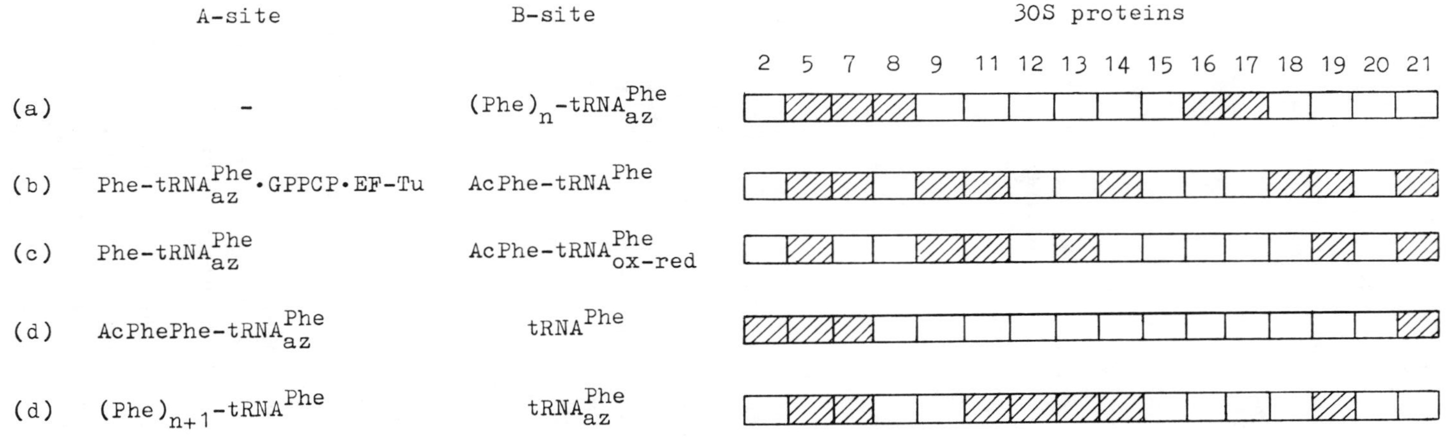

FIGURE 5. 30S ribosomal proteins modified by photoreactive tRNA derivative $tRNA_{az}^{Phe}$ in 70S ribosomes under conditions imitating different elongation steps. (From Vladimirov, S. N., Graifer, D. M., Karpova, G. G., Semenkov, Yu. P., Makhno, V. I., and Kirillov, S. V., *FEBS Lett.*, 181, 367, 1985.)

Table 1

**DISTRIBUTION OF THE LABEL BETWEEN RIBOSOMAL
COMPONENTS OBTAINED IN THE EXPERIMENTS ON
AFFINITY MODIFICATION OF *E. coli* RIBOSOMES IN THE
TERNARY COMPLEX WITH tRNA^{Phe} AND REACTIVE
OLIGOURIDILATE ANALOGUES**

$$CH_3 \diagdown N - \bigcirc - CH_2 - NH(pU)_n$$
$$ClCH_2CH_2 \diagup$$

Number	Moles of covalently attached reagent per mole of subunit		rRNA %	Protein %	Proteins modified
	30S	50S			
4	0.133	0.03	70	30	S3, S4; L7/L12
5	0.176	0.104	90	10	—
6	0.125	0.114	58	42	S3, S9, S11, S13; L23, L25, L32
7	0.130	0.120	14	86	S5, S11, S13; L2, L7/ L12, L19

Among further investigations, most systematic studies with oligouridylate derivatives bearing the above-mentioned RCl residue attached either to the 3′ or to 5′ end: $p(U_p)_{n-1}$ U > CHRCl (n = 5 ÷ 8)[319] and ClRCH$_2$NH(pU)$_n$ (n = 4 ÷ 7).[466] In the experiments under consideration, the ternary complex of *E. coli* ribosomes with reagent and *E. coli* tRNA^{Phe} was formed in the presence of the excess of reagent and the unbound reagent was removed by gel filtration. After incubation corresponding to at least half-time of the formation of ethyleneimmonium cation, the level of labeling was estimated as well as the distribution of the label between 30S and 50S subunits, between rRNA and proteins, and between different ribosomal proteins. The results obtained with the reagent with alkylating group bound to the 5′ end of oligo(U) are given in Table 1. It is seen that in all cases both subunits are modified. According to the consideration given in Section A, this means that the mRNA binding area is close to the subunit interface. It is clearly seen that in a number of cases the results change strongly with the change of the addressing part of the reagent by one nucleotide residue. The same was found to be with the reagents bearing the reactive group at the 3′ end. This could be due to a number of factors: the presence of a rather large spacer separating addressing and reactive moieties, dynamics of the system favoring some occasional contacts between the reactive group and some parts of ribosome advantageous for chemical reaction (see Chapter 4, Section II.C), and sliding of oligo(U) moiety along ribosome. The role of dynamics of the system could be especially significant due to the absence of tRNA in the A site.

At present, only first attempts are described which try to lower the multiplicity of modification and labeling of occasional proteins. To escape large spacers, direct cross-linking of 5′-terminal phosphate of (pU)$_7$ with *E. coli* ribosomes was carried out using water-soluble carbodiimide (see Chapter 2, Section II.B). This modification was performed in the presence of tRNA^{Phe} in the P site with the A site being either vacant or occupied by Phe-tRNA^{Phe}. Only in the latter case was selective labeling of one definite protein, namely S3, observed.[467] Similar results were obtained with ApUpGp(Up)$_5$U > CHRCl as the alkylating template analogue. The codon ApUpG recognizes initiator tRNA$_f^{Met}$ and, therefore, no sliding of the reactive template analogue should be expected. As the correct initiation proceeds only in the presence of special initiator factors, the complete system containing these factors was used. However, in this case up to nine proteins were found to be labeled. This number strongly decreases when, besides initiation factors and fMet-tRNA$_f^{Met}$ (fMet — N-formyl-methionine), the ternary complex EF-Tu·GTP·Phe-tRNA^{Phe} containing AA-tRNA recognized by the second trinucleotide of the analogue was added. Stabilization of the codon-anticodon interaction resulted in the significant decrease of the number of labeled proteins, only proteins S13 and L10 being modified.[468]

It is interesting that with rat liver ribosomes rather unequivocal modification with similar alkylating mRNA analogues proceeded without special introduction of ternary complex EF-1·GTP·AA-tRNA. Oligouridylate with a ClR-residue at the 3' end alkylated mainly protein S3/S3a and with a similar residue at the 5' end alkylated protein S26.[469-471]

The results presented clearly demonstrate that affinity modification may become a powerful tool of the study of functional topography of ribosomes. However, rather complicated investigations are necessary, including the use of heterogeneous templates stabilized in definite positions by cognate tRNAs, both in P and A sites, and of a number of reagents differing in the points of attachment and in the chemical nature of reactive groups.

Among other functional regions of ribosomes, the sites of binding of various antibiotics were subjected to rather numerous investigations. However, these sites are not involved directly in the main ribosomal function and, therefore, we shall restrict ourselves by providing references to some recent review papers.[472,474]

A successful attempt to cross-link one of the initiation factors IF-3 to *E. coli* ribosome was described recently.[475] Two groups of authors tried to use reactive GTP derivatives to elucidate the area responsible for GTPase activity of ribosomes.[476,477] The results of these two groups look contradictory and their consideration may be found in original papers.

E. Complementary Addressed Modification of Nucleic Acids

As mentioned in Section I, the general approach to sequence-specific affinity modification of nucleic acids with reactive derivatives of oligonucleotides called complementary addressed modification was formulated and realized by Grineva et al.[2,3] Due to many interesting possibilities opened by the approach, including generation of the nucleic acid-targeted drugs as well as the development of efficient methods of oligonucleotide synthesis, complementary addressed modification is receiving growing attention in the last years.[91,478]

The first experiments dealing with this approach were carried out in the early 1970s. At that time nucleic acids of known sequences as well as heterogeneous oligonucleotides were hardly available. First experiments were performed with alkylating derivatives of oligoadenylate with p-(N-2-chloroethyl-N-methylaminophenyl) (ClR) residue attached to the 2',3'-*cis*-diol group via acetal fragment (Ap)$_5$ACHRCl and rRNA as a target nucleic acid.[3] The main result of this investigation was the demonstration of high efficiency of the reaction. At moderate excess of reagent, up to 90% of ethyleneimmonium cation formed in the rate-limiting step was consumed for alkylation of rRNA. Such high efficiency (see Chapter 4, Section II.B) is typical of an intracomplex reaction.

Sequence specificity of complementary addressed alkylation was demonstrated for the first time in the reaction of yeast tRNA$_1^{Val}$ with the oligonucleotide derivative dCpdGp-ACHRCl. It was found that this tRNA is preferentially alkylated at three points, all of them being in the same position towards CpG sequences of tRNA$_1^{Val}$ complementary to dCpdG fragment of the address, namely at the fourth nucleotide from the region of complementarity.[479]

All main recent investigations were performed with single-stranded DNA fragments of the known sequence.[280,338,480] The most thoroughly studied model is the 303-nucleotides-long fragment corresponding to a part of the tick-borne encephalitis virus RNA sequence.

```
           260                    270
            ↓                      ↓
5'...-T-T-G-A-A-T-G-G-G-A-A-G-A-G-G-G-T-C-A-G-G-...3'
                 ╲                          ↑
    3'{     rA-C-C-C-T-T-C-T-C-C-C-A-G-TpX₁ }
                                              5'
          X₂rA-C-C-C-T-T-C-T-C-C-C-A-G-Tp
```

$$X_1 = \begin{matrix} ClCH_2CH_2 \\ \\ CH_3 \end{matrix} \!\!\! N\!\!-\!\!\bigcirc\!\!-\!\!CH \qquad X_2 = \, ^-NHCH_2\!\!-\!\!\bigcirc\!\!-\!\!N \begin{matrix} CH_2CH_2Cl \\ \\ CH_3 \end{matrix}$$

FIGURE 6. The part of nucleotide sequence of DNA fragment subjected to complementary addressed alkylation and the reagent used. Hyphens are substituted for phosphate symbols and prefixes "d" (deoxy) are omitted. Upper arrows indicate the numbers of nucleotide residues of the template counting from 5' end. Lower arrows indicate preferentially alkylated residues of the target DNA. (From Vlassov, V. V., Zarytova, V. F., Kutiavin, I. V., Mamaev, S. V., and Podyminogin, M. A., *Nucleic Acids Res.*, 14, 4065, 1986.)

The sequence of the part of the fragment subjected to complementary addressed alkylation is presented in Figure 6 together with some reagents used. It was found that at low concentrations of both fragment $(1.5 \times 10^{-8} \, M)$ and reagents $(5 \times 10^{-7} \, M)$ a high level of alkylation was achieved. Alkylation points were determined by subsequent splitting at this point using the procedure described in Chapter 3, Section IV.B. Reagent bearing alkylating RCl-residue at the 5' end modified nucleotide residue adjacent to the last base-paired unit of the template. The position of alkylation with reagent bearing RCl-residue at the 3' end was similar to that determined in the above-mentioned studies of alkylation of tRNA$_1^{Val}$. Neighboring nucleotides are scarcely affected by these reagents, especially by the second one. Since appropriate treatment of alkylated DNA results in the cleavage of the polynucleotide chain, complementary addressed modification opens the possibility of site-specific chemical cleavage of single-stranded DNAs.[481]

However, in one last case besides alkylation of the target nucleotide G-259, a significant level of modification of G-179 far from the complementarity region was observed. A similar result was reported for addressed modification of DNA fragment obtained from single-stranded DNA of phage M13. The reagent with the oligonucleotide sequence complementary to the sequence 39 to 47 of the fragment bearing the reactive group attached to 5'-terminal thiophosphate (see Chapter 2, Section III.D) alkylated this fragment in the region 80 to 90 and around the nucleotide residue in position 150.[280]

At least two explanations may be given for specific modification of the target nucleic acid with the reactive oligonucleotide derivative in the region removed from the region of complementarity along the polynucleotide chain. The first one is the spatial proximity of this region to the modification point due to the specific tertiary structure of nucleic acid. The second one is the possibility of modification due to the formation of some unperfect complex with only partial base pairing of the address and nucleic acid subjected to modification.

The last possibility was analyzed in the studies of alkylation of 16S rRNA from *E. coli* with four derivatives addressed to unique sequences of nucleic acid shown in the parentheses:

d(pCpCpTpApCpGpGpTpT)rACHRCl (1506–1515)

d(pApCpCpTpTpGpTpT)rACHRCl (1498–1506)

d(pTpTpApCpGpApCpT)rUCHRCl (1492–1500)

d(pTpTpTpGpCpTpCpCpCpC)rACHRCl (771–781)

The dependences of the limit-modification yield achieved after several half-times of ionization of the 2-chloroethylamino group on the initial concentration of the reagents were determined and extrapolated to infinite concentration. Only the first reagent gave the extrapolated value close to unity. For three other nucleotides these values were estimated at 20°C as 2.0, 8.6, and 12.0, respectively. Interestingly, these numbers correlated well with those calculated from analysis of the primary 16S rRNA structure as they were able to give unperfect complexes with the addressing oligonucleotide, namely 2, 9, and 12. For example, the second reagent is complementary to the sequence 1498 to 1506 pUpApApCpApApGpGpA of 16S rRNA. Sequence 69 to 76 of the same rRNA pUpApApCpApGpGpGpA may form a complex with 7 nucleotide units of the addressing fragment, 6 of which are base pairs being normal Watson Crick ones and 1 being a G · U pair.[482]

To increase the affinity of oligonucleotides to complementary sequences, Letsinger and Schott[483] have proposed to attach intercalating dye to oligonucleotides. It was demonstrated that compound

forms a complex with poly(A) with a melting point near 47°C whereas thymidine oligonucleotides smaller than the pentamer fail to bind with poly(A) at 0°C. This approach was further developed by Asseline and co-workers[484] who have prepared a series of oligothymidylates bearing acridine dye attached to 3′-terminal phosphate via oligomethylene bridge. It was found that acridine enhances the melting point of the complex with poly(A) by more than 20°C.

This method of stabilization of double helices was applied to enhance the efficiency of complementary addressed modification. Oligonucleotide derivatives with phenazinium residue at the 5′ end and alkylating group at the 3′ end were prepared.[485] The derivative

with the oligonucleotide moiety complementary to the sequence 261 to 268 of the previously mentioned DNA fragment (Figure 6) was used for alkylation of this fragment. Efficiency of alkylation by this reagent at concentration $4 \times 10^{-6}\ M$ exceeded significantly that with the derivative of the same oligonucleotide lacking phenazinium residue at $1 \times 10^{-4}\ M$.[338] A significant enhancement of efficiency of the poly(dA) cleavage with EDTA derivative of oligothymidylate in the presence of Fe(II), H_2O_2, and dithiothreitol was observed after attachment of acridine residue to 5′-phosphate of the oligonucleotide.[486]

Interesting observations were done in experiments on alkylation of chromatin and isolated metaphase chromosomes with alkylating oligonucleotide derivatives. It is known that the

Table 2
COMPLEMENTARY ADDRESSED MODIFICATION OF DNA IN CHROMATIN AND METAPHASE CHROMOSOMES OF MOUSE FIBROBLASTS WITH ALKYLATING OLIGONUCLEOTIDE DERIVATIVES

| | | Alkylation extent (moles of reagent per moles of DNA nucleotide $\times 10^6$) | | |
| | | | | |
Reagent	In the presence of	Chromatin	Metaphase chromosomes	Isolated DNA
ClRCH$_2$NH(pdT)$_{16}$	—	42	17	0.2
	Excess of (pdT)$_{15}$	10[a]	2[b]	—
	Tenfold excess of (pdN)$_{16}$[c]	40	18	—
ClRCH$_2$NH(pdApdC)$_6$	—	—	100	0.4
	Fourfold excess of (pdApdC)$_6$	—	25	—
	Excess of (pdN)$_{16}$	—	87[a]	—

Note: 25°C, 18 hr, 0.01 *M* Hepes, pH 7.05; reagent concentration 10^{-6} *M*.

[a]   Fourfold excess of free oligodeoxynucleotide.
[b]   Tenfold excess of free oligodeoxynucleotide.
[c]   p(dN)$_{16}$ = pd(CATGCAAAACCTTCCC).

state of DNA in these structures differs significantly from that inherent to free DNA. Thus, the ratio of the yields of N7-methylguanine to N1-methyladenine in alkylation with dimethylsulfate is 6.6. This value is significantly lower than that for native DNA (40) and is closer to the value found for denatured DNA (2.2).[487] The possibility of complementary addressed modification of chromatin DNA and DNA in metaphase chromosomes was tested in experiments with chromatin isolated from human placenta and mouse fibroblasts grown in culture and with chromosomes isolated from the fibroblasts. As targets, poly(A) and (pdGpdT)$_n$ sequences were chosen since they are present in eukaryotic DNA in large amounts. The results are presented in Table 2. It is seen that with both types of addressing oligonucleotides, specific alkylation takes place which is suppressed by free oligonucleotide of the same structure and is not influenced by heterogeneous oligonucleotide of arbitrary structure.[488]

All above-mentioned results dealt with reactive derivatives containing a reactive group at one terminal nucleotide residue. Recently, complementary addressed modification was described with derivatives containing reactive ethyleneimino residue in the middle of a 21-membered oligonucleotide chain

$$\text{d(T–A–I–T–A–C–C–T–A–Y–Z–T–T–A–C–T–T–C–T–C–T)}$$

where I is deoxyinosine residue, Y is either C or T, and Z contains derivatized 5-methylcytosine (XXI in Chapter 2). When these reagents were used to modify a complementary 21-membered oligonucleotide, cross-linking was observed as revealed by gel electrophoresis.[489]

In conclusion, it should be mentioned that modification of nucleic acids at definite positions may also be achieved with reagents supplied with other types of addresses, first with antibiotics interacting with DNA. Thus, sequence-specific cleavage was demonstrated in the experiment with the bromacetyl derivative of distamycin (X in Chapter 2). This derivative alkylated the 167 base-pair fragment of DNA of plasmid pBR 322 at a single point, namely adenine residue 49, and, therefore, could be selectively split at this point.[490]

III. BIOLOGICAL, PHARMACOLOGICAL, AND THERAPEUTIC APPLICATIONS OF AFFINITY MODIFICATION

Main applications of affinity modification in biological and pharmacological fields are related to the two fundamental properties of affinity reagents, namely the ability to bind covalently to the target biopolymers and to inactivate them. The first property allows tagging of specific biopolymers providing the possibility to identify and characterize specific bio-polymers present in complex systems and, therefore, greatly facilitating isolation. Specific inactivation of certain biopolymers under physiological conditions allows the investigation of the role in various biochemical processes and provides the background for rational design of drugs which correct certain biochemical disorders by decreasing the level of specific enzymatic activities (enzyme pattern-targeted chemotherapy) or inactivate infectious agents by reacting with biopolymers which are specific for these agents. Diverse applications of affinity modification for the characterization of various cellular biopolymers capable of specific interactions and rapidly developing pharmacological approaches based on the using of affinity reagents make the attempt to cover systematically all these applications in this book impractical. We can only illustrate by single examples the main types of experiments and directions of development of the technique in order to give the reader ideas about the current state of research and emerging methods in this field. Biological and pharmacological applications of affinity modifications, in particular using this technique for investigation of cellular receptors, were extensively reviewed.[491-499]

A. Localization of the Ligand Binding Sites

The covalent attachment of appropriately labeled affinity reagents to the binding sites allows investigating the localization of these sites in biological systems at various levels of resolution. As an example of visualization of a specific binding site on the protein surface using affinity modification, one can consider experiments on localization of ATP binding site in myosin. This force-generating protein has a special binding site, the ATPase site, which binds ATP molecules supplying energy for the protein functioning. Spatial location of this site was investigated using affinity modification with a photoreactive ADP analogue bearing biotin moiety.

After modification, myosin was treated with the platinum-derivatized protein avidin which possesses extremely high affinity to biotin. Electron microscopy of the modified protein with ATPase sites visualized by attached electron-dense avidin derivatives allowed the location of the site at the position 140 Å apart from the head-rod junction, on the head of the protein.[500]

Another example of localization of the ligand binding sites on the target is the experiment on visualization of ecdysterone binding sites on polytene chromosomes of *Drosophila melanogaster*. The hormone ecdysterone stimulates specific genes on polytene chromosomes of *Diptera* and causes changes in the pattern of specific structural modifications, so-called puffs, present in chromsomes. In order to learn if there are specific chromosomal sites for the hormone binding, direct UV cross-linking of endogenous ecdysterone to chromosomes in salivary glands of *D. melanogaster* was performed. The chromosome preparations were treated with antiecdysterone rabbit antibodies and the latter were developed with fluorescein isothiocyanate-coupled goat antirabbit immunoglobulins in order to visualize the ecdysterone binding sites by the fluorescence microscopy. The microscopic investigation revealed the specific pattern of ecdysterone binding to various chromosomal loci that are known to respond to the hormone.[501] It should be emphasized that analysis of the ecdysterone binding sites was only possible when the hormone could be covalently linked to the binding sites, because preparation of the chromosome samples requires denaturating treatments which remove the noncovalently bound hormone.

Using radiolabeled affinity reagents allows localizing the binding sites in cell, tissues, and organs by means of autoradiography, as it will be considered in the following sections.

B. Affinity Modification of Enzymes in Biological Systems

Affinity modification of enzymes with radiolabeled reagents followed by electrophoretic analysis and quantitation of the labeled proteins can be used for the investigation of spectra of enzymes interacting with the same ligand in crude cellular extracts and various tissues. Thus, modification with [^{3}H]-diisopropyl phosphorofluoridate which is specific for serine proteinases (see Chapter 1, Section II, Equation 9) was used for comparative studies of serine proteases present in plant tissues.[502]

A unique specificity of affinity reagents allows titration of specific enzymes in crude tissue homogenates and even in vivo. Indeed, very specific and efficient reagents are needed for such experiments, such as suicide reagents. Thus, the ^{14}C-labeled suicide reagent α-difluoromethylornithine (see Chapter 2, Section II.G) was used for investigation of the amount of ornithine decarboxylase present in rat liver under various conditions. Treatment of liver extracts with this reagent resulted in labeling of the only protein, the target enzyme. It was concluded that titration with α-difluoromethylornithine provides a valuable method for quantification of the amount of active ornithine decarboxylase present in mammalian tissues.[503] Autoradiographic examination of organs from mice treated with [^{14}C] difluoromethylornithine allowed the distribution of ornithine decarboxylase in the organism to be determined.[504]

The following example illustrates the possibility of using affinity modification for titration of specific enzymes in vivo. It presents also an approach which allows the achieving of selective labeling of an enzyme using a nonabsolutely specific affinity reagent. The enzyme to be determined was mouse brain glutamate decarboxylase (EC 4.1.1.5). This very important pyridoxal phosphate-linked enzyme is responsible for the biosynthesis of the major inhibitory neurotransmitter γ-aminobutyric acid (see Chapter 2, Section II.F). Monitoring of the level of this enzyme in vivo is of interest for neuropharmacological and behavioral investigations. The suicide affinity reagent for this enzyme is 4-amino-1-hexynoic acid (CXLVIII).

$$CH{\equiv}C-\underset{\underset{NH_2}{|}}{CH}-CH_2-CH_2-CO_2H$$

(CXLVIII)

(CXLIX)

This inhibitor is not absolutely specific and reacts also with some other enzymes such as γ-aminobutyric acid transaminase. Therefore, it was used in combination with another suicide reagent, natural neurotoxin gabaculine (CXLIX) which is a potent inactivator of the same enzymes except for glutamate decarboxylase. In the experiments, consecutive administration of the unlabeled gabaculine and [α-³H] acetylenic inhibitor CXLVIII allowed the target enzyme to be labeled in a highly specific way.[505]

The susceptibility of affinity modification with certain reagents allows discrimination between different enzymes present in complex biological systems and catalyzing similar reactions. Thus, serine proteases (see Chapter 1, Section II, Equation 9) are readily inactivated by diisopropyl phosphorofluoridate in contrast to proteases which contain an essential thiol group in the active sites, such as papain (thiol proteases). On the contrary, the latter enzymes are sensitive toward thiol-specific reagents. A representative of the class-specific inhibitors is L-*trans*-epoxysuccinyl-leucylamido-(4-guanidino)butane which reacts efficiently with thiol enzymes and does not modify serine proteases. The reagent was used for identification of thiol proteinases such as papain, ficin, and bromelains in plants.[506] Enzymes using similar catalytic mechanisms but showing different specificity toward substrates can be discriminated using corresponding reactive derivatives of the substrates. Thus, methylene chlorides of peptides with C-terminal lysine and arginine show more high reactivity toward trypsin-like enzymes while those with C-terminal aromatic amino acid residues react more readily with chymotrypsin-like enzymes.

An example of characterization of enzymes with similar specificity is represented by experiments with blood plasma proteases that function in blood coagulation and fibrinolysis. These enzymes show trypsin-like specificity with respect to substrates. They usually cleave peptide bonds formed by arginine residues, and show some specificity toward the neighboring amino acid residues. Therefore, oligopeptides corresponding to these specific sequences were used as addressing structures of the affinity reagents designed for modification of the enzymes. Thus, factor X_a, which is a protease involved in a blood coagulation pathway, hydrolyzes two bonds in the conversion of prothrombin to thrombin after sequences Glu-Gly-Arg. Thrombin, which is the last protease in the blood coagulation cascade, catalyzes the conversion of fibrinogen to fibrin by cleaving the polypeptide after sequence Gly-Val-Arg. It was demonstrated that methylene chlorides corresponding to these sequences are efficient and specific inactivators of the enzymes.[507]

Characterization of proteases involved in the processing of insulin represents an example of experiments where affinity modification was used to identify enzymatic activity involved in certain biochemical process. Hormone insulin is synthesized in so-called Langerhans islets in the pancreas as the larger precursor, preproinsulin, which is cleaved to proinsulin and further to insulin. One of the last steps of the insulin processing takes place in the Golgi apparatus and maturing secretory granule. At this step, cleavage of the precursor between amino acid residues Arg-32 and Glu-33 in a sequence Ala-Arg-Arg-Glu occurs, suggesting the existence of an enzyme with trypsin-like specificity. To characterize this activity, affinity modification of the enzyme in a lysed crude secretory granule fraction from rat islets of Langerhans was investigated. It was found that serine protease inhibitors such as diisopropyl fluorophosphate did not substantially inhibit the enzyme while it was readily inactivated by the thiol-specific reagents. It was concluded, therefore, that the protease was a thiol type one. A radiolabeled alkylating affinity reagent L-alanyl-L-lysyl-L-arginyl methylene chloride imitating the amino acid sequence attacked by the enzyme was used for further characterization of the protein. This reagent efficiently inactivated the enzyme, and gel electrophoresis of the modification products has revealed the major band corresponding to protein of 31.5 kdaltons.[508]

The possibility to inactivate certain biopolymers allows the biochemical functions to be investigated. Affinity modification with benzyloxycarbonylphenylalanylalanyldiazomethane,

Z-Phe-Ala-CHN$_2$, an inhibitor of thiol proteases, was used for elucidation of the role of lysosomal proteolysis in the intracellular protein degradation. In experiments with cultured macrophages, it was shown that this reagent enters the cell by a pinocytosis mechanism which gives access to the lysosomal system and efficiently inactivates the thiol proteases therein. Treatment of macrophages with this reagent resulted in considerable inhibition of protein degradation in the cells, thus demonstrating the important role of the lysosomal thiol proteases in the degradation of the intracellular proteins.[509]

C. Affinity Modification of Cell Receptors

Affinity modification has become an invaluable tool for investigating cell receptors, since it allows to identify them in complex systems, to control the integrity of receptors at any step of purification processes, and to prove that the isolated protein is identical to the receptor under investigation.

When involvement of a specific receptor in an interaction with a certain ligand is suggested, affinity modification with a radiolabeled reactive derivative of the ligand represents the most straightforward way to learn whether the receptor exists. A typical experiment on labeling and isolation of a putative receptor is the experiment on affinity modification of vinblastine binding protein from the multiple drug-resistant tumor cells. Emergence of a multiple drug-resistant tumor cell population in the course of cancer therapy allows the tumor to survive treatments with various drugs and represents a common problem of oncology. Investigations of the resistant cells have demonstrated that membrane vesicles from these cells bind more anticancer drugs such as the efficient anticancer alkaloid vinblastine as compared to the vesicles from the sensitive cells. It was suggested that specific biopolymers may be involved in the increased drug binding. To identify the putative receptor, the membrane vesicles were labeled with photoreactive vinblastine analogues, N-(p-azido-[3,5-^{3}H]benzoyl)-N'-(β-aminoethyl)vindesine and N-(p-azido-[3-^{125}I]salicyl)-N'-(β-aminoethyl)vindesine. Both analogues were shown to have similar pharmacologic properties to vinblastine and were expected to bind to the same cellular targets. When membrane vesicles from the multidrug-resistant cells were treated with these reagents, labeling of specific 150- to 170-kdalton protein was observed. In parallel experiments, no labeling was found in vesicles from the drug-sensitive cells parental to the resistant cells tested. The obtained results proved involvement of a specific protein in drug binding in the multidrug-resistant cells which can be involved in mechanisms providing the cell resistance.[510]

Besides the general purpose to label and isolate specific receptors, affinity modification can be used to characterize receptors present in various tissues and to study specificity of the interaction with different ligands. As a typical example of such studies one can consider experiments on affinity modification of receptors for phencyclidine (angel dust), CLa, with a photoreactive analog N-(1-[3-azidophenyl]cyclohexyl)piperidine, CLb.

$$X = H \quad a$$
$$X = N_3 \quad b$$

(CL)

Phencyclidine was introduced primarily as a general anesthetic and was withdrawn from clinical use because of psychotomimetic effects. In recent years, interest in biochemical pharmacology of phencyclidine has increased because of frequent abuse of the drug with harmful consequences. In humans, the drug produces bizarre dissociative behavioral effects ranging from schizophrenia-like states to self-mutilation and aggressive behavior. Phencyclidine was shown to interfere with functions of several different receptors including opiate ones. In order to check if phencyclidine and opiates interact with a common receptor, affinity modification of rat hippocampal phencyclidine receptors with radiolabeled reagent CLb was investigated. SDS-polyacrylamide gel electrophoresis has revealed five major labeled 90-, 62-, 49-, 40-, and 33-kdalton peptides. The most intensively labeled ones were peptides 90 and 33 kdaltons. When the modification was performed in the presence of various ligands, it was found that opiates interacting with so-called sigma receptor binding opioids of benzomorphan series suppress modification of the above peptides. At the same time, morphine, D-Ala-D-Leu-enkephalin and some other ligands which do not bind to sigma receptor show no influence on the reaction. Various sigma ligands decreased the labeling of the five peptides to a different extent. The most selective sigma ligands influenced the most efficient labeling of the 90-kdalton polypeptide. It was concluded that this polypeptide represents a component of the major phencyclidine/sigma receptor.[511] In further experiments, uneven distribution of various phencyclidine receptors in the brain was demonstrated. Synaptosomal membranes from various brain regions were modified. In all brain regions except for the cerebellum, the same five labeled polypeptides were observed although in different proportion. In the cerebellum, only two polypeptides, namely 90- and 33-kdalton polypeptides, were detected. Clear differences between brain regions with respect to the polypeptides interacting with phencyclidine demonstrated heteterogeneity of the brain phencyclidine receptors.[512]

Modification with ^{125}I- or ^{3}H-labeled affinity reagents allows to investigate easily localization of receptors in cells and tissues at high resolution level. Since the label is bound covalently, the problems of denaturing treatments required for analysis are circumvented. Thus, to study localization of receptors for benzodiazepine minor tranquilizers in brain tissue, the receptors were modified with [^{3}H]-flunitrazepam (see Chapter 2, Section II.F) and distribution of the label in tissue was investigated by electron microscopic autoradiography. Using this technique it was found that the labeled receptors are localized predominantly in regions of synaptic contacts.[513,514]

The possibility of labeling specific receptors by means of affinity modification allows some dynamic aspects of the receptor functioning to be investigated. The example is given by experiments with affinity-modified insulin receptors.[515-517]

When hormones bind to the cellular receptors, the latter are internalized and may be sequestered within the cell, degraded, or recycled back to the cell surface. Since insulin is always present in plasma under normal conditions, internalization of insulin receptors proceeds constantly and it was reasonable to think that the receptors should be recycled. The recycling may also provide one more regulatory component for the adjustment of the receptor number at the cell surface. To learn whether the insulin receptor recycling takes place, affinity modification with ^{125}I-labeled photoreactive derivatives of insulin such as insulin with the 2-nitro-4-azidophenacetyl group substituted for the N-terminal Phe residue in the insulin B chain was used. The derivative was produced by acylation of the amino group of B2 after removal of B1-phenylalanine. The time course of distribution of the receptors subjected to photoaffinity labeling in hepatocytes and adipocytes was investigated using two approaches. One was the determination of the labeled receptors on the cell surface taking the advantage of the accessibility to trypsin degradation. The second approach was using quantitative electron microscopic autoradiography which allows quantitative determination of the number of grains produced by the radioisotope decays in cell autoradiographs and, hence, the localization of radiolabeled compounds within the cell. Photomodification was performed at 15°C, where internalization processes are slowed down. It was found that the major modified polypeptide is an α-subunit of the receptor, and that 80% of the labeled receptors are localized initially at the cell surface. When the labeled hepatocytes were incubated at 37°C, trypsin-accessible receptors disappeared within 30 to 60 min due to the internalization of the receptor-insulin complex, and by 60-min incubation 80% of the labeled receptors were internalized. At the same time, progressive labeling of two cellular compartments, endosomes and lysosome-like structures, was observed. This finding was in accordance with the adsorptive endocytosis internalization mechanism including the concentration of the receptor-ligand complex in the coated pits, coated vesicles formation, and further processing of the complex in the multivesicular bodies and lyposomes. After 4-hr incubation, progressive reappearance of the labeled receptor complex at the hepatocyte surface was observed. Labeling of endosomes demonstrated inverse time dependence. The results obtained evidence that the journey of the receptor-insulin complex ends at the cell surface membrane and hence the receptor is recycled. The interesting finding was that the dissociation of the internalized complex, which was believed to be crucial for receptor recycling, is not the obligatory event, since the labeled complex studied could not dissociate by definition.

D. Affinity Modification as an Approach to Produce Locally Damaged Biopolymers

The ability of affinity reagents to introduce damages within specific areas of biopolymer molecules provides unique possibilities for the preparation of important modified biopolymer derivatives. One example of this application of the technique is related to preparation of the protein toxoids. Toxoids are modified toxins which are more poisonous but retain the most characteristic antigenic determinants of the parent toxins, and can be used for vaccination against toxin-mediated diseases. The usual way to prepare toxoids consists in nonspecific chemical modification, for example, with cross-linking reagents such as formaldehyde which is damaging for both the toxin function and immunological properties. In the case of toxins expressing enzymatic activity, affinity modification suggests the most rational way for preparation of toxoids. Local damage introduced by the affinity reagents directed to the toxoid binding sites should destroy the activity and eliminate the toxic effect. At the same time, most of the antigenic determinants important for immunological recognition should be unaffected in this case. As an example, preparation of toxoids of *Pseudomonas aeruginosa* exotoxin A can be considered. The usual approach based on formaldehyde treatment was found to be inefficient in this case; the produced toxoids were reversibly inactivated or antigenically altered. The mechanism of the toxin A action consists in the inhibition of protein synthesis in cells by the catalytic transfer of ADP-ribose from NAD to elongation factor 2 (similarly to diphtheria toxin, see Chapter 2, Section II.F). Therefore, it was proposed to modify the NAD binding site of the protein. When the toxin was subjected to photoaffinity modification with 8-azidoadenosine, irreversibly inactivated, antigenically intact toxoid of this toxin was prepared.[518]

Very interesting possibilities are suggested by the development of complementary addressed modification of nucleic acids (see Section E) which can be used for accomplishing local modifications in polynucleotides. This approach provides unique possibilities for the development of new methods for gene-directed mutagenesis. The first reported experiments on gene-directed mutagenesis in viruses and plasmids with affinity reagents were performed using reactive derivatives of RNA and DNA fragments. Early transcripts of the bacteriophage T7 carrying alkylating groupings introduced by modification with a heterobifunctional reagent according to Chapter 2, Equation 80 were used for modification of the corresponding genes in the T7 DNA. It was found that modification with reactive transcript derivatives resulted in the generation of mutations in the genes related to the transcripts and did not influence other genes.[284] Thus, in one of the experiments, a transcript of early gene 1.3 which codes for T7 DNA ligase, an enzyme important for the DNA replication which forms phosphodiester bonds at the single-stranded breaks in the double stranded DNA, was used as a carrier of the reactive groups. It was found that the reactive derivative of this transcript induces mutations in this gene with a high yield without any changes in the other genes.

Using the same approach, reactive derivatives of single-stranded DNA fragments complementary to the plasmid pBR 322 gene Tc^r responsible for tetracyclin resistance were prepared. It was shown that modification of the plasmid with these derivatives induces mutation specifically in the gene Tc^r.[519]

Another set of experiments was performed with 16Sr RNA bearing photoreactive psoralen monoadducts (see Chapter 2, Section II.F). Using this derivative, specific modification of the genes coding for the rRNA sequence present in a supercoiled plasmid was accomplished and it was found that the modification resulted in specific mutations in this sequence.[520]

Indeed, the described version of directed mutagenesis with reactive polynucleotide derivatives is not as precise as the most popular method based on the priming of DNA replication with synthetic oligonucleotides containing desired point mutations.[521] However, the techniques, although applied today only to in vitro experiments, may be transformed to in vivo gene-directed mutagenesis.

E. Pharmacological and Therapeutic Applications of Affinity Modification

In this section, we shall overview the applications of affinity modification in pharmacology and therapy. The main physicochemical background of therapy is selective toxicity.[522] The most obvious cases where selective toxicity should play the key role are infectious and oncogenic diseases. In both cases the problem can be solved when one can damage vital systems inherent to an infectious agent or malignant cell by using the means which have a minor effect on host cells and tissues. In principle, this can be achieved if the enzyme is identified as being unique to the infectious agent or malignant cells or the latter possess some unique receptors at the surface. These specific biopolymers can be used as molecular targets for specific drugs. The idea of selective toxicity was formulated first by Paul Ehrlich[523] who, nearly a century ago, suggested the concept of the chemotherapeutic "magic bullet", the bifunctional compound which can recognize the specific target structure and by doing this deliver some damaging group to this structure. Affinity modification was born in an attempt to develop enzyme-targeted inhibitors for therapeutic purposes, and now it is clear that the potential of affinity reagents as drugs is tremendous. Application in therapy represents the main direction of development of the affinity modification.

It should be noted that using chemically reactive compounds is only one of the possibilities to realize the principle of selective toxicity. Some very tightly binding enzyme inhibitors interfere with the enzyme function[524] and, from the point of view of therapy, are not very different from the affinity reagents.

Antibodies specific to unique receptors present at the surface of certain cells can be used for targeting of some damaging constructions to these cells.[525-535] These constructions covalently bound to the antibodies can be reactive compounds[525-528] or toxin molecules, such as the toxic subunit of diphtheria toxin (see Chapter 2, Section II.F) or the toxic subunit of the efficient plant toxin ricin. The latter type of targeted cell-damaging compounds has received great attention since using such immunotoxins demonstrates promising results in specific cell eradication. Thus, it was shown that using a specific immunotoxin it is possible to kill neoplastic human B cells present in a mixture of bone marrow cells without any diminution in viability of the latter. A prolonged regression of the mice tumor induced by administration of tumor BCL_1 cells was observed when the mice were treated with immunotoxin targeted to these cells.

Another example of using antibody targeting potential is given by the applications in the liposome techniques. Lipid vesicle liposomes can be charged with cytostatic compounds and delivered to certain cells by specific autobodies anchored to the liposomes.[536-539] This approach may be a method of choice for selective cell killing. It can be used also for delivery of various compounds including affinity reagents to certain cells.

The treatment with specific affinity reagents represents a straightforward and efficient approach to regulate the level of corresponding enzymatic activities in cells and to cause specific damages to an organism by blocking certain biochemical pathways. In fact, many of the known natural and synthetic compounds used in chemotherapy and many of the most efficient poisons are enzyme-targeted affinity reagents. We have already mentioned the affinity modification-like character of the inhibition of acetylcholinesterase by nerve agents (Chapter 2, Section II.B). A number of examples of natural toxic compounds, suicide enzyme inhibitors can be found in Table 2 in the Appendix. Mechanisms of action can be found in special reviews.[156-158] Enzyme pattern-targeted chemotherapy as an approach to the correction of biochemical disorders is receiving considerable attention. Discussion of this topic and numerous examples can be found in special books.[540-542] To illustrate the approach we can mention several typical examples. Allopurinol is an important drug which is used in therapy of the family disease termed gout. In patients with gout, excess uric acid is produced via xanthine due to enhanced purine turnover. The acid crystallizes in the joints causing inflammation and pain. Xanthine oxidase (EC 1.2.3.2), the enzyme converting xanthine (CLI) to uric acid (CLII) according to Equation 7, can be suicidally inhibited by allopurinol (CLIII) which serves as a substrate to the enzyme and is transformed into alloxanthine (CLIV). The latter binds exceedingly tightly to the enzyme forming a complex with the molybdenium ion present in the enzyme structure[543]

Another example of specific enzyme inactivation allowing to block a certain unwanted biochemical pathway is given by norethisterone (17α-ethynyl-19-nortestosterone). This major ingredient of oral contraceptives was found to be a suicide inhibitor of enzyme estrogen synthetase (aromatase) which converts androgens to estrogens by aromatizing the A ring of androstendione.[544]

It should be noted that in most cases inactivation of an enzyme influences several biochemical processes. A typical case is the therapeutic use of inhibitors of monoamine oxidases.[545-547] The physiological role of these enzymes is the inactivation of transmitters such as noradrenaline, dopamine, and 5-hydroxytryptamine. The enzymes are also responsible for the inactivation of toxic monoamines in blood and pressor amines of foodstuffs. It was found that inhibition of the enzymes can serve as an approach to treat certain forms of depression and some inhibitors of monoamine oxidases have been used clinically. A representative of the acetylenic suicide reagents, inhibitors of monoamine oxidase which bind covalently to the N5 atom of the flavin of the enzyme, is deprenyl (CLV).

Wide use of these drugs is limited by the side effects produced due to suppression of the activities responsible for elimination of pressor amines and toxic monoamines.

Since biosynthesis of nucleic acids is crucial for growth and multiplication of living organisms, in particular for the rapidly multiplying viruses and malignant cells, it represents the primary target for chemotherapy. Thus, as mentioned in Chapter 2, Section II.G, 5-fluorouracil, a suicide inhibitor of thymidylate synthetase responsible for providing an important component for DNA biosynthesis, is used chemically for certain types of epithelial cell cancer. Another approach to limit cell proliferation based on deprivation of cells of the components necessary for adequate functioning of nucleic acids consists in the inhibition of polyamine biosynthesis.[548,549] Polyamines play an important role in the structural arrangement of DNA and functions of nucleic acids. Cells require an adequate supply of polyamines in order to grow at an optimal rate and exhibit the highest polyamine biosynthetic activity during periods of rapid growth and proliferation. This dependence is general and it seems to be possible to reduce the growth rate of any kind of cell, fast growing tumor and bacterial cells in particular, by inhibiting the polyamine synthesis. The conversion of ornithine into putrescine and CO_2 catalyzed by ornithine decarboxylase is the limiting step of polyamine synthesis. This is the only means of putrescine formation in mammalian cells and consequently an obligatory step in the polyamine biosynthetic pathway. A potent suicide inhibitor of ornithine decarboxylase is α-difluoromethylornithine (DFMO) (see Chapter 2, Section II.G). After administration of DFMO, the levels of putrescine and spermidine fall rapidly, and the polyamine depletion causes marked suppression of growth and proliferation of various cells. In clinical studies, patients with various types of cancer have been treated with DFMO in combination with another chemotherapeutic agent, methylglyoxal-bis(guanylhydrazone) (CLVI) which inhibits another enzyme of polyamine biosynthesis, adenosylmethionine decarboxylase.

$$NH_2-\overset{\overset{NH}{\|}}{C}-NH-N=CH-\overset{\overset{CH_3}{|}}{C}=N-NH-\overset{\overset{NH}{\|}}{C}-NH_2 \qquad (CLVI)$$

In the studies, tumor regression has been recorded occasionally and rapid therapeutic responses in childhood leukemia was observed.

Since putrescine and spermidine are absolute requirements for the growth of some bacteria and fungi, attempts have been made to develop approaches to control bacterial growth by specific interference with the polyamine biosynthesis. Since most bacteria can synthesize polyamines from several precursors, several enzymes should be inhibited to achieve efficient polyamine deprivation. It was demonstrated that growth of bacteria *E. coli* and *P. aeruginosa* can be markedly slowed by a combination of D,L-α-monofluoromethylornithine (suicide inhibitor of ornithine decarboxylase), D,L-α-difluoromethylarginine (suicide inhibitor of arginine decarboxylase), and dicyclohexylammonium sulfate (competitive inhibitor of spermidine synthase). The results obtained suggest that the approach based on the inhibition of polyamine biosynthesis may be a viable approach toward the control of certain bacterial infections.[550]

A special direction in applications of affinity reagents in chemotherapy is related to the use to overcome drug resistance of infectious agents. Defensive mechanisms developed by infectious agents are often based on enzymatic inactivation of drugs. A straightforward approach to overcome this problem is suggested by using reactive derivatives of corresponding drugs which can inactivate these enzymes and hence destroy the defensive mechanism of the infectious agent. The classic mentioned example of realization of this principle is related to the bacterial enzyme β-lactamase (see Chapter 2, Section II.G) which destroys β-lactam antibiotics and makes bacteria resistant to these drugs. This main defensive mechanism of resistant bacteria was overcome using suicide reagents, the natural compound clavulanic acid (LXIII in Chapter 2) isolated from *Streptomyces clavigerus,* and a group of synthetic derivatives.[551,552] Being poor antibiotics, the compounds act in synergy with the known β-lactam antibiotics and drastically lower the required inhibitor antibiotic concentration toward resistant bacteria. β-Lactamase inhibitors have come into clinical use in combinations of drugs Timentin and Augmentin which represent mixtures of clavulanic acid with the β-lactam antibiotics ticarcillin and amoxicillin, respectively.

Nucleic acid-targeted affinity reagents, although they are not as well refined as compared to the enzyme-targeted inhibitors, now represent a very promising type of potential therapeutic agent. Currently used nucleic acid-targeted drugs are represented by anticancer antibiotics such as bleomycin, adriamycin, and some others. Although details of the mechanism of the specific antiproliferative activity remain to be elucidated, it was shown that the main biological effect consists in specific damage of DNA which is accomplished with some specificity with respect to the DNA sequences.[553]

A rational approach to nucleic acid-targeted chemotherapy may be provided by the use of complementary addressed reagents (see Section II.E) which are capable of attacking certain nucleotide sequences and hence to accomplish selective modification of nucleic acids containing unique elements of primary structure. In contrast to approaches based on inhibition of specific enzymes, complementary addressed modification suggests the principal possibility of permanent suppression of certain biochemical pathways by damaging corresponding genes. A discussion of potential applications of complementary addressed modification in therapy can be found in the paper by Summerton.[285]   Binding of oligonucleotides to single-stranded nucleic acids can influence functional activity of these methods. The possibility of specific arrest of mRNA translation by oligonucleotides complementary to these mRNAs has been demonstrated in in vitro translation experiments and in experiments where synthesis of specific proteins in cultured cells was suppressed.[554-557] It was reported that the treatment of cultured cells with oligonucleotides complementary to functionally important regions of the Rous sarcoma virus RNA and to RNA of virus HIV-1, which is the etiological agent of acquired immunodeficiency syndrome (AIDS), inhibits multiplication of these viruses in the cells.[558-560]

In order to facilitate entering of oligonucleotides into cells and protect them from nuclease attack, nonionic oligonucleotide derivatives with blocked phosphate groups were prepared. These derivatives, deoxyribooligonucleoside methylphosphonates and oligonucleotide ethylphosphotriesters, were found to enter animal cells and to arrest translation of the target mRNAs and splicing of viral RNAs.[561-563]

Recently, it was found that mRNA functions can be inhibited by complementary RNAs transcribed from plasmids or genes introduced into the cellular genome using the methods of genetic engineering.[564] These RNAs were termed antisense RNAs since the sequences are complementary to the sense nucleic acid sequences.

The possibility of complementary addressed modification of cellular nucleic acids was demonstrated in experiments with alkylating oligothymidylate derivatives.[478,565] It was shown that the reagents can penetrate mammalian cells and selectively react with poly(A) sequences of cellular mRNA. Nonionic oligothymidylate ethylphosphotriester derivatives entered the cells easier as compared to oligothymidylate derivatives with unprotected negatively charged phosphates.

Alkylating derivatives of oligonucleotides complementary to immunoglobulin G κ light chain mRNA were shown to inhibit translation of this mRNA in mouse myeloma cells MOPC 21.[478]

APPENDIX

Table 1
HETEROBIFUNCTIONAL REAGENTS FOR THE PREPARATION OF REACTIVE DERIVATIVES OF BIOPOLYMERS

Reagent number	Code	Ref.
	Photoreactive cleavable heterobifunctional reagents	
1	p2-(s24-c1)-r4	67
2	r3-(s2-c1-s24)-p2	60
3	p2-(s24-c1-s1)-r1	67
4	p2-(s24-c1-s2)-r1	258
5	p2-(s24-c1-s5)-r1	566
6	p2-(s24-c1-s2)-r3	67
7	p2-(s24-c1-s3)-r3	67
8	p2-(s24-c1-s3)-r4	67
9	p2-(s38-s10-s2-c1-s2)-r1	60
10	p2-(s47-s10-s2-s12-s2-c1-s2-s10-s9-s2-s10-s47)-p2	567
11	r2-(c1-s12-s28)-p2	568
12	p2-(s28-s12-s2-c1-s2)-r2	67
13	p2-(s37-s12-s2-c1-s2)-r3	61
14	p2-(s24-c2-s1)-r3	569
15	p2-(s30-s12-c3-s2)-r1	570
16	p2-(s46-s12-s1-c3)-r7	62
17	p2-(s46-s12-s4-c3)-r7	62
18	p2-(s46-s1-s12-s4-c3)-r7	62
19	p2-(s28-s12-s1-c3-s1)-r1	570
20	p2-(s24-c4-s24)-r5	571
21	p2-(s24-c4-s24-s12-s2)-r1	67
22	p2-(s24-c4-s24-s12-s5)-r1	67
23	p2-(s24-c4-s24-s12-s5-s5)-r1	67
24	p2-(s27-c4-s41-s12-s2)-r1	62
25	p2-(s38-s10-s2-c5-s2-s10-s38)-p2, or p2-(s38-s10-s2-r12-s2-s10-s38)-p2	67
	Photoreactive noncleavable heterobifunctional reagents	
26	p1-(s17-s12-s15-s1)-r14	572
27	p1-(s24)-p2	573
28	p2-(s24)-r1	258
29	p2-(s24)-r3	61
30	p2-(s24)-r4	61
31	p2-(s24)-r5	62
32	p2-(s24)-r7	61
33	p2-(s24)-r10	234
34	p2-(s24)-r13	67
35	p2-(s24)-r15	574
36	p2-(s24)-r21	575
37	p2-(s24)-r27	67
38	p2-(s28)-r1	258
39	p2-(s37)-r1	61
40	p2-(s37)-r21	576
41	r13-(s38)-p2	67
42	p2-(s38)-r16	67
43	p2-(s38)-r18	67
44	p2-(s38)-r23	60

Table 1 (continued)
HETEROBIFUNCTIONAL REAGENTS FOR THE
PREPARATION OF REACTIVE DERIVATIVES OF
BIOPOLYMERS

Reagent number	Code	Ref.
45	p2-(s39)-r16	67
46	p2-(s42)-r16	577
47	p2-(s43)-r16	577
48	p2-(s45)-r5	60
49	p2-(s24-s14)-r3	569
50	p2-(s24-s1-s16)-r1	569
51	p2-(s24-s9-s1)-r22	578
52	p2-(s24-s9-s13-s10-s1)-r1	61
53	p2-(s24-s9-s13-s10-s15-s1)-r8	234
54	p2-(s24-s12-s1)-r1	67
55	p2-(s24-s12-s1)-r4	76
56	p2-(s24-s12-s5)-r1	67
57	p2-(s24-s12-s1-s12-s19)-r1	67
58	r21-(s10-s20)-p2	579
59	r7-(s28-s9)-p2	67
60	p2-(s28-s9-s13-s10-s19)-r1	67
61	p2-(s37-s12-s1)-r4	61
62	p2-(s37-s12-s1)-r1	62
63	p2-(s38-s10-s3)-r1	320
64	p2-(s38-s10-s5)-r1	67
65	p2-(s38-s10-s6-s5)-r1	67
66	p3-r24	580
67	p3-r26	123
68	p3-(s1)-r24	131
69	r3-(s3-s8-s35)-p2	60
70	p4-(s8-s1-s26)-r8	67
71	p5-(s1-s11-s2-s7-s2-s11)-r21	581

Cleavable heterobifunctional reagents

72	r8-(c1-s2)-r1	67
73	r27-(s24-s12-s2-c1-s2)-r7	582
74	r7-(s25-c4-s25)-r1	583
75	r7-(s26-c4-s25)-r1	583
76	r7-(s29-c4-s25)-r1	583
77	r7-(s30-c4-s25)-r1	583
78	r7-(s30-c4-s24)-r1	583
79	r7-(s26-c4-s29)-r1	583
80	r7-(s30-c4-s24)-r2	583
81	r7-(s29-c4-s25)-r2	583
82	r14-(c6)-r24	584
83	r9-(s1-s9-s8-c6)-r24	584

Noncleavable heterobifunctional reagents

84	r21-r4	60
85	r7-(s1)-r1	585
86	r7-(s2)-r1	585
87	r7-(s5)-r1	585
88	r8-(s2)-r1	67

Table 1 (continued)
HETEROBIFUNCTIONAL REAGENTS FOR THE
PREPARATION OF REACTIVE DERIVATIVES OF
BIOPOLYMERS

Reagent number	Code	Ref.
89	r1-(s41)-r7	585
90	r7-(s24)-r1	585
91	r7-(s31)-r1	585
92	r7-(s25)-r1	585
93	r7-(s33)-r1	585
94	r1-(s26)-r7	67
95	r7-(s1-s24)-r1	67
96	r1-(s3-s24)-r7	585
97	r1-(s18)-r9	586
98	r22-(s24)-r1	587
99	r22-(s24)-r2	587
100	r2-(s41)-r7	585
101	r2-(s26)-r7	585
102	r7-(s24)-r22	67
103	r7-(s24)-r25	588
104	r24-(s5)-r7	589
105	r28-(s24-s12-s5)-r7	589
106	r17-(s44)-r17	590
107	r16-(s30)-r15	574
108	r5-(s32)-r6	65
109	r6-(s24-s1-s34)-r6	65
110	r21-(s10-s24)-r24	591
111	r21-(s10-s24)-r11	591
112	r19-(s1-s23-s2)-r26	581
113	r19-(s1-s23-s2-s55)	592
114	r20-r29	593

Note: Reactive groups and spacer structures are given using abbreviations introduced in Chapter 2, Tables 4 and 5. Where possible, references to general reviews on bifunctional reagents are given.

Nucleosides, nucleotides, and amino acids are abbreviated as shown in Table 2.

SU	subunit
C-SU	catalytic subunit
R-SU	regulatory subunit
reg.	regulatory
L-SU	light subunit
ph.m.	photoaffinity modification
ss	suicide substrate
AdoHcy	*S*-homocysteinyl adenosine
o	oxidized by $NaBO_4$
NAP	2-nitro-4-azido-phenyl
Py	pyridine
PyAD	pyridine adenine dinucleotide
PLP	pyridoxal-5′-phosphate

ClR	N-2-chloroethyl-N-methylaminophenyl
Z	carbobenzyl oxycarbonyl
ϵ	$1,N^6$-etheno
ϵA	ethenoadenosine

Peptide sequences containing modified residue (marked by [a]) are given in one letter notation. Z is Gln or Glu and B is Asp or Asn.

Table 2
SURVEY OF AFFINITY MODIFICATION DATA FOR ENZYMES

Source	Reagent	Modification point	Ref.
1.1.1.1 Alcohol dehydrogenase			
Horse liver, yeast	Cl-$(CH_2)_n$OpAdo n = 2,3,4,6;		
	Br-$(CH_2)_n$OpAdo n = 2,3,6		594
	$HC\equiv C$-CH_2-CH_2OH (ss)		595
Horse liver	S-2-Chloro-3(imidazol-5-yl)propionate (R enantiomer inactive)	Cys-46	596
	2-Bromo-3(imidazol-5-yl)propionate		597
	ICH_2COO^-, $BrCH_2COO^-$, $CH_3CHBrCOO^-$, $BrCH_2CH_2COO^-$, and $CH_3CH_2CHBrCOO^-$	Cys-46	598
Human liver	ICH_2COO^-	Cys-174	599
Horse liver, yeast	$BrCH_2COCH_2Br$	Cys-46	597
Equine liver	9-Azidoacridine (ph.m.)		214
Horse liver	Diazonium-1H-tetrazole	Cys-174	600
Yeast	p-N_3-C_6H_4-$COCH_2OCOCH_2I$ (ph.m.) and 3-N_3PyAD (ph.m.)		252
	3-$BrCH_2COPy^+$-$(CH_2)_3$-ppAdo and 4-Br-CH_2COPy^+-$(CH_2)_3$-ppAdo	Cys-43	255
Horse liver	3-$BrCH_2COPy^+$-$(CH_2)_3$-ppAdo and 4-Br-CH_2COPy^+-$(CH_2)_3$-ppAdo	Cys-174	255
Horse liver, yeast	3-$BrCH_2COPy^+$-$(CH_2)_4$-ppAdo	Cys-43, Cys-46	601
Yeast	PLP-pAdo	Lys	602
Horse liver	3-$ClCH_2CO$-PyAD		254, 603
Yeast	A3'-O-NAP-β-Ala-NAD^+		604
Horse liver	N^6[N-(6-aminohexyl)carbamoylmethyl]-NAD^+ + CDI		380, 605
	Nicotinamide-5-bromoacetyl-4-methyl-imidazole dinucleotide	Cys-46, Cys-174	606
Yeast	Nicotinamide-5-bromoacetyl-4-methyl-imidazole dinucleotide	Cys-43	606
	3-N_2^+-PyAD	Cys	251
1.1.1.23 Histidinol dehydrogenase			
Salmonella typhimurium	7-Chloro-4-nitro-2,1,3-benzoxadiazole	Cys-116, Cys-377	607
1.1.1.27 Lactate dehydrogenase			
Lactobacillus casei	3-Bromopyruvate	His	608
Pig heart	9-Azidoacridine (ph.m.)		214
Peptostreptococcus elsdenii	$HC\equiv C$–OCH–COO^- (ss) OH	FAD	609
Mouse C_4	PLP		610
Rabbit muscle	3-(3H-diazirino)-PyAD (ph.m.)		253
	4-(Iodoacetamido)-salicylic acid		95

Table 2 (continued)
SURVEY OF AFFINITY MODIFICATION DATA FOR ENZYMES

Source	Reagent	Modification point	Ref.
1.1.1.30 3-Hydroxybutyrate de-hydrogenase			
Beef heart mitochondria	A3'-O-NAP-β-Ala-NAD+ (ph.m.)		611
	A3'-O-NAP-β-Ala-NAD+ (ph.m.)	MESYC[a]	
		TSGSTD	
		TSPVIK	612
1.1.1.37 Malate dehydrogenase			
Pig heart mitochondria	FSO$_2$BzAdo	Lys	613
1.1.1.40 Malate dehydrogenase decarboxylating (NADP+)			
Pigeon liver	oNADP	Lys	417
	Bromopyruvate	Cys	614
1.1.1.41 Isocitrate dehydro-genase (NAD+)			
Pig heart	oADP	Lys	615
	3-Bromo-2-ketoglutarate	Cys	616
	3,4-Didehydro-2-ketoglutarate	Cys	617
	6-(4-Bromo-2,3-dioxobutyl)thioadenosine 5'-diphosphate		618
1.1.1.42 Isocitrate dehydro-genase (NADP+)			
Pig heart	2-[(4-Bromo-2,3-dioxobutyl)thio]-1,N^6-ethen-oadenosine 2',5'-bis-phosphate		304
	3-Bromo-2-ketoglutarate		619
1.1.1.49 Glucose 6-phosphate dehydrogenase			
Candida utilis	oNADP+	Lys	620
Leuconostoc mesenteroides	PLP	Lys	621
1.1.1.50 3α-Hydroxysteroid dehydrogenase			
Pseudomonas testosteroni	5-β-Pregnane-3,20-dione-12-α-iodoacetate and 5α-androstane-3-one-17β-bromoacetate	His Met	622
1.1.1.53 3α,20β-Hydroxysteroid dehydrogenase			
Streptomyces hydrogen-ans	17β-(1-Oxo-2-propynyl)androst-4-ene-3-one		404
	2α-Bromoacetoxyprogesterone and 11α-bro-moacetoxyprogesterone		623
	17(Bromoacetoxy)progesterone and 5α-dihy-drotestosterone 17-bromo-acetate		405
	5'-Br-CH$_2$CONH-5'-Ado		624
1.1.1.62 Estradiol 17β-dehydro-genase			
Human placenta	3-(NAP-β-Ala)estrone (ph.m.) and 17β-(NAP-β-Ala)estradiol 3-methyl ether (ph.m.)		379
	15-Estradiol-17β analogues bearing –COCH$_2$Br or –COCH$_2$I residue in eight different posi-tions of steroid (1,2,3,4,6,7,12,16); details in Chapter 2, Section III.B		232
	3-Methoxyestriol 16-bromoacetate and 12β-Hydroxy-4-estrene-3,17-dione-12-bro-moacetate	His (TDIH[a] TFH[a]R)	625
	3-Bromoacetoxyestrone		626

Table 2 (continued)
SURVEY OF AFFINITY MODIFICATION DATA FOR ENZYMES

Source	Reagent	Modification point	Ref.
	3-Hydroxy-16-methylene estra-1,3,5(10)-triene-17-one (16-methylene estrone) (ss)		627
	4-Bromoacetamidoestrone methyl ether	Lys, Cys	628
	6β-Bromoacetoxyprogesterone	Cys, His	629
	12β-(Bromoacetoxy)-4-estrene-3,17-dione and 16α-(bromoacetoxy)-1,3,5(10)-estratriene-3,17β-diol 3-methyl ether	His	630
	16α-Bromoacetoxyprogesterone	His	407
	17β-(Oxo-2-propynyl)androst-4-en-3-one		406
	Iodo[2′-^{14}C]acetoxy-3-oestrone and iodo[2′-^{14}C]acetoxy-3-estradiol	His	631
	3-ClCH$_2$CO-PyADP	Cys	632
	5′-FSO$_2$BzAdo		408, 633
1.1.3.12 Alcohol oxidase			
Yeast	Propargylalcohol and 1,4-butynediol		
C. boidinii			634
1.2.1.11 L-Aspartate-β-semi-aldehyde dehydrogenase			
Escherichia coli	L-2-Amino-4-oxo-5-chloropentanoic acid	His (FVGGD H^{a}TVS)	635
	3-ClCH$_2$CO-PyADP$^+$		636
1.2.1.12 Glyceraldehyde 3-phosphate dehydrogenase			
Rabbit muscle	9-Azidoacridine (ph.m.)		214
	p-N$_3$C$_6$H$_4$COCH$_2$Br (ph.m.)	Cys-149	252
Yeast	A3′-NAP-β-Ala-NAD$^+$ (ph.m.)		637
Bacillus stearo-thermophilus, sturgeon, rabbit muscle	3-ClCH$_2$CO-PyAD	Cys-149	638
	3-BrCH$_2$COPy$^+$-(CH$_2$)$_3$-ppAdo		639
1.2.1.14 Inosine 5-monophosphate dehydrogenase			
E. coli	6-Chloro-9β-D-ribofuranosylpurine 5′-phosphate		640
1.2.3.2 Xanthine oxidase			
Buttermilk	1-Methyl-2(bromomethyl)-4,7-dimethoxybenzimidazole		393
1.3.99.2 Butyryl-CoA dehydrogenase			
M. elsedenii	3-Butynoylpantetheine (ss) and 3-pentynoylpantetheine (ss)		641
	3-Chloro-3-butynoylpantetheine (ss)	Glu	642
1.3.99.3 General acyl-CoA dehydrogenase			
Pig kidney	Methylenecyclopropylacetyl-CoA (ss)		643
	3,4-Pentadienoyl-CoA		392
Pig liver	3-Butynoyl-CoA and 3-octynoyl-CoA (ss)		644
1.3.99.3 Short-chain acyl-CoA dehydrogenase			
Beef liver	Propionyl-CoA		645
1.3.99.5 Steroid 5α-reductase			
Rat liver microsomes	21-Diazo-4-methyl-4-aza-5α-pregnane-3,20-dione (ph.m.)		646

Table 2 (continued)
SURVEY OF AFFINITY MODIFICATION DATA FOR ENZYMES

Source	Reagent	Modification point	Ref.
1.3.99.7 Glutaryl-CoA dehydrogenase			
Pseudomonas fluorescens	3-Butynoylpantetheine and 3-pentynoylpantetheine		641
1.4.1.1 Alanine dehydrogenase			
B. subtilis	9-Azidoacridine		214
1.4.1.3 Glutamate dehydrogenase			
Beef liver	FSO₂BzAdo	reg. site	423
	FSO₂Bz εAdo	reg. site, Tyr-262	647
	8-N₃-ADP (ph.m.)		648
	Iodoacetyldiethylstilbestrol	reg. site, Cys-89	649
	6-[(4-Bromo-2,3-dioxobutyl)thio]-6-deaminoadenosine 5'-diphosphate	Cys-319	650
	4-(Iodoacetamido)-salicylic acid		342
Mammalian liver	4-(Iodoacetamido)-salicylic acid		95
1.4.1.13 Glutamate synthase			
E. coli K 12	L-2-Amino-4-oxo-5-chloropentanoic acid and iodoacetamide	Cys	651
1.4.3.3 D-Amino acid oxidase			
Pig kidney	*N*-Chloro-D-leucine	Tyr (DLER GIYᵃ)	652
1.4.3.4. Monoamine oxidase			
Beef liver	D,L-*trans*-Phenylcyclopropylamine (ss)		653
	1-Phenylcyclopropylamine (ss)		165
—	1-Phenylcyclobutylamine (ss)		654
Pig liver	*N*-Cyclopropyl-*N*-arylalkyl amines (ss)		655
	N-(1-Methyl)cyclopropylbenzylamine (ss)		656
Beef heart mitochondria	3-{4[(3-chlorophenyl)methoxy]-phenyl}-5[(methylamino)methyl]-2-oxazolidinone methanesulfonate (MD 780236)		657
Rat heart, human placenta (A-form)	[³H]*N*-methyl-*N*(2-propynyl)benzylamine (pargyline) (ss)	Cys	
		(SGGCᵃY)	658, 659
Pig, beef liver (B-form)	[³H]*N*-methyl-*N*(2-propynyl)benzylamine (pargyline) (ss)		658, 659
Beef kidney	[³H]*N*-methyl-*N*(2-propynyl)benzylamine (pargyline) (ss)		660
Plasma	2-Chloroallylamine, 2-butynylamine, 2-propynylamine		661
Beef kidney	R–NH–NH₂ (R=CH₃,Ph,α-naphtyl)		660
1.5.1.3 Dihydrofolate reductase			
Chicken liver	4,6-Diaminodihydrotriazine	Tyr-31	662
E. coli			

(NSC 127755)

Table 2 (continued)
SURVEY OF AFFINITY MODIFICATION DATA FOR ENZYMES

Source	Reagent	Modification point	Ref.
L.actobacillus casei	(NSC 127755)	Arg-52	326
	(NSC 127755)	Arg-31	326
1.5.1.5 Methylenetetrahydro-folate dehydrogenase			
Pig liver	^{3}H-folate + CDI		105
	8-N$_3$-AdoMet (ph.m.)		663
1.5.1.11 Octopine dehydro-genase			
Pecten maximus	Bromopyruvate		664
1.6.1.1 NAD(P$^+$)-Transhydro-genase			
Beef heart mitochondria	FSO$_2$BzAdo		665
1.6.99.2 NAD(P)H Dehydro-genase			
Rat liver	3(α-Acetonyl-*p*-azidobenzyl)-4-hydroxycou-marin (ph.m.)		666
1.6.99.3 NADH Dehydro-genase			
Beef heart mitochondria	Arylazidoamorphigenin (hydroxy analogue of rotenone) (ph.m.)		667
	A3'-O-NAP-β-Ala-NAD$^+$ (ph.m.)		668, 669
1.9.3.1 Cytochrome *c* oxidase			
Beef heart	NAP-Lys(13)-cytochrome *c*		235, 670
Pigeon breast, mitochon-dria	2,4-Dinitroazidophenyl		
	Cytochrome *c*,[^{3}H]-*p*-azidophenacylbromide-(methyl-4-mercaptobutyrimidate)-cytochrome *c* (ph.m.)		381
			671
Yeast	102-*S*-(2-Thiopyridyl)-iso-1-cytochrome and 102-*S*-(4-azidophenacyl)-iso-1-cytochrome (ph.m.)		575
	5,5'-Dithiobis(2-nitrobenzoate)-cytochrome *c*		170
Beef heart	8-N$_3$-ATP (ph.m.)		672
1.10.2.2 Ubiquinol-cytochrome *c* reductase			
Beef heart mitochondria	2,3-Dimethoxy-5-methyl-6{10-(NAP-β-Ala)-decyl}-1,4-benzoquinone		673
Beef heart	Deformamidoazidoantimycin A (ph.m.)		674
1.11.1.5 Cytochrome *c* peroxidase			
Yeast	NAP-Lys(13)-cytochrome *c*		675
1.11.1.8 Iodothyronine deiodinase			
Rat liver	*N*-Bromoacetyl-3,3',5-triiodothyronine		676
1.13.11.12 Lipoxidase			
Soybean	5,8,11,14-Eicosatetranoic acid		677
1.13.12.4 Lactate oxidase			
Mycobacterium smegmatis	2-Hydroxy-3-butynoic acid (ss)	FMN	678
1.14.11.2 Prolyl 4-hydroxylase			
Chick embryos	*N*-(NAP)glycyl-(Pro-Pro-Gly)$_5$	α SU	679
1.14.14.1 Benzo(a)pyrene hydroxylase			

Table 2 (continued)
SURVEY OF AFFINITY MODIFICATION DATA FOR ENZYMES

Source	Reagent	Modification point	Ref.
Rat liver microsomes	1-Ethynylpyrene (ss); trans-1-(2-bromovi-nyl)pyrene (ss); and methyl-1-pyrenyl acety-lene		680
1.14.17.1.Dopamine-β-hydroxylase			
	X–C$_6$H$_4$–CH$_2$–CH=CH$_2$ (X– p–HO–, p–CH$_3$O–, m–HO–, m–CH$_3$O–, p–Br–,p–CN–)		681
Beef adrenals	1-Phenyl-1-aminomethylethene		682
1.15.1.1 Cu,Zn-superoxide dismutase			
Yeast	H$_2$O$_2$ at alkaline pH	His	201
1.17.4.1 Ribonucleoside di-phosphate reductase			
Corinebacterium nephridii	oADP and oCDP		683
E. coli	Radioactive dTTP, dATP (UV cross-linking)	Cys-289	684
Mouse T-lymphosarama	dATP (UV cross-linking)		685
E. coli, L. leichmannii	2′-Cl-2′-dUDP and 2′-Cl-2′-dUTP (ss)		686
E. coli	2′-Cl′-dCDP and 2′-N$_3$-dCDP		687
1.18.1.2 Oxidoreductase ferredoxin-NADP$^+$			
Spinach	oNADP$^+$		688
Ammonia monooxygenase			
Nitrosomonas europaea	HC≡CH (ss)		689
Cytochrome P-450			
Rat liver	2-Isopropyl-4-pentenamide (ss); 1-Ethynylcy-clopentanol (ss); 17α-propadienyl-19-nortes-tosterone (ss); fluoroxene (ss); and 5,6-dichloro-1,2,3-benzothiadiazole (ss); 1-ami-nobenzotriazole (ss) (isozyme induced by phenobarbital); fluoroxene (ss); 1-aminoben-zotriazole (ss); and 5,6-dichloro-1,2,3-benzo-thiadiazole (ss) (isozyme induced by methylcholanthrene)		690
	Chloramphenicol (ss)	Lys	691
Fatty acid synthase			
Rat mammary gland	3-Chloropropionyl coenzyme A	Cys	692
Lactoperoxidase			
Beef	Sulfide ion; 2-mercaptobenzimidazole; and 1-methyl-2-mercaptoimidazole (ss)		391
15-Lipoxygenase of arachidonic acid			
Soybean	14,15-Dehydroarachidonic acid (ss)		166
Luciferase			
V. harveyi	1-Diazo-2-oxoundecane (ph.m.)		693
Firefly	F-SO$_2$BzAdo	Lys, Lys[a]-Gly-Glx-Asx-Ser-Lys	694
Lysyl oxidase			
Beef aorta	BrCH$_2$CH$_2$NH$_2$; ClCH$_2$CH$_2$NH$_2$; NCCH$_2$CH$_2$NH$_2$; and NO$_2$CH$_2$CH$_2$NH$_2$ (ss)		695
Pyruvate dehydrogenase			
Beef kidney and heart	PLP	Lys	696
Testosterone 5α-reductase			

Table 2 (continued)

SURVEY OF AFFINITY MODIFICATION DATA FOR ENZYMES

Source	Reagent	Modification point	Ref.
Rat prostate	(5α,20–R)-4-diazo-21-hydroxy-20-methylpreg-nan-3-one		697
Thyroid peroxidase			
Pig thyroid	L-Thyroxine		698
			699
Aromatase			
Human placenta	10-Propargylestr-4-ene-3,17-dione (ss); 10-[(1S)-1-hydroxy-2-propynyl]estr-4-ene-3,17-dione (ss); and 10-(1-oxo-2-propynyl)estr-4-ene-3,17-dione (ss)		
	10β-Difluoromethyl-estr-4-ene-3,17-diene (ss); 10β-propargyl-estr-4-ene-3,17-dione; and 10β-allenyl-estr-4-ene-3,17-dione		700
	19,19-Difluoroandrost-4-ene-3,17-dione		701
	6-α,β-Hydroperoxyandrostenedione		702
	19(2-Propan-2,3-dien-androst-4-ene-3,17-dione) and 19(2-propynyl-androst-4-ene-3,17-dione)		703
	6α-Bromo-androst-4-ene-3,17-dione		704
	17α-Ethynyl-19-nortestosterone		705
2.1.1.1 Serine transhydroxy-methylase			
Sheep, rabbit liver	F-Ala	Cys	706
2.1.1.6 Catechol O-methyl-transferase			
E. coli	8-N₃-AdoMet		707
Rat liver	4-Methyl-5,6-dihydroxyindol + O_2; 7-methyl-5,6-dihydroxyindol + O_2; 4,7-dimethyl-5,6-dihydroxyindol + O_2		296
	5-Hydroxy-3-mercapto-4-methoxybenzoic acid + O_2		119
	2-[6-Amino-3,4-dihydroxyphenylethyl]-amine (6-aminodopamine) + O_2		708
	6-Hydroxydopamine + O_2; 6-hydroxydopa + O_2; 5-hydroxydopamine + O_2; N^1,-D,L-seryl-N^2(2,3,4-trihydroxybenzyl)hydrazine + O_2; pyrogallol + O_2; 5,6-dihydroxyindole + O_2; and 6-hydroxydopamine-p-quinone		709
	N-Haloacetyl-3,5-dimethoxy-4-hydroxy phenylalkylamines:		710

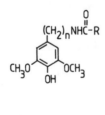

n	R
2	—CH₂Br
2	—CH₂I
3	—CH₂I
1	—CH₂I

Table 2 (continued)
SURVEY OF AFFINITY MODIFICATION DATA FOR ENZYMES

Source	Reagent	Modification point	Ref.
2.1.1.8 Histamine *N*-methyl-transferase			
Pig brain	oAdoHcy; oAdoMet; and oAdo		711
2.1.1.20 Glycine methyltransferase			
Rat liver	FSO₂BzAdo	Cys, Tyr	712
2.1.1.24 Protein-*O*-methyltransferase			
2.1.1.28 Phenylethanolamin *N*-methyltransferase			
Human erythrocytes	AdoMet (UV-cross-linking)		713
2.1.1.45 Thymidylate synthase			
L. casei	5(α-Bromoacetyl)-2'-dUMP (ss)		248
	5-Nitro-2'-dUMP (ss)		714, 715
	1-(β-ᴅ-2'-Deoxyribofuranosyl)-8-azapurin-2-one-5'-monophosphate (ss)		716
	5(E)-(3-azidostyryl)-2'-dUMP (ss)		717
L 1210 cells	5'-Bromoacetamido-5'-dThd; 5'-iodacetamido-5'-dThd; and 5'-chloroacetamido-5'-dThd		718
2.1.3.1 Methylmalonyl-CoA: pyruvate carboxyltransferase			
—	*p*-N₃-BzCoA (ph.m.)		719
2.1.3.2 Aspartate transcarbamylase			
E. coli	ᴅ- and ʟ-bromosuccinate	C-SU (Lys-83, -84, -224)	720
2.3.1.7 Carnitine acetyltransferase			
Pigeon breast	*p*-Azidophenacyl-ᴅ,ʟ-thiocarnitine (ph.m.)		721
2.3.1.9 Thiolase			
Pig heart	3-Pentynoyl-CoA (ss); 3-butynoyl-CoA (ss); and 4-bromocrotonyl-CoA (ss)		722
Bovine liver	5-(4-Bromo-2,3-dioxobutyl)-CoA		723
2.3.1.13 Acyl-CoA:Glycine *N*-acyltransferase			
	p-N₃-BzCoA (ph.m.)		724
	S-Benzoyl(3'-dephospho-8-azido)-CoA (ph.m.)		725
2.3.1.28 Chloramphenicol acetyltransferase			
Gram-negative bacteria	3-(Bromoacetyl)chloramphenicol	His-189	726
2.3.2.2 γ-Glutamyl transpeptidase			
Rat kidney	6-Diazo-6-oxo-ᴅ-norleucine		727
Human kidney	6-Diazo-6-oxo-ʟ-norleucine	L-SU	
	ʟ-azaserine		728
	ʟ-(αS,5S)-α-amino-3-chloro-4,6-dihydro-5-isoxazoleacetic acid (AT-125)	L-SU	104
Human fetal lung fibroblasts WI-38	6-Diazo-5-oxo-L-norleucine and 5-iodoacetamidofluorescein	L-SU	729
Rat kidney	PheCH₂SO₂F		730
Hydra attenuata	ʟ-Azaserin		731
2.4.1.1 Glycogen phosphorylase			

Table 2 (continued)
SURVEY OF AFFINITY MODIFICATION DATA FOR ENZYMES

Source	Reagent	Modification point	Ref.
Rabbit muscle	bis-PLP	Lys-573	732
	8-N₃-AMP	reg. site	733
	Cyanogen bromide activated maltooligacchar-ides 4 ≤ n ≤ 10; maltoheptaose; and 4-*O*-methylmaltoheptaose		425
	K₂FeO₄		199
	8-R-adenine R = − SCH₂C₆H₄ −NHCONH–C₆H₄SO₂F (p,m-isomer), R = −C₆H₄CH₂NHCOC₆H₄SO₂F (m,m-isomer)		734, 735
2.4.1.11 Glycogen synthase			
Rabbit muscle	PLP-pUrd	Lys (NVADKₐ VGGIY)	736
2.4.1.22 Lactose synthase			
Bovine milk	NAP-ppUrd	Galactosyl transferase α SU	737
	N-Diazoacetylglucosamine (ph.m.)	Galactosyl transferase α SU	738
2.4.1.25 4-α-D-Glucanotrans-ferase			
Rabbit muscle	1-*S*-Dimethylarsino-1-thio-β-D-glucanopyrano-side		410
2.4.1.38 UDPgalactose-glyco-protein galactosyltransferase (β-4-galactosyltransferase)			
Bovine colostrum	oUDP		739
2.4.2.1 Purine nucleoside phos-phorylase			
Human erythrocytes	6-Chloro-9-(3,4-dioxopentyl)purine and 9-(3,4-dioxopentyl)hypoxanthine	Arg	117
2.4.2.14 Amidophosphoribosyl-transferase			
E. coli	Diazo-5-oxo-norleucine		740
Asparagine *N*-glucosyltrans-ferase			
—	Arg-Asn-Gly-epoxyethylglycine-Ala-Val-OMe		741
2.5.1.1 Prenyltransferase			
Avian liver	*o*-Azidophenethyl pyrophosphate (ph.m.); *p*-azidophenethyl pyrophosphate (ph.m.); and 3-azido-1-butyl pyrophosphate (ph.m.)		742
Chicken liver	*o*-Azidophenylethyl pyrophosphate (ph.m.)		743
L. plantarum	2-Diazo-3-trifluoropropionyloxygeranyl pyro-phosphate		744
E. coli	2-Diazo-3-trifluoropropiohyloxygeranyl pyro-phosphate		411
2.5.1.18 Glutathione *S*-trans-ferase			
Sheep liver	*S*-(*p*-Azidophenacyl)-glutathione (ph.m.)		745
2.6.1.1 Aspartate aminotrans-ferase			
Pig heart	Propargylglycine (ss)		746
	CH₂=CHCH(NH₂)COOH (ss)	Lys-258	747
	N-(Bromoacetyl)pyridoxamine	Lys-258	748

Table 2 (continued)
SURVEY OF AFFINITY MODIFICATION DATA FOR ENZYMES

Source	Reagent	Modification point	Ref.
Pig heart (cytosolic)	α-N-(2,4-Dinitro-5-fluorophenyl)-β-N-phos-phopyridoxyldiaminopropionate	Lys-258, Cys-45	102
Pig heart (mitochondrial)	α-N-(2,4-Dinitro-5-fluorophenyl)-β-N-phos-phopyridoxyldiaminopropionate	Lys-258, Lys-343	102
	β-Bromopropionate	Lys	749
Pig heart (cytosolic, mitochondrial)	4'-N-(2,4-Dinitro-5-fluoro-phenyl)pyridoxamine-5-phosphate	Lys-258	750
2.6.1.2 Alanine aminotrans-ferase			
Pig heart	3-Cl-Ala (ss)		751
	L-Propargylglycine (ss)		752
	β-CN-Ala (ss)		753
2.6.1.13 Ornithin-oxoacid aminotransferase			
Rat liver	4-Aminohex-5-ynoic acid (RMI 71645) (ss) and 5-amino-1,3-cyclohexadienyl-carboxylic acid (gabaculine) (ss)		754
	3-Amino-1,5-cyclohexadienyl carboxylic acid (isogabaculin)		755
2.6.1.19 γ-Aminobutyrate amino-transferase			
Pig brain	4-Amino-5-halopentanoic acids (F,Cl,Br)		756
	3-Amino-1,5-cyclohexadienyl carboxylic acid (isogabaculin)		755
Rat, mouse brain	5-Fluoro-4-oxo-pentanoic acid		757
Rat brain	4-Amino-hex-5-enoic acid		758
2.7.1.1 Hexokinase			
Brain	6-Mercapto-9-β-D-ribofuranosyl-purine 5'-tri-phosphate	Cys	759
Yeast	PLP-pAdo		602
	ClRCH$_2$NHpppAdo and ClRCH$_2$NHpAdo		240
2.7.1.11 Phosphofructokinase			
Sheep heart	FSO$_2$BzAdo	reg. site Lys (DFATK[a] MGAK)	760
Rabbit muscle	FSO$_2$Bz-2-aza-εAdo	reg. site	761
	cAMP (UV cross-linking) 2'-O-(COCN$_2$CO$_2$Et)-cAMP; N^6,2'-O-(COCN$_2$CO$_2$Et)$_2$-cAMP; and N^6-(COCN$_2$CO$_2$Et)-cAMP		762 763
2.7.1.21 Thymidine kinase			
E. coli	5-I-dUMP and 5-I-6-aza-dUrd		764
2.7.1.37 Protein kinase			
Human blood and rat brain Mouse embryo fibro-blasts	8-N$_3$-cAMP (ph.m.)		765
3T3-L1	8-N$_3$-cAMP (ph.m.)		766
Bovine brain	8-N$_3$-cAMP (ph.m.)		767
Pig heart (RSU)	8-N$_3$-cAMP (ph.m.)	Tyr (KREISH Y[a]NNQLV KM)	768
Corpinus macrorhirus	8-N$_3$-cAMP (ph.m.)	R.SU	769
Bovine heart	8-N$_3$-cAMP (ph.m.)	R.SU	770, 771
Bovine lung (cGMP-dependent)	8-N$_3$-cGMP (ph.m.)		772

Table 2 (continued)
SURVEY OF AFFINITY MODIFICATION DATA FOR ENZYMES

Source	Reagent	Modification point	Ref.
Vascular smooth muscle (cGMP-dependent)	8-N$_3$-cIMP (ph.m.)		773
Bovine heart (C.SU)	FSO$_2$BzAdo		774
Rabbit reticulocytes	FSO$_2$ BzAdo	α SU	775
Pig heart (C.SU)	FSO$_2$ BzAdo	Lys-71	776
Bovine lung (cGMP-dependent)	FSO$_2$BzAdo		777
	FSO$_2$BzAdo	Lys (VNL VQLKSNNS KTFAMKIU KKRHIVDT RQQNHIRS GK)	778
Human tumor cells A 431	FSO$_2$BzAdo		779
Bovine muscle	O-Phthalaldehyde	Lys-72, Cys-199	780
Bovine lung (cGMP-binding, bovine heart	N$^\alpha$-Tosyl-LysCH$_2$Cl		413
Bovine brain (C.SU)	Cibacron blue F3GA		781
—	8-N$_3$-ATP (ph.m.)	C.SU	782
Pigeon breast (C.SU)	ClCH$_2$-ppAdo; FSO$_2$C$_6$H$_4$-pAdo; p-FSO$_2$BzAdo; m-FSO$_2$BzAdo; 2,4,6-(CH$_3$)$_3$C$_6$H$_2$-COOppAdo; HPO$_3$-CH(NHCOCH$_2$Br)-ppAdo; and HPO$_3$-CH(NHCOCH$_2$Br)-pAdo		259
2.7.1.38 Phosphorylase kinase			
Rabbit muscle	FSO$_2$BzAdo	β SU	783
	8-N$_3$-ATP and O-8-N$_3$-ATP	β SU	784
2.7.1.40 Pyruvate kinase			
Yeast	Bromopyruvate	Cys	785
	FSO$_2$BzAdo		786
Rabbit muscle	FSO$_2$BzAdo and FSO$_2$BzGuo	Cys,Tyr	787
	oADP		788
	FSO$_2$BzAdo	Cys,His	108
	2-[(4-Bromo-2,3-dioxobutyl)-thio]-AMP		789
	PLP-pAdo		602
2.7.1.70 Histone kinase			
Pig brain (C.SU)	oATP	Lys	790
2.7.2.3 Phosphoglycerate kinase			
Yeast, B.stearothermophilus	N^6-(p-Bromoacetaminobenzyl)-AMP		246
Yeast	PLP-pADO		602
2.7.3.2 Creatine kinase			
Rabbit muscle	N-(2,3-Epoxypropyl)-N-amidinoglycine		791
	p-N$_3$-C$_6$H$_4$NHpppAdo (ph.m.)		297, 792
	ClRCH$_2$NHpppAdo		793
	p-N$_3$-C$_6$H$_4$NHppp εAdo (ph.m.) and p-N$_3$-C$_6$H$_4$NHpp εAdo (ph.m.)		388
Rat heart	oADP and oATP		794
Mitochondrial rabbit muscle	oADP and oATP		312
	ClCH$_2$CH$_2$N(CH$_3$)-pppAdo		324
2.7.3.3 Arginine kinase			
Homarus vulgaris muscle	p-N$_3$-C$_6$H$_4$NHpppAdo (ph.m.)		297

Table 2 (continued)
SURVEY OF AFFINITY MODIFICATION DATA FOR ENZYMES

Source	Reagent	Modification point	Ref.
2.7.4.3 Adenylate kinase			
Rabbit muscle	PLP-pAdo		602
Pig muscle	K_2FeO_4	Tyr-95, Cys-25	795
Myosin light chain kinase			
Rabbit skeletal muscle	p-Azidobenzamidinium-calmodulin (ph.m.)		796
2.7.7.6 RNA polymerase			
E. coli	Oligo 4-thiothymidilate (UV cross-linking)	β′,β SU	428
	$(pT)_5p$-5-Br-U$(pT)_4$ (UV cross-linking)	β SU (GG VVQYVDA-SR), β′ SU (LSNVKLV ITSRLIDE FGR)	431
	Poly(A,8-N_3-A) (ph.m.)	β′,β,σ SU	283
	Poly d(A-T) (UV cross-linking)		426
	Poly d(A-T) (UV cross-linking) and E. coli DNA (UV cross-linking)	β′,β,σ SU	427
	T7 phage DNA (UV cross-linking)	β′,β,σ SU	433
	5-Br-U substituted DNA fragment containing lac UV5 promoter (UV cross-linking)	β,σ SU	434
	5-Br-U substituted G2 promoter fragment from phage fd (UV cross-linking)	β′,β SU	435
	Partially depurinated DNA fragment containing lac UV5 promoter	β′,β,σ SU	436
	5-Formyl-UTP	β SU	249
	oUTP	β′,β,σ SU	797
	oNTP	α SU(Lys)	798
	Periodate oxidized 6-methylthiopurine riboside	β SU	799
	Periodate oxidized N-(acetylaminoethyl)-1-naphthylamine-5-sulfonate of 6-thiopurine riboside	β SU	800
	4-Thio-UTP	β′,β SU	428
	9-β-Arabinofuranosyl-6-thiopurine	β SU(Cys)	801
	5′-(4-Azidophenacylthio)pApU (ph.m.)	β′,β SU	441
	8-N_3-ATP (ph.m.)	β′,σ SU	802
	3′-Azido-3′-deoxy-β-D-xylofuranosyl-AMP (ph.m.)	σ SU	803
	3′-Azido-3′-deoxyxylofuranosyl-ATP (ph.m.) and 3′-azido-3′-deoxyxylofuranosyl-GTP (ph.m.)	β SU	804
	β-(4-Azidophenyl)ppApU (ph.m.) and β-(4-azidophenyl)ppAdo (ph.m.)		805
	p-N_3-C_6H_4NHpppAdo (ph.m.) and p-N_3-C_6H_4NHpppGuo (ph.m.)	β′,β, σ SU	439
	Imidazolides of AMP, ADP, GMP, GDP; p- and o-formylphenyl esters of ATP; Cl-RNHpppAdo; ClRCH₂NHpppAdo; Cl-RCH₂NHppAdo; ClCH₂CH₂N(CH₃)-pppAdo; p-OCH-C_6H_4-N(CH₃)CH₂CH₂pAdo; p-OCH-C_6H_4-N(CH₃)CH₂CH₂ppAdo; p-OCH-C_6H_4-N(CH₃)CH₂CH₂pppAdo; p-OCH-C_6H_4N(CH₂CH₂Cl)CH₂CH₂OpAdo; and cyclic trimethaphosphate of Ado and Guo		324
Wheat germ	4-[N-β-Hydroxyethyl-N-methyl] benzaldehyde esters of AMP, ADP, and ATP	140 kDa SU	445

Table 2 (continued)
SURVEY OF AFFINITY MODIFICATION DATA FOR ENZYMES

Source	Reagent	Modification point	Ref.
Reovirus	PLP		806
Calf thymus	8-N$_3$-ATP (ph.m.) and 8-N$_3$-GTP (ph.m.)	37 kDa SU	
			446
B. cubtilus	Plasmid pUC8 with insert containing promoter (UV cross-linking)	β,σ SU	807
Drosophila	Adenovirus 2DNA (UV cross-linking); nascent RNA transcripts from adenovirus 2 DNA (UV cross-linking)		808
2.7.7.7 DNA polymerase			
E. coli	Epoxy ATP (ss)		809
	Periodate oxidized 6-methylthiopurine riboside	Lys	810
	ClRCH$_2$NHpppAdo and ClRCH$_2$NHpppdThd		811
	5-N$_3$-dUTP (ph.m.)		250
	Imidazolides of dATP, dCTP, dGTP, and dTTP		449
	dTTP (UV cross-linking) and 8-N$_3$-ATP (ph.m.)		812
	ATP; dATP; dTTP (UV cross-linking)	α,τ,γ,δ SU	813
	PLP	Lys	814
	PLP		815
Human placenta	(pT)$_2$pC[Pt^{2+}(NH$_3$)$_2$OH](pT)$_7$		448
E. coli	*E. coli* DNA (UV cross-linking) and poly(dA-dT) (UV cross-linking)		816
2.7.7.27 Glucose-1-phosphate adenyltransferase			
	8-N$_3$-ADP-Glc (ph.m.)	Lys-108 Arg-114	331
	8-N$_3$-AMP (ph.m.)		817
2.7.7.31 Terminal deoxynucleotidyltransferase			
Calf thymus glands	8-N$_3$-ATP (ph.m.)		818
2.7.9.1 Pyruvate, phosphate dikinase			
Bacteriodes symbiosus	PLP	Lys	819
2.8.2.1 Phenol sulfotransferase			
Rat liver	oADP		820
3.1.1.4 Phospholipase A$_2$			
Crotalus atrox	1,2-Di-*O*-hexylglycero-3-(ethyldiazomalonamidoethyl phosphate) (ph.m.)		821
3.1.1.7 Acetylcholinesterase			
Bovine erythrocyte	*N,N*-Dimethyl-2-phenyl-aziridinium ion		355
	2-Chloro-*N*-(chloroethyl)-*N*-methyl-2-phenylethylamine		822
	p-Dimethylaminobenzene diazonium fluoroborate (ph.m.)		174
Electric eel	7-(Diethoxyphosphinyloxy)-*N*-methylquinolinium fluorosulfate		823
Electrophorus electricus	RSO$_2$CH$_3$ (a-c); R = m—(CH$_3$)$_3$N$^+$C$_6$H$_4$O— (a), 3–N–CH$_3$–Py$^+$–O– (b), and N–Py$^+$–(CH$_2$)$_5$–N–Py$^+$–O(3)– (c); 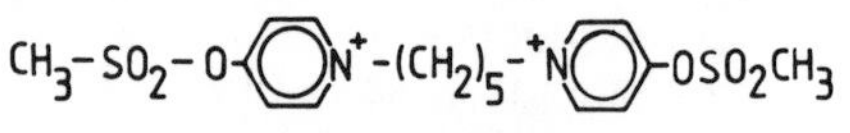		341

$$CH_3-SO_2-O-\langle\bigcirc\rangle N^+-(CH_2)_5-^+N\langle\bigcirc\rangle-OSO_2CH_3$$

Electric eel	Methyl(acetoxymethyl)nitrosamine (ph.m.) and methyl(butyroxymethyl)nitrosamine (ph.m.)		175

Table 2 (continued)
SURVEY OF AFFINITY MODIFICATION DATA FOR ENZYMES

Source	Reagent	Modification point	Ref.
3.1.1.8 Acylcholine acyl-hydrolase			
Horse serum	*N,N*-Dimethyl-2-phenylaziridinium ion		358
Horse plasma	2-Chloro-*N*-(chloroethyl)-*N*-methyl-2-phenyl-ethylamine		822
3.1.2.1 Acyl-CoA hydrolase			
Bovine kidney cytosol	*S*-(4-Bromo-2,3-dioxobutyl)-CoA		723
3.1.3.1 Alkaline phosphatase			
Human placenta	oAMP		824
Calf intestine	9-azidoacridine (ph.m.)		214
3.1.3.2 Acid phosphatase			
Rat liver, human prostate, potato, wheat germ, rabbit erythrocytes	FeO_4^{2-}		825
3.1.3.11 Fructose-bisphosphatase			
Pig kidney	oAMP	reg. site	826
	2-N_3-AMP (ph.m.)	reg. site	827
	8-N_3-AMP (ph.m.)	reg. site	828
3.1.4.1 Phosphodiesterase			
Bovine heart muscle	*p*-Azidobenzamidinium-calmodulin (ph.m.)		796
3.1.27.5 Pancreatic ribonuclease A			
Bovine pancreas	6-Chloropurine 9-β-ribofuranosyl 5′-mono-phosphate	Lys-1	829
	9-Azidoacridine (ph.m.)		214
	ClRCH$_2$NHpdTpdA; (N → P)*N*-methylimida-zolide d(pTpA)		115
	Ferrate ion	His-119	830
	6-Chloropurine riboside	Lys-1, -37, -41, -91	831
3.2.1.1 α-Amylase			
Pig pancreas	9-Azidoacridine (ph.m.)		214
3.2.1.2 β-Amylase			
Soybean	2′,3′-Epoxypropyl-α-D-glucopyranoside		100
3.2.1.23 β-Galactosidase			
Escherichia coli	β-D-Galactopyranosylmethyl-*p*-nitrophenyltri-azene		832
3.2.1.33 Amylo-1,6-glucosidase			
Rabbit muscle	1-*S*-Dimethylarsino-1-thio-β-D-glucopyranoside		410
3.3.1.1 Adenosylhomocystein-ase			
Human placenta	8-N_3-Ado (ph.m.) and 8-N_3-cAMP (ph.m.)		833
Calf liver	2′-Ado (ss)		834
Human red cells	5′-Deoxy-5′-methylthio-adenosine		835
Bovine liver	Nucleoside analogues: 2-chloro-Ado, Ara-Ado, aristeromycin, pyrazomycin, 2-amino-6-chloropurine riboside, purine riboside, 2-chloro-Ara-Ado, 2′-dAdo; N^6-methyl-Ado, 7-deaza-Ado, 8-aza-Ado, 8-amino-Ado, 5-iodo-5′-dAdo, 3-deaza-Ado, 9-(3′-azido-3′-deoxy-2′-ethylthio-β-D-arabinofuranosyl) adenine, 8-bromo-Ado, xylosyladenine, 3′-dAdo, inosine, Guo, 5′-dAdo, 8-aza-9-deaza-Ado, Ado		836

Table 2 (continued)
SURVEY OF AFFINITY MODIFICATION DATA FOR ENZYMES

Source	Reagent	Modification point	Ref.
3.4.11.1 Leucine aminopeptidase			
Pig kidney	pN$_2^+$-Phe-Phe		325
3.4.11.2 Aminopeptidase (microsomal)			
Pig kidney microsomes	*p*-N$_3$-Phe (ph.m.) and *p*-N$_3$-Phe-Ala-Gly-Gly (ph.m.)		837
3.4.11.8 Pyroglutamyl peptidase			
B. amyloliquefaciens	*N*-Carbobenzoxy-L-pyroglutamyl diazomethane	Cys	838
	L-Pyroglutamyl chloromethane		839
	N$^\alpha$-Carbobenzoxy-L-pyroglutamyl chloromethane		840
3.4.14.1 Cathepsin C			
Bovine spleen	Gly-Phe-CHN$_2$		841
3.4.15.1 Dipeptidyl carboxy peptidase			
Bovine lung	Chlorambucil-(ClCH$_2$CH$_2$)$_2$NC$_6$H$_4$-(CH$_2$)$_3$COOH; chlorambucilyl-Pro		842
3.4.17.1 Carboxypeptidase A			
Bovine pancreas	2-Benzyl-3-mercaptopropanoic acid		395
	N-Bromoacetyl-*N*-methyl-Phe		843
3.4.17.2 Carboxypeptidase B			
Pig pancreas	2-Mercaptomethyl-5-guanidinopentanoic acid		395
3.4.21.1 Chymotrypsin[b]			
—	*N*-Bromoacetylarylamides (24 compounds)	Met-192 Ser-195	845
—	Bromoacetyl-(R)-(+)-1-aminoindan; bromoacetyl-(S)-(−)-1-aminoindan; bromoacetyl-3-(aminomethyl)indol; and bromoacetyl-3-[(methylamino)methyl]-indol	Ser-195	846
Bovine pancreas	9-Azidoacridine		214
	Butyl isocyanate	Ser-195	847
	Z-Ala-Ala-Phe; Z-Ala-Ala-Phe(p-NO$_2$); Z-Ala-Ala-Phe(p-N$_3$); Z-Ala-Ala-Phe(m-N$_3$); and Z-Ala-Ala-Phe(p-N$_3$,o-NO$_2$)		848
	Dansyl-Leu-Phe-CH$_2$Cl		849
	3- and 5-Benzyl-6-chloro-2-pyrone and 3- and 5-methyl-6-chloro-2-pyrone		850
	Ac-Leu-Phe-CH$_2$F		851
	D-*N*-Nitroso-*N*-benzyl-*N*'-isobutyryl-Ala-amide; D-*N*-nitroso-*N*-benzyl-*N*'-isobutyryl-Phe-amide; *N*-nitroso-1,4-dihydro-6,7-dimethoxy-3(2H)-isoquinolinone		168
	Z-Gly-Leu-Phe-CH$_2$Cl and Boc-Gly-Leu-Phe-CH$_2$Cl		852
	N-Nitroso-2-oxo-3-aza-6,7-dimethoxy(1,2,3,4-tetrahydronaphthalene)		853
	CH$_3$(CH$_2$)$_n$NHCOOC$_6$H$_4$NO$_2$ (n = 1—7)		854
3.4.21.4 Trypsin[b]			
Bovine pancreas	D-Val-Leu-Lys-CH$_2$Cl; Ala-Phe-Lys-CH$_2$Cl; D-Val-Phe-Lys-CH$_2$Cl; pyro-Glu-Phe-Lys-CH$_2$Cl; and respective *p*-nitrophenyl anilides		855
	NAP-sulfenyl-Trp(93)-soybean trypsin inhibitor (ph.m.)		856

Table 2 (continued)
SURVEY OF AFFINITY MODIFICATION DATA FOR ENZYMES

Source	Reagent	Modification point	Ref.
	Dansyl-Ala-Lys-CH$_2$Cl	His-46	857
	(*p*-Amidinophenyl)methanesulfonylfluoride		171
	Peptides with C-terminal Lys-CH$_2$Cl		858
	4-Amidinobenzolsulfofluoride and 4-(2-amino-ethyl)-benzolsulfofluoride		859
	H$_3$N$^+$(CH$_2$)$_n$NHCOOC$_6$H$_4$NO$_2$ (n = 2,5,7)		854
	p-[*m*(*m*-Fluorosulfonylphenyl-ureido)phenoxyethoxy]benzamidine		860
3.4.21.5 Thrombin			
Human blood	D-Val-Leu-Lys-CH$_2$Cl		855
	4-Amidinobenzolsulfofluoride and 4-(2-amino-ethyl)-benzolsulfofluoride		859
Bovine plasma	Dansyl-Ala-Lys-CH$_2$Cl		849
Bovine blood	Peptides with C-terminal Arg-CH$_2$Cl		861
Human blood	PLP		862
	2-Phenyl-4-(3-*N*-nitroguanidinopropyl)-5-pivaloyloxyoxazole		863
	(*p*-Amidinophenyl)methanesulfonyl fluoride		171
	Peptides with C-terminal Lys-CH$_2$Cl		858
	m-[*O*-(2-Chloro-5-fluorosulfonylphenyl-ureido)phenoxybutoxy]benzamidine		864
	D-Phe-Pro-Arg-CH$_2$Cl; D-Tyr-Pro-Arg-CH$_2$Cl; 4-NH$_2$-D-Phe-Pro-Arg-CH$_2$Cl; ε-Bz-D-Lys-Pro-Arg-CH$_2$Cl; Bz-Ala-D-Phe-Pro-Arg-CH$_2$Cl; D-Phe-Pro-Phe(4-Gn)-CH$_2$Cl; and D-Phe-Phe-Arg-CH$_2$Cl		310
Bovine blood	*N*-Tosyl-Lys-CH$_2$Cl and n^α-(*p*-nitro-Z)Arg-CH$_2$Cl	His-43	865
3.4.21.7 Plasmin			
Human plasminogen	D-Val-Leu-Lys-CH$_2$Cl; Ala-Phe-Lys-CH$_2$Cl; D-Val-Phe-Lys-CH$_2$Cl; Glu-Phe-Lys-CH$_2$Cl; and respective *p*-nitrophenylanilides		855
Human plasma	Peptides with C-terminal Arg-CH$_2$Cl		861
	(*p*-Amidinophenyl)methane-sulfonyl fluoride		171
	4-N$_3$-Bz-Gly-Lys (ph.m.)		866
3.4.21.8 Kallikrein			
Human plasma	Peptides with C-terminal Arg-CH$_2$Cl		861
	Peptides with C-terminal Lys-CH$_2$Cl		858
Pig pancreas	*p*-[*m*(*m*-Fluorosulfonylphenylureido)phenoxyethoxy]-benzamidine		860
3.4.21.11 Elastase			
—	Butyl isocyanate	Ser-188	847
Pig pancreas	Dansyl-Ala-Pro-Ala-CH$_2$Cl; dansyl-Pro-Ala-CH$_2$Cl; and dansyl-Pro-Ala-Pro-Ala-CH$_2$Cl		849
	Peptides with C-terminal Ala-CH$_2$Cl		867
	CH$_3$(CH$_2$)$_n$NHCOOC$_6$H$_4$NO$_2$ (n = 1—7)		854
	Ethyl *p*-nitrophenyl alkylphosphonates (alkyl = *n*-propyl, *n*-butyl, *n*-pentyl, *n*-hexyl)		868
Pig liver	3- and 5-Benzyl-6-chloro-2-pyrone and 3- and 5-methyl-6-chloro-2-pyrone		850
Human pancreas	Ac-Ala-Ala-Pro-Ala-CH$_2$Cl; Ac-Ala-Ala-Ala-Ala-CH$_2$Cl; Ac-Ala-Ala-Pro-Ile-CH$_2$Cl; Ac-Ala-Ala-Pro-Val-CH$_2$Cl; $^-$OOC–CH$_2$–CH$_2$–CO–Ala-Ala-Pro-Leu-CH$_2$Cl; $^-$OOC–CH$_2$–CH$_2$–CO–Ala-Ala-Pro-Val-CH$_2$Cl; CH$_3$OOC–CH$_2$–CH$_2$–CO–Ala-Ala-Pro-Val-CH$_2$Cl; NH$_2$-Ala-Ala-Pro-Leu-CH$_2$Cl; and Ac-Ala-Ala-Pro-Leu-CH$_2$Cl		869

Table 2 (continued)
SURVEY OF AFFINITY MODIFICATION DATA FOR ENZYMES

Source	Reagent	Modification point	Ref.
—	1-Bromo-4-(2,4-dinitrophenyl)butan-2-one	Glu-6	870
3.4.21.14 Proteinase K			
Fungus *Tritirachium album*	Z-Ala-Ala-CH$_2$Cl	His-68 Ser-221	334
3.4.21.30 Crotalase			
C. adamanteus venom	Pro-Phe-Arg-CH$_2$Cl		871
3.4.21.31 Urokinase			
Human urine	Dansyl-chloride; N$^\alpha$-tosyl-Lys-CH$_2$Cl		872
3.4.22.1 Cathepsin B$_1$			
Bovine spleen	Z-Phe-Ala-CHN$_2$; Z-Ala-Phe-Ala-CHN$_2$; Z-Phe-Gly-CHN$_2$		841
	Z-Phe-Phe-CHN$_2$		873
Bovine spleen, pig liver	(Gly-Phe-NHCH$_2$CH$_2$S)$_2$; (Ala-Ala-Phe-NH-CH$_2$CH$_2$S)$_2$		874
Human liver	L-*trans*-Epoxysuccinyl-Leu-amido(4-guanidino)butane		875
Bovine spleen	Z-Phe-Ala-CHN$_2$; Z-Ala-Phe-Ala-CHN$_2$; Z-Phe-Gly-CHN$_2$; and Z-Phe-Phe-CHN$_2$		876
3.4.22.2 Papain			
Papaya latex	Z-Phe-Phe-CHN$_2$		877
	L-*trans*-Epoxysuccinyl-Leu-amido-(4-guanidino)butane (E-64)		875
3.4.22.3 Ficin			
—	N-Tosyl-Lys-CH$_2$Cl and N-tosyl-Phe-CH$_2$Cl		878
3.4.22.4 Bromelain			
Pineapple stem juice	N-Tosyl-Phe and N-tosyl-Lys		879
3.4.22.8 Clostripain			
—	Z-Lys-CHN$_2$; Z-Phe-CHN$_2$; and Z-Phe-Phe-CHN$_2$		841
Clostridium histolyticum	α-N-Tosyl-Lys-CH$_2$Cl		880
3.4.22.15 Cathepsin L			
Rat liver	L-*trans*-Epoxysuccinyl-Leu-amido-(4-guanidino)butane and L-*trans*-epoxysuccinyl-Leu-amido(3-methyl)butane		875
	Z-Phe-Phe-CHN$_2$; and Z-Phe-Ala-CHN$_2$		881
3.4.23.1 Pepsin			
Pig stomach mucosa	p-Diazobenzenesulfonic acid (ph.m.); N-α-(p-N$_3$-Phe)-Leu-Val-Lys-Pro-Leu-Val-Arg-Lys-Lys-Ser-Leu-OH (ph.m.); pNO$_2$Phe (ph.m.); and N-α-(p-NO$_2$Phe)-Leu-Val-Lys-Val-Pro-Leu-Val-Arg-Lys-Lys-Ser-Leu-OH (ph.m.)		882
Pig pepsinogen	1,2-Epoxy-3-NAP-oxypropane; p-azidophenacyl bromide (ph.m.); p-azido-α-diazoacetophenone (ph.m.)		883
	Z-Phe-CHN$_2$		884
	1-Diazo-4-phenyl-2-butanone	Asp (IVD^a	885
	Tosyl-Phe-CHN$_2$	T(G,T) SL)	886
3.5.1.4 Amidase			
P. aeruginosa	Chloroacetone		345
3.5.2.6 β-Lactamase			
E. coli	Clavulanic acid and 9-deoxyclavulanic acid		887
S. albus G, actinomadura R 39	Clavulanic acid		888
B. cereus, E. coli, Staphylococcus aureus, B. licheniformis	6-β-Bromopenicillanic acid		889

Table 2 (continued)
SURVEY OF AFFINITY MODIFICATION DATA FOR ENZYMES

Source	Reagent	Modification point	Ref.
B. cereus	6-β-Bromopenicillanic acid	Ser-44	890
	Diazo-6-aminopenicillanic acid	Ser-44	891
E. coli	Penicillanic acid sulfone		892
569/H, *E. coli* RTEM	6-β-X-penicillanic acid sulfones: X = H, CF₃SO₂NH-, CH₃SO₂NH-, p-CH₃C₆H₄SO₂NH-, p-CH₂OC₆H₄SO₂NH-; quinacillin sulfonamide		893
E. coli	Derivative of olivanic acid		894

$$^-O_3S-O-\overset{\underset{|}{CH_3}}{CH} \cdots S-CH=CH-NH-\overset{O}{\overset{\|}{C}}-CH_3 \quad (COO^-)$$

Source	Reagent	Modification point	Ref.
3.5.3.6 Arginine deaminase			
Mycoplasma arthritidis	Formamidinium ion		895
3.5.4.4 Adenosine deaminase			
Calf intestinal mucosa	9-(p-Bromoacetamidobenzyl)Ade and 9-(o-bromoacetamidobenzyl)Ade		896
3.5.4.6 AMP-deaminase			
Rat muscle	oGTP		897
	oATP		898
Pig liver	Carbodiimide-activated folate		105
3.6.1.1 Inorganic pyrophosphatase			
Yeast	Methylphosphate; O-phosphopropanolamine; O-phosphoethanolamine; N-acetylphosphoserine; and phosphoglycollic acid		899
	N-Chloroacetylphosphoethanolamine		900
E. coli	Methylphosphate; N-acetylphosphoserine; O-phosphoethanolamine; and O-phosphopropanolamine		901
3.6.1.3 F₁ ATPase			
Bovine heart mitochondria	3'-O-(NAP-NH(CH₂)₃-CO)-ADP (ph.m.) and 3'-O-(NAP-NH(CH₂)₃-CO)-ATP (ph.m.)	α,β SU	902
	3'-O-[5-Azidonaphthoyl]-ADP (ph.m.)	α,β SU	400
	oATP		903
	p-N₃-C₆H₄COCH₂-OSCP (oligomycin sensitivity conferring protein) (ph.m.)	α,β SU	904
	3'-O-(NAP-NH-CH₂CH₂CO)-ATP (ph.m.)		905
Bovine heart mitochondria	3'-O-(NAP-NH-CH₂CH₂CO)-8-N₃-ATP (ph.m.)	Cross-link between α and β SU	263
Thermophilic bacterium PS3	3'-O-(NAP-NH-CH₂CH₂CO)-8-N₃-ATP (ph.m.)		402
Bovine heart mitochondria	8-N₃-ATP (ph.m.)	β SU	906, 907
	8-N₃-ATP (ph.m.) and 8-N₃-ATP (ph.m.)	β SU	401
	2-N₃-ATP (ph.m.)	β SU	243
	2-N₃-ATP (ph.m.) and 2-N₃-ADP (ph.m.)	β SU	908
Bovine liver mitochondria	o-εATP		909
Pig heart mitochondria	3'-O-(NAP)CH₂CH₂CO-ADP (ph.m.)	α,β SU	398
	3'-O-(NAP-NH-CH₂CH₂CO)ADP (ph.m.)	β SU	910
	FSO₂BzAdo	β SU, Tyr (IMDPNIV GSEHYᵃDV AR)	911

Table 2 (continued)
SURVEY OF AFFINITY MODIFICATION DATA FOR ENZYMES

Source	Reagent	Modification point	Ref.
Rat liver mitochondria	3'-O-(4-Benzoyl)benzoyl-ATP (ph.m.)	β SU	912
Chloroplast	2-N$_3$-ATP (ph.m.)	β SU	913
Thermophilic bacterium PS3	FSO$_2$BzAdo	Tyr(ALAP EIV-GEEH Y^{a}QVAR)	914
	3'-O-(4-Benzoyl)benzoyl-ADP(ph.m.)	β SU	915
Micrococcus luteus	8-N$_3$-ATP (ph.m.)	β SU	416
E. coli	8-N$_3$-ATP (ph.m.) and 4-chloro-7-nitrobenzofurazan	α,β SU	916
Yeast mitochondria	8-N$_3$-εATP (ph.m.)	β SU	917
Yeast *Saccharomyces cerevisiae*	FSO$_2$BzAdo	Tyr(HY^{a}D VASKVQE)	918
E. coli, thermophilic bacterium PS3	NAP-phosphate (ph.m.)	β SU	919
Mycobacterium phlei	oATP	α SU	920
	oADP	α,β SU	
3.6.1.3 Na,K-ATPase			
Pig kidney	2-Nitro-5-azidobenzoyl-ouabain (ph.m.)	α SU	921
	(ClCH$_2$)pAdo; (ClCH$_2$CH$_2$O)pAdo; (BrCH$_2$CO)pAdo; (BrCH$_2$CH$_2$)ppAdo; and (Ms)pAdo		242

$$\text{Ms} = \quad CH_3 - \overset{\displaystyle CH_3}{\underset{\displaystyle CH_3}{\bigcirc}} - C(O)O-$$

Source	Reagent	Modification point	Ref.
	(Ms)ppAdo; pppAdo-8-NH(CH$_2$)$_6$N-HCOCH$_2$Cl; p(CH$_2$)ppAdo-8-NH(CH$_2$)$_6$NHCOCH$_2$Cl;(p-FO$_2$SC$_6$H$_4$O)-pAdo; (p-FO$_2$SC$_6$H$_4$CO)Ado; and (m-FO$_2$SC$_6$H$_4$CO)Ado (3β,5β,14β,20E)-24-azido-3-[(2,6-dideoxy-β-D-ribo-hexopyranosyl)oxy]-14-hydroxy-21-norchol-20(22)-en-23-one (ph.m.)		
Canine kidney	Cr(III)-3'-O-(NAP-NH-CH$_2$CH$_2$CO)-ATP (ph.m.)		923
	oATP	β SU	924
	FSO$_2$BzAdo	α SU	925
Pig kidney and *Electrophorus electricus*	1-Myristoyl,2-N-NAP-dodecanoyl-sn-glycero-3-phosphocholine (ph.m.);1-palmitoyl,2-(2-azido-4-nitro)benzoyl-sn-glycero-3-phosphocholine (ph.m.); N'-NAP-12-amino-N-dodecanoyl-glucosamine; N'-NAP-12-amino-N-dodecanoyl-glycine	Hydrophobic domain α,β SU	926
E. electricus and canine kidney	Ouabain p-aminobenzenediazonium derivative (ph.m.)	α SU	927
E. electricus	N-(Ouabain)-N'-(NAP)ethylenediamine	12,8 kdalton SU 93,3 kdalton SU	
	N-(Strophanthidin)-N'-(NAP)-ethylenediamine	93,3 kdalton SU	928
	4'''-Diazomalonyldigitoxin (ph.m.); 3'''-diazomalonyldigitoxin (ph.m.)	α,β SU	929
Rat kidney	6-[(3-Carboxy-4-nitrophenyl)-thiol]-9-β-D-ribofuranosylpurine 5'-triphosphate		351

Table 2 (continued)
SURVEY OF AFFINITY MODIFICATION DATA FOR ENZYMES

Source	Reagent	Modification point	Ref.
Pig kidney	Disulfide of thioinosine triphosphate		351
Pig mucosa parietal cells	Fluorescein 5'-isothiocyanate	Lys(HVLM K[a]-GAPEQL SIR)	212
Guinea pig brain	Hellebrigenin 3-iodoacetate and hellebrigenin 3-bromoacetate		930
3.6.1.3 Ca^{2+} ATPase			
Rabbit muscle sarcoplasmatic reticulum	Fluorescein isothiocyanate		351, 931
	3'-O-(NAP-NH-CH$_2$CH$_2$CO)ATP(ph.m.)		932
	3'-O-(4-benzoyl)benzoylATP (ph.m.)		933
Escherichia coli GTPase	oADP and oATP		934
Bovine rod outer segments	8-N$_3$-GTP (ph.m.)	37 kdalton SU	935
3.6.1.5 ATP diphosphohydrolase			
Pig pancreas	8-N$_3$-ATP		936
Microsomal cholesterol oxide hydrolase			
Rat liver microsomes	7-Dehydrocholesterol 5,6-β-oxide		937
4.1.1.1 Pyruvate decarboxylase			
Brewer's yeast	4-(4-Chlorophenyl)-2-oxo-3-butenoate (ss)		390
4.1.1.17 Ornithine decarboxylase			
Rat liver	α-(Difluoromethyl)ornithine		938, 939
Hepatoma cells	α-(Difluoromethyl)ornithine		940
4.1.1.19 Arginine decarboxylase			
E. coli, P. aeruginosa, Klebsiella pneumoniae	D,L-α-(Difluoromethyl)-arginine (RMI 71897)		941
4.1.1.22 Histidine decarboxylase			
Lactobacillus 30a	3-Bromoacetylpyridine	Cys-5	211
4.1.1.26 Aromatic L-amino acid decarboxylase			
Pig kidney	D,L-α-Difluoromethyl-L-3,4-dihydroxyphenyl alanine (ss)		942
4.1.1.28 3-(3,4-Dihydroxyphenyl)alanine decarboxylase			
Pig kidney	2-(Fluoromethyl)-3-(3,4-Dihydroxyphenyl)alanine (ss)		943
4.1.1.31 Phosphoenolpyruvate carboxylase			
E. coli	Bromopyruvate	Cys	944
4.1.1.32 Phosphoenolpyruvate carboxykinase			
Pig liver	N-(Iodoacetylaminoethyl)-5-naphthylamine-1-sulfonic acid		945
Pig liver mitochondria	Bromopyruvate		946
4.1.1.39 Ribulose bisphosphate carboxylase/oxygenase			
Rhodospirillum rubrum	2-(4-Bromoacetamido)anilino-2-deoxy-pentitol 1,5-bisphosphate	Cys,His AGYGYV ATAAH[a]-FAAESSTGTNVE	947

Table 2 (continued)
SURVEY OF AFFINITY MODIFICATION DATA FOR ENZYMES

Source	Reagent	Modification point	Ref.
Spinach	N-Bromoacetylethanolamine phosphate	VCᵃTTDDF TR Lys,Cys	948
	3-Bromo-1,4-dihydroxy-2-butanone 1,4-bis-phosphate	Lys YGRPLLGC TIKᵃPK,L SGGDHIHS GTVVGKᵃL EGER	331
	2-Bromoacetylaminopentitol 1,5-bisphosphate	Met LQGASGIH TGTMGFGK MᵃEGESSDR	949
R. rubrum, spinach	2-Bromo-1,5-dihydroxy-3-pentanone 1,5-bis-phosphate		950
Spinach	PO_3^{2-}-OCH_2-CH_2-Br-CO-$CH_2OPO_3^{2-}$	Lys	951
4.1.2.13 Fructose 1,6-bisphos-phate aldolase			
Rabbit muscle	Fructose 1,6-bisphosphate and $K_3[Fe(CN)_6]$ (paracatalytic modification)	Cross-linking of Lys-146-Lys-227	952
	$BrCH_2CONHC_2H_4OPO_3^{2-}$	His-359, Lys-146	953
Rabbit muscle	2-Keto-4,4,4-trifluorobutyl phosphate		954
4.1.2.14 2-Keto-3-deoxy-6-phosphogluconic aldolase			
P. putida strain A3.12	Bromopyruvate		955
4.1.3.53 Hydroxy-3-methyl-glutaryl-CoA synthase			
Avian liver	3-Chloropropionyl-CoA	Cys	956
4.1.3.6 Citrate lyase			
K. aerogenes	p-N_3-Bz-CoA (ph.m.)	β SU	957
4.1.3.7 Citrate synthase			
Pig heart	S-(4-Bromo-2,3-dioxobutyl)-CoA		723
4.1.3.27 Antranilate synthase			
Serratia marcescens	PLP + NaCN	Lys	958
Salmonella typhimurium	6-Diazo-5-oxo-L-norleucine	Cys	959
Serratia marcescens	L-(αS,5S)-α-amino-3-chloro-4,5-dihydro-5-isoxazoleacetic acid (AT-125)	Cys-83	960
4.1.99.1 Tryptophanase			
E. coli B/1t 7-A	PLP	His	961
	3-Bromopyruvate	Cys	962
4.2.1.1 Carbonic anhydrase			
Bovine erythrocytes	Cyanogen	Cross-linking Glu-His	963
Bovine, rabbit, human, and dog erythrocytes	Cyanogen		200
Spinach leaves	Bromopyruvate	Cys	964
	Bromoacetazolamide	Cys	
Bovine erythrocytes	N_3-$C_6H_4SO_2NH_2$ (ph.m.)		252
4.2.1.49 Urocanase			
P. putida	Urocanic acid + O_2	Cys	965
4.2.99.9 Cystathionine γ-syn-thase			
Salmonella typhimurium	2-Amino-4-chloro-5-(p-nitrophenylsulfi-nyl)pentanoic acid (ss)		163
	L-Propargylglycine (ss)		966
4.4.1.1 Cystathionase			
Rat liver	Propargylglycine (ss)		967
	β,β,β-Trifluoroalanine	Lys, CSATKᵃYM	968

Table 2 (continued)
SURVEY OF AFFINITY MODIFICATION DATA FOR ENZYMES

Source	Reagent	Modification point	Ref.
	β-Trifluoromethyl-D,L-alanine and β-trifluoro-vinyl-D,L-alanine		969
4.4.1.5 Glyoxalase I			
Yeast	S-(p-Bromobenzyl)-glutathione		970
4.4.1.11 Methionine γ-lyase			
P. ovalis	2-Amino-4-chloro-5-(p-nitrophenylsulfi-nyl)pentanoic acid (ss)		163
	L-Propargylglycine (ss)		966
4.6.1.1 Adenylate cyclase			
Bordetella pertussis	p-$N_3C_6H_4C$(=N)-calmodulin (ph.m.)		971
Ns protein	oGTP		972
Bovine brain and rat liver	oATP		292
Rabbit heart membranes	oATP		973
Chick embryo heart	8-N_3-GTP		974
β-Hydroxydecanoyl thioester-dehydrase			
E. coli	3-Decynoyl-N-acetylcysteamine (ss); 2,3-deca-dienoyl-N-acetylcysteamine; and 2,3-dodeca-dienoic acid (dextrorotatory isomer)		975
5.1.1.1 Alanine racemase			
E. coli B	β,β-Difluoroalanine (ss); β,β,β-trifluoroalan-ine (ss)		976
5.1.3.2 Uridine diphosphate galactose 4-epimerase			
E. coli B	p-(Bromoacetamido)phenyl uridyl pyrophos-phate	NAD	394
5.3.1.1 Triose phosphate iso-merase			
Rabbit muscle	HalCH₂COCH₂OPO₃H₂(Hal-Br,I,Cl) and glyci-dol-P₁	Glu-165 Glu,AYE[a] PVW	953, 977, 978
Mammals, chicken mus-cle, fish, and yeast	Ferrate anion	Tyr-164, Cys-165, Trp-168	979
5.3.1.9 Glucosephosphate iso-merase			
Human placenta, rabbit muscle	N-Bromoacetylethanolamine phosphate	His	205
5.3.1.19 Glucosamine phosphate isomerase			
Pig muscle	PLP	LGK[a]Q, IASK[a]T	980
Rabbit muscle	PLP	Lys	953
Rabbit muscle, yeast	Mixture: 1,2-anhydro-D-mannitol 6-phosphate and 1,2-anhydro-L-gulitol 6-phosphate	Glu	953
E. coli, P. aeruginosa, Arthrobacter aurescens	Anticapsin	Cys	981
Salmonella typhimurium	Nᵝ-Fumarylcarboxyamido-L-2,3-diaminopro-pionic acid		982
5.3.3.1 Δ⁵-3-Ketosteroid iso-merase			
P. testosteroni	O-Carboxymethylagarose-ethylene-diaminesuc-cinyl-17β-O-19-nertestosterone and O-carbox-ymethylagarose-ethylene-diaminesuccinyl-17β-O-4,6-androstadien-3-one (ph.m.)		983
	Δ⁶-Testosterone-agarose; Δ⁶-testosterone-oxi-ranylestra-1,3,5(10),6,8-pentaene-3-ol; and Δ⁶-testosterone-Δ⁴-androsten-3-one (ph.m.)	Asp-38	329

Table 2 (continued)
SURVEY OF AFFINITY MODIFICATION DATA FOR ENZYMES

Source	Reagent	Modification point	Ref.
	β-Spiro-3-oxiranyl-5α-androstan-17β-ol		984
	Spiro-17β-oxiranyl-5α-androstan-3β-ol and spiro-17β-oxiranyl-Δ⁵-androsten-3β-ol		985
	Spiro-17β-oxiranylestra-1,3,5(10),6,8-pentaene-3-ol and spiro-17β-oxiranyl-Δ⁴-androsten-3-one		986
	(3S)-Spiro[5α-androstane-3,2'-oxirane]-17-ol		987
	(17S)-Spiro[estra-1,3,5(10),6,8-pentaene-17,2'-oxiran]-3-ol	Asp-38	988
	5,10-Secestr-5-yne-3,10,17-trione (ss)	Asp-57 YAEᵃS	989
	5,10-Seco-19-norpregn-5-yne-3,10,20-trione (ss)		990
	10β-(1-Oxoprop-2-ynyl)oestr-4-ene-3,17-dione; 5,10-secoestr-4-yne-3,10,17-trione; 17β-hydroxy-5,10-secoestr-4-yne-3,10-dione and 17β-(1-oxoprop-2-ynyl)androst-4-en-3-one		991
	3-Oxo-4-estren-17β-yl acetate(ph.m.)		135
	17β-Hydroxy-4-estren-3-one; and 17β-hydroxy-4-androsten-3-one (ph.m.)		992
P. putida	1,4,6-Androstatrien-3-one-17β-ol		993
	2α-Cyanoprogesterone and 2-hydroxy-methyleneprogesterone		994
6.1.1.1 Tyrosyl-tRNA synthetase			
Bacillus stearothermophilus	oATP		995
E. coli	tRNA^Tyr (UV cross-linking)		996
	otRNA^Tyr	Lys-229, 234, 237	997
	tRNA^Tyr_{1-2}	U64 of tRNA	998
6.1.1.2 Tryptophanyl-tRNA synthetase			
Bovine liver, yeast	(ClCH₂CH₂)₂NC₆H₄(CH₃)₂CO-Trp-tRNA^Trp		418
Bovine pancreas	Trp-CH₂Cl		999
	2,4,6-(CH₃)₃C₆H₂COOppAdo ClCH₂ppAdo; ClCH₂CH₂OpAdo; and 2,4,6-(CH₃)₃C₆H₂COOpAdo		1000, 1001
E. coli	oATP		995
Bovine pancreas	p-N₃-C₆H₄NHpppAdo (ph.m.)		239, 1002
B. stearothermophilus	Brown MX-5BR		1003
6.1.1.3 Threonyl-tRNA synthetase			
E. coli MRE600	p-N₃C₆H₄NHpppAdo (ph.m.)		239
6.1.1.4 Leucyl-tRNA synthetase			
E. coli MRE600	p-N₃-C₆H₄NHpppAdo (ph.m.)		239
Euglena gracilis	2,4,6-(CH₃)₃C₆H₂COOpppAdo; ClCH₂ppAdo; ClCH₂CH₂-OpAdo		1004
Escherichia coli	6-Amino-7-chloro-5,8-dioxyquinoline		1005
6.1.1.5 Isoleucyl-tRNA synthetase			
E. coli	IleCH₂Br		1006
	oATP		995
E. coli B	oATP		1007
E. coli	p-N₃-C₆H₄NHpppAdo (ph.m.)		239

Table 2 (continued)
SURVEY OF AFFINITY MODIFICATION DATA FOR ENZYMES

Source	Reagent	Modification point	Ref.
MRE-600			
	L-IleCH$_2$Br	Cys	1008
E. coli B	N-BrCH$_2$CO-Ile-tRNA^{Ile}		1009
E. coli	ATP (UV cross-linking)		1010
	tRNA^{Phe} (UV cross-linking)		336
6.1.1.9 Valyl-tRNA synthetase			
E. coli	p-N$_3$-C$_6$H$_4$NHpppAdo (ph.m.)		239
MRE-600			
E. coli	ValCH$_2$Cl		1011
	tRNA^{Phe} (UV cross-linking)		336
Bovine liver	ValCH$_2$Cl		1012
6.1.1.10 Methyonyl-tRNA synthetase			
B. stearothermophilus, E. coli	oATP		995
E. coli	otRNA$_f^{Met}$	Lys-61, 142, 147, 149, 335	1013
	tRNA^{Met}-(s^{4}U8)-CH$_2$CONHC$_6$H$_4$N$_3$ (ph.m.); otRNA$_f^{Met}$ treated with p-NH$_2$NHCOC$_6$H$_4$N$_3$ (ph.m.)		1014

Source	Reagent	Modification point	Ref.
	(statistically modified)	Lys-402, 439, 465, 640	1015
E. coli EM20031	otRNA$_f^{Met}$ treated with NaBH$_3$CN tRNA^{Met} (UV cross-linking)		266 / 1016
E. coli	tRNA$_i^{Met}$	C74,A76 of tRNA	998
6.1.1.19 Arginyl-tRNA synthetase			
E. coli MRE-600	p-N$_3$-C$_6$H$_4$NHpppAdo (ph.m.)		239
6.1.1.20 Phenylalanyl-tRNA synthetase			
E. coli	oATP		995
	Adenosine-5′-trimetaphosphate		389, 397
E. coli MRE-600	Ethenoadenosine-5′-trimetaphosphate		247, 389
	Ethenoadenosine-5′-trimetaphosphate (cross-linking)		1017
	(ClCH$_2$CH$_2$)$_2$NC$_6$H$_4$(CH$_3$)$_2$CO-Phe-tRNA^{Phe}	α SU	397,
	tRNA^{Phe}-(G)CH$_2$CH$_2$N(CH$_3$)C$_6$H$_4$CH$_2$NH-C$_6$H$_4$(NO$_2$)$_2$N$_3$ (statistically modified) (ph.m.); tRNA^{Phe}-(s^{4}U8)-CH$_2$CONHC$_6$H$_4$N$_3$ (ph.m.);		1018, 1019
	N-BrCH$_2$CO-Phe-tRNA^{Phe}; *p*-azidoanilidate of phenylalanine; and *N*-bromoacetyl-L-phenyla-laninyladenylate		198, 397, 1020, 1021
E. coli	p-N$_3$-C$_6$H$_4$NHpppAdo (ph.m.)		1020
	p-N$_3$-C$_6$H$_4$-N(CH$_3$)CH$_2$-ATP		1022
	Ac(Br)-Phe-tRNA^{Phe}	β SU	371, 396
	N$_3$C$_6$H$_4$NHppCH$_2$CH$_2$-CH$_2$-C$_6$H$_5$		198

Table 2 (continued)
SURVEY OF AFFINITY MODIFICATION DATA FOR ENZYMES

Source	Reagent	Modification point	Ref.
Yeast	3′-Oxidized tRNA^{Phe}	YNWKPEEQ CK^{a}GN- (A,L),D- (G,H),D	1023
	tRNA^{Phe} (UV cross-linking) and otRNA^{Phe}	Lys	1024
	tRNA^{Phe} (UV cross-linking)		336
6.2.1.4 Succinyl-CoA synthetase (GDP)			
Pig heart	S-Methanesulfonyl-CoA		1025
E. coli pig heart	CoA disulfide oxidized with peracetic acid		1026
Rat liver	oGDP		1027
6.2.1.5 Succinyl-CoA synthetase (ADP)			
E. coli	oADP		1028
6.3.1.2 Glutamic synthetase			
E. coli	FSO$_2$BzAdo; PLP; thiourea dioxide	Lys	206
	β, γ (CrIII)(H$_2$O$_4$)ATP		178
6.3.4.3 Formyltetrahydrofolate synthetase			
Pig liver	Carbodiimide-activated folate		105
6.3.4.16 Carbamyl phosphate synthetase (NH$_3$)			
Rat liver	FSO$_2$BzAdo	Cys	1029
	FSO$_2$BzAdo		1030
6.3.5.3 2-Formamido-N-ribosylacetamide 5′-phosphate: L-glutamine amido-ligase			
S. typhimurium	Azaserine	Cys	1031
6.3.5.5 Carbamyl phosphate synthetase (Gln)			
E. coli	FSO$_2$BzAdo		1032
6.4.1.1 Pyruvate carboxylase			
Sheep liver mitochondria	Complex (Mg-oATP)$^{2-}$	Lys	1033
	Bromopyruvate	Cys	1034
6.4.1.2 Acetyl-CoA carboxylase			
Rat liver	FSO$_2$BzAdo		1035
Cytidine triphosphate synthetase			
E. coli	6-Diazo-5-oxo-1-norleucine	Cys	1036

Note: The data are presented in the order of the enzyme classification numbers. Unclassified enzymes are given tentatively at the end of each class. When the same process is described in several papers by nearly the same group of authors, only the last reference is given.

[b]   A set of reagents for chymotrypsin and trypsin elaborated before 1967 are reviewed by Shaw.[844]

REFERENCES

1. **Baker, B. R.,** *Design of Active Site Directed Irreversible Inhibitors,* John Wiley & Sons, New York, 1967.
2. **Belikova, A. M., Zarytova, V. F., and Grineva, N. I.,** Synthesis of ribonucleosides and diribonucleoside phosphates containing 2-chloroethylamine and nitrogen mustard residues, *Tetrahedron Lett.,* 3557, 1967.
3. **Grineva, N. I. and Karpova, G. G.,** Complementarily addressed modification of rRNA with p-(chloroethylmethylamino)-benzylidene hexanucleotides, *FEBS Lett.,* 32, 351, 1973.
4. **Jakoby, W. B. and Wilcheck, M., Eds.,** *Methods in Enzymology,* Vol. 46, Academic Press, New York, 1977.
5. **Knorre, D. G., Ed.,** *Affinity Modification of Biopolymers,* Nauka, Novosibirsk, U.S.S.R., 1983.
6. **Lipscomb, W. N., Reeke, G. N., Jr., Hartsuck, J. A., Quiocho, F. A., and Bethge, P. H.,** The structure of carboxypeptidase A. VIII. Atomic interpretation at 0.2 nm resolution, a new study of the complex of glycyl-L-tyrosine with CPA, and mechanistic deductions, *Philos. Trans. R. Soc. London, Ser. B,* 257, 177, 1970.
7. **Pribnow, D.,** Bacteriophage T7 Early promoters: nucleotide sequences of two RNA polymerase binding sites, *J. Mol. Biol.,* 99, 419, 1975.
8. **Sloan, D. L., Loeb, L. A., Mildvan, A. S., and Feldman, R. J.,** Conformation of deoxynucleoside triphosphate substrates on DNA polymerase I from *Escherichia coli* as determined by nuclear magnetic relaxation, *J. Biol. Chem.,* 250, 8913, 1975.
9. **Robertus, J. D., Ladner, J. E., Finch, J. T., Rhodes, D., Brown, R. S., Clark, B. F. C., and Klug, A.,** Structure of yeast phenylalanine tRNA at 3 Å resolution, *Nature,* 250, 546, 1974.
10. **Suddath, F. L., Quigley, G. J., McPherson, A., Sneden, D., Kim, J. J., Kim, S. H., and Rich, A.,** Three-dimensional structure of yeast phenylalanine transfer RNA at 3.0 Å resolution, *Nature,* 248, 20, 1974.
11. **RajBhandary, U. L. and Chang, S. H.,** Studies on polynucleotides. LXXXII. Yeast phenylalanine transfer ribonucleic acid: partial digestion with ribonuclease T_1 and derivation of the total primary structure, *J. Biol. Chem.,* 243, 598, 1962.
12. **Rich, A. and Rajbhandary, U. L.,** Transfer RNA: molecular structure, sequence and properties, *Annu. Rev. Biochem.,* 45, 805, 1976.
13. **Kurganov, B. I., Sugrobova, N. P., and Mil'man, L. S.,** Supramolecular organisation of glycolytic enzymes, *Mol. Biol. (USSR),* 20, 53, 1986.
14. **Lamb, R. A. and Choppin, P. W.,** The gene structure and replication of influenza virus, *Annu. Rev. Biochem.,* 52, 467, 1983.
15. **Robertson, A. S.,** 5′ and 3′ terminal sequences of the RNA genome segments of influenza virus, *Nucleic Acids Res.,* 6, 3745, 1979.
16. **Nierhaus, K. H.,** The assembly of the prokaryotic ribosome, *Biosystems,* 12, 273, 1980.
17. **Langone, J. J. and Vunakis, H. V., Eds.,** *Methods in Enzymology,* Vol. 74 (Part C), Academic Press, New York, 1981.
18. **Lehninger, A. L.,** *Principles of Biochemistry,* Worth, New York, 1982.
19. **Cantor, C. R. and Schimmel, P. R.,** *Biophysical Chemistry,* W.H. Freeman, San Francisco, 1980.
20. **Alberts, B., Bray, D., Lewis, J., Raff, M., Roberts, K., and Watson, J. D.,** *Molecular Biology of the Cell,* Garland, New York, 1983.
21. **Gilbert, H. F. and O'Leary, M. H.,** Modification of arginine and lysine in proteins with 2,4-pentandione, *Biochemistry,* 14, 5194, 1975.
22. **Kochetkov, N. K., Budowsky, E. I., and Shibaeva, R. P.,** The selective reaction of O-methylhydroxylamine with cytidine nucleus, *Biochim. Biophys. Acta,* 68, 493, 1963.
23. **Kochetkov, N. K., Budowsky, E. I., Sverdlov, E. D., Shibaeva, R. P., Shibaev, V. N., and Monastirskaya, G. S.,** The mechanism of the reaction of hydroxylamine and O-methylhydroxylamine with cytidine, *Tetrahedron Lett.,* 3253, 1967.
24. **Woody, R. W., Friedman, M. E., and Scheraga, H. A.,** Structural studies of ribonuclease. XXII. Location of the third buried tyrosyl residue in ribonuclease, *Biochemistry,* 5, 2034, 1966.
25. **Dixon, M. and Webb, E. C.,** *Enzymes,* 2nd ed., Academic Press, New York, 1964, 465.
26. **Glazer, A. N., Delange, R. J., and Sigman, D. S.,** Laboratory techniques in biochemistry and molecular biology, in *Chemical Modification of Proteins,* 2nd ed., Work, T. S. and Burdon, R. H., Eds., Elsevier, Amsterdam, 1985.
27. **Mirzabekov, A. D., Krutilina, A. I., Reshetov, P. D., Sandakhchiyev, L. S., Knorre, D. G., Khokhlov, A. S., and Bayev, A. A.,** Preparative scale production of the enriched valine accepting transfer RNA from baker's yeast, *Dokl. Akad. Nauk SSSR,* 160, 1200, 1965.
28. **Gillam, I., Millward, S., Blew, D., von Tigerstrom, M., Wimmer, E., and Tener, G. M.,** The separation of soluble ribonucleic acids on benzoylated diethylaminoethylcellulose, *Biochemistry,* 6, 3043, 1967.
29. **Woodward, J., Ed.,** *Immobilized Cells and Enzymes. A Practical Approach,* IRL Press, Oxford, 1982.

30. **Sconten, W. H.,** *Affinity Chromatography,* (Chemical Analysis Series, Vol. 59), John Wiley-Sons, New York, 1981.
31. **Dalchau, R. and Fabre, J. W.,** The purification of antigens and other studies with monoclonal antibody affinity columns: the complementary new dimension of monoclonal antibodies, in *Monoclonal Antibodies in Clinical Medicine,* McMichael, A. J. and Fabre, J. W., Eds., Academic Press, London, 1982.
32. **Commerford, S. L.,** In vitro iodination of nucleic acids, in *Methods in Enzymology,* Vol. 70 (Part A), Langone, J. J. and Vunakis, H. V., Eds., Academic Press, New York, 1980, 247.
33. **Bolton, A. E. and Hunter, W. M.,** The labelling of proteins to high specific radioactivities by conjugation to ^{125}I-containing acylating agent. Application to radioimmunoassay, *Biochem. J.,* 133, 529, 1973.
34. **Means, G. E. and Feeney, R. E.,** Reductive alkylation of amino groups in proteins, *Biochemistry,* 7, 2192, 1968.
35. **Rice, R. H. and Means, G. E.,** Radioactive labeling of proteins in vitro, *J. Biol. Chem.,* 246, 831, 1971.
36. **Goldberger, R. F. and Anfinsen, C. B.,** Reversible masking of amino groups in ribonuclease and its possible usefulness in the synthesis of the protein, *Biochemistry,* 1, 401, 1962.
37. **Dixon, H. B. F. and Perham, R. N.,** Reversible blocking of amino groups with citraconic anhydride, *Biochem. J.,* 109, 312, 1968.
38. **Cole, D.,** S-aminoethylation, in *Enzyme Structure,* (Methods in Enzymology Series, Vol. 11), Hirs, C. H. W., Ed., Academic Press, New York, 1967, 315.
39. **Edman, P.,** Method for the determination of the amino acid sequences in peptides, *Acta Chem. Scand.,* 4, 283, 1950.
40. **Gross, E. and Witkop, B.,** Nonenzymatic cleavage of peptide bonds: the methionine residues in bovine pancreatic ribonuclease, *J. Biol. Chem.,* 237, 1856, 1962.
41. **Maxam, A. M. and Gilbert, W.,** Sequencing end-labeled DNA with base-specific chemical cleavages, in *Methods in Enzymology,* Vol. 65, Langone, J. J. and Vunakis, H. V., Eds., Academic Press, New York, 1980, 499.
42. **Blackburn, P. and Moore, S.,** Pancreatic ribonuclease, in *Nucleic Acids,* (Part B), 3rd ed., (The Enzymes Series, Vol. 15), Boyer, P. D., Ed., Academic Press, New York, 1982, 317.
43. **Wyckoff, H. W., Hardman, K. D., Allewell, N. M., Inagami, T., Johnson, L. N., and Richards, F. M.,** The structure of ribonuclease-S at 3.5Å resolution, *J. Biol. Chem.,* 242, 3984, 1967.
44. **Crestfield, A. M., Stein, W. H., and Moore, S.,** Alkylation and identification of the histidine residues at the active site of ribonuclease, *J. Biol. Chem.,* 238, 2413, 1963.
45. **Crestfield, A. M., Stein, W. H., and Moore, S.,** Properties and conformation of the histidine residues at the active site of ribonuclease, *J. Biol. Chem.,* 238, 2421, 1963.
46. **Lukashin, A. V., Vologodskii, A. V., Frank-Kamenetskii, M. D., and Lyubchenko, Y. L.,** Fluctuational opening of the double helix as revealed by theoretical and experimental study of DNA interaction with formaldehyde, *J. Mol. Biol.,* 108, 665, 1976.
47. **KiKuchi, H., Goto, Y., and Hamaguchi, K.,** Reduction of the buried intrachain disulfide bond of the constant fragment of the immunoglobulin light chain: global unfolding under physiological conditions, *Biochemistry,* 25, 2009, 1986.
48. **Vlassov, V. V., Puzyriov, A. T., Ebel, J.-P., and Giege, R.,** Chemical modification studies on the structure of tRNA^{Phe} in solution, *Bioorg. Khim. (USSR),* 7, 1487, 1981.
49. **Vlassov, V. V., Grineva, N. I., Knorre, D. G., and Pavlova, R. M.,** Relative reactivities of some guanosine residues in the "halves" of tRNA$_1^{Val}$ (yeast) and in their complex, *FEBS Lett.,* 28, 322, 1972.
50. **Ross, W. C. J.,** *Biological Alkylating Agents,* Butterworths, London, 1962.
51. **Lavery, R., Pullman, A., Pullman, B., and Oliveira, M.,** The electrostatic molecular potential of tRNA^{Phe}. The potentials and steric accessibilities of sites associated with the bases, *Nucleic Acids Res.,* 8, 5095, 1980.
52. **Romby, P., Moras, D., Bergdoll, M., Dumas, Ph., Vlassov, V. V., Westhof, E., Ebel, J. P., and Giege, R.,** Yeast tRNA^{Asp} tertiary structure in solution and areas of interaction of the tRNA with aspartyl-tRNA synthetase. A comparative study of the yeast phenylalanine system by phosphate alkylation experiments with ethylnitrosourea, *J. Mol. Biol.,* 184, 455, 1985.
53. **Galas, D. J. and Schmitz, A.,** DNAase footprinting: a simple method for detection of protein-DNA binding specificity, *Nucleic Acids Res.,* 5, 3157, 1978.
54. **Schultz, P. G. and Dervan, P. B.,** Distamycin and penta-N-methylpyrrolecarboamide binding sites on native DNA. A comparison of methidiumpropyl-EDTA-Fe (II) footprinting and DNA affinity cleaving, *J. Biomolec. Struct. Dyn.,* 1, 1133, 1984.
55. **Kopka, M. L., Yoon, C., Goodsell, D., Pjura, P. and Dickerson, R. E.,** The molecular origin of DNA-drug specificity in netropsin and distamycin, *Proc. Natl. Acad. Sci. U.S.A.,* 82, 1376, 1985.
56. **Reisler, E., Burke, M., and Harrington, W. F.,** Cooperative role of two sulfhydryl groups in myosin adenosine triphosphatase, *Biochemistry,* 13, 2014, 1974.
57. **Birchmeier, W., Wilson, K. J., and Christen, P.,** Cytoplasmic aspartate aminotransferase: syncatalytic sulfhydryl group modification, *J. Biol. Chem.,* 248, 1751, 1973.

58. **Kalomiris, E. L. and Coller, B. S.,** Thiol-specific probes indicate that the β-chain of platelet glycoprotein Ib is a transmembrane protein with a reactive endofacial sulfhydryl group, *Biochemistry,* 24, 5430, 1985.

59. **Marfey, P. S., Uziel, M., and Little, J.,** Reactions of bovine pancreatic ribonuclease A with 1,5-difluoro-2,4-dinitro-benzene. II. Structure of an intramolecularly bridged derivative, *J. Biol. Chem.,* 240, 3270, 1965.

60. **Han, K. -K., Richard, C., and Delacourte, A.,** Chemical cross-links of proteins by using bifunctional reagents, *Int. J. Biochem.,* 16, 129, 1984.

61. **Middaugh, C. R., Vanin, E. F., and Ji, T. H.,** Chemical cross-linking of cell membranes, *Mol. Cell. Biochem.,* 50, 115, 1983.

62. **Gaffney, B. J.,** Chemical and biochemical crosslinking of membrane components, *Biochim. Biophys. Acta,* 822, 289, 1985.

63. **Fasold, H., Klappenberger, J., Meyer, C., and Remold, H.,** Bifunctional reagents for the crosslinking of proteins, *Angew. Chem. Int. Ed. Engl.,* 10, 795, 1971.

64. **Ji, T. H.,** The application of chemical crosslinking for studies on cell membranes and the identification of surface reporters, *Biochim. Biophys. Acta,* 559, 39, 1979.

65. **Wold, F.,** *Bifunctional Reagents,* (Methods in Enzymology Series, Vol. 25B), Langone, J. J. and Vunakis, H. V., Eds., Academic Press, New York, 1972, 623.

66. **Peters, K. and Richards, F. M.,** Chemical cross-linking: reagents and problems in studies of membrane structure, *Annu. Rev. Biochem.,* 46, 523, 1977.

67. **Ji, T. H.,** *Bifunctional Reagents,* (Methods in Enzymology Series, Vol. 91), Langone, J. J. and Vunakis, H. V., Eds., Academic Press, New York, 1983, 580.

68. **Kunkel, G. R., Mehrabian, M., and Martinson, H. G.,** Contact-site cross-linking agents, *Mol. Cell. Biochem.,* 34, 3, 1981.

69. **Burke, M. and Reisler, E.,** Effect of nucleotide binding on the proximity of the essential sulfhydryl groups of myosin. Chemical probing of movement of residues during conformational transition, *Biochemistry,* 16, 5559, 1977.

70. **Haidu, J., Bartha, F., and Friedrich, P.,** Crosslinking with bifunctional reagents as a means for studying the symmetry of oligomeric proteins, *Eur. J. Biochem.,* 68, 373, 1976.

71. **Hucho, F., Müllner, H., and Sund, H.,** Investigation of the symmetry of oligomeric enzymes with bifunctional reagents, *Eur. J. Biochem.,* 59, 79, 1975.

72. **Carpenter, F. H. and Harrington, K. T.,** Intermolecular cross-linking of monomeric proteins and cross-linking of oligomeric proteins as a probe of quarternary structure, *J. Biol. Chem.,* 247, 5580, 1972.

73. **Wittmann, H. G.,** Architecture of procaryotic ribosome, *Annu. Rev. Biochem.,* 52, 35, 1983.

74. **Brimacombe, R. and Steige, W.,** Structure and function of ribosomal RNA, *Biochem. J.,* 229, 1, 1985.

75. **Sköld, S. -E.,** RNA-protein complexes identified by cross-linking of polysomes, *Biochimie,* 63, 53, 1981.

76. **Millon, R., Olomucki, M., Le Gall, J. -Y., Golinska, B., Ebel, J. -P., and Ehresmann, B.,** Synthesis of a new reagent, ethyl 4-azidobenzoylaminoacetimidate, and its use for RNA-protein cross-linking within *Escherichia coli* ribosomal 30 S subunits, *Eur. J. Biochem.,* 110, 485, 1980.

77. **Sigman, D. S. and Mooser, G.,** Chemical studies of enzyme active sites, *Annu. Rev. Biochem.,* 44, 890, 1975.

78. **Kaiser, E. T., Lawrence, D. S., and Rokita, S. E.,** The chemical modification of enzymatic specificity, *Annu. Rev. Biochem.,* 54, 565, 1985.

79. **Cohen, L. A.,** Chemical modification as a probe of structure and function, in *The Enzymes,* 3rd ed., Boyer, P. D., Ed., Academic Press, New York, 1970, 148.

80. **Chapeville, F., Lipmann, F., von Ehrenstein, G., Weisblum, B., Ray, W. J., and Benzer, S.,** On the role of soluble ribonucleic acid in coding for amino acids, *Proc. Natl. Acad. Sci. U.S.A.,* 48, 1086, 1962.

81. **Sluyterman, L. A. Ae.,** The rate-limiting reaction in papain action as derived from the reaction of the enzyme with chloroacetic acid, *Biochim. Biophys. Acta,* 151, 178, 1968.

82. **Gorshkova, I. I., Dacy, I. I., Lavrik, O. I., and Mamaev, S. V.,** The role of the lysine residues in the phenylalanyl-tRNA synthetase substrates interaction, *Mol. Biol. (USSR),* 12, 1096, 1978.

83. **Gorshkova, I. I., Dacy, I. I., and Lavrik, O. I.,** The role of arginine residues in phenylalanyl-tRNA synthetase substrates interaction, *Mol. Biol. (USSR),* 14, 118, 1980.

84. **Schulman, L. H. and Chambers, R. W.,** Transfer RNA. II. A structural basis for alanine acceptor activity, *Proc. Natl. Acad. Sci. U.S.A.,* 61, 308, 1968.

85. **Schulman, L. H., Pelka, H., and Sundari, R. M.,** Structural requirements for recognition of *Escherichia coli* initiator and non-initiator transfer ribonucleic acid by bacterial T factor, *J. Biol. Chem.,* 249, 7102, 1974.

86. **Siebenlist, U. and Gilbert, W.,** Contacts between *Escherichia coli* RNA polymerase and an early promoter of phage T7, *Proc. Natl. Acad. Sci. U.S.A.,* 77, 122, 1980.

87. **Vlassov, V. V., Knorre, D. G., and Skobeltsyna, L. M.,** On the role of N7 atoms of guanosine in tRNA[Phe] (*E. coli*) in interaction with ribosomes, *Biokhimija (USSR),* 42, 784, 1977.

88. **Kochetkov, N. K. and Budovskii, E. I.**, *Organic Chemistry of Nucleic Acids*, Plenum Press, New York, 1972.

89. **Dugas, H. and Penny, C.**, *Bioorganic Chemistry. A Chemical Approach to Enzyme Action*, Cantor, R., Ed., Springer-Verlag, New York, 1981.

90. **Grineva, N. I. and Karpova, G. G.**, Alkylation of ribosomal RNA complementary complexed with 2'-3'-O-4-(N-2-chloro-ethyl-N-methylamino)-benzylidene oligonucleotides. Chemical selectivity and positional specificity, *Bioorg. Khim. (USSR)*, 1, 588, 1975.

91. **Knorre, D. G. and Vlassov, V. V.**, Complementary-addressed (sequence-specific) modification of nucleic acids, *Prog. Nucleic Acid Res. Mol. Biol.*, 32, 291, 1985.

92. **Bayley, H.**, *Photogenerated Reagents in Biochemistry and Molecular Biology*, Elsevier, Amsterdam, 1983.

93. **Plapp, B. V.**, Application of affinity labeling for studying structure and function of enzymes, in *Methods in Enzymology*, Vol. 87 (Part C), Purich, D. L., Ed., Academic Press, New York, 1982, 469.

94. **Wilman, D. E. V. and Connors, T. A.**, Molecular structure and antitumor activity of alkylating agents, in *Molecular Aspects of Anti-Cancer Drug Action*, Neide, S. and Waring, M. J., Eds., Verlag Chemie, Weinheim, West Germany, 1983.

95. **Baker, B. R., Lee, W. W., and Tong, E.**, Non-classical antimetabolites. VI. 4-(Iodoacetamido)-salicylic acid, an exo-alkylating irreversible inhibitor, *J. Theor. Biol.*, 3, 459, 1962.

96. **Gundlach, G. and Turba, F.**, Wechselwirkung zwischen Chymotrypsin und Halogenacylderivaten bei Benutzung von Substratanalogen als Reaktionsvehikel, *Biochem. Z.*, 335, 573, 1962.

97. **Shaw, E.**, Chemical modification by active-site-directed reagents, in *The Enzymes*, Vol. 1, 3rd ed., 1970, 91.

98. **Knorre, D. G. and Kisselev, L. L.**, Aminoacyl-tRNA-synthetases: approaches to the structure and function of the active centers, in *Proc. Int. Symp. Frontiers of Bioorganic Chemistry and Molecular Biology*, Ananchenko, S. N., Ed., Pergamon Press, Oxford, 1980, 315.

99. **Webb, T. R. and Matteucci, M. D.**, Sequence-specific cross-linking of deoxyoligonucleotides via hybridization-triggered alkylation, *J. Am. Chem. Soc.*, 108, 2764, 1986.

100. **Isoda, Y. and Nitta, Y.**, Affinity labeling of soybean β-amylase with 2',3'-epoxypropyl α-D-glucopyranoside, *J. Biochem.*, 99, 1631, 1986.

101. **Pinkus, L. M.**, Glutamine binding sites, in *Methods in Enzymology*, Vol. 46, Jakoby, W. B. and Wilchek, M., Eds., Academic Press, New York, 1977, 414.

102. **Carotti, D., Andria, F., Giartosio, A., Turano, C., and Riva, F.**, Specific labeling of cytosolic and mitochondrial aspartate aminotransferases, *Eur. J. Biochem.*, 146, 619, 1985.

103. **Lowe, C. R., Small, D. A. P., and Atkinson, A.**, Some preparative and analytical applications of triazine dyes, *Int. J. Biochem.*, 13, 33, 1981.

104. **Gardell, S. Y. and Tate, S. S.**, Affinity labeling of γ-glutamyl transpeptidase by glutamine antagonists, *FEBS Lett.*, 122, 171, 1980.

105. **Smith, D. D. S. and MacKenzie, R. E.**, Methylenetetrahydrofolate dehydrogenase-methenyltetrahydrofolate-cyclohydrolase-formyltetrahydrofolate synthetase. Affinity labeling of the dehydrogenase-cyclohydrolase active site, *Biochem. Biophys. Res. Commun.*, 128, 418, 1985.

106. **Pal, P. K., Wechter, W. J., and Colman, R. F.**, Affinity labeling of the inhibitor DNPH site of bovine liver glutamate dehydrogenase by 5'-fluorosulfonylbenzoyl adenosine, *J. Biol. Chem.*, 250, 8140, 1975.

107. **Colman, R. F.**, Affinity labeling of purine nucleotide sites in proteins, *Annu. Rev. Biochem.*, 52, 67, 1983.

108. **Tomich, J. M. and Colman, R. F.**, Reaction of 5'-p-fluorosulfonylbenzoyl-1, N_6-ethenoadenosine with histidine and cysteine residues in the active site of rabbit muscle pyruvate kinase, *Biochem. Biophys. Acta*, 827, 344, 1985.

109. **Drutsa, V. L., Kozlov, I. A., Milgrom, Y. M., Shabarova, Z. A., and Sokolova, N. I.**, An active-site-directed adenosine triphosphate analogue binds to the β-subunit of factor F_1 mitochondrial adenosine triphosphatase with its triphosphate moiety, *Biochem. J.*, 182, 617, 1979.

110. **Sokolova, N. I. and Tretyakova, S. S.**, Reactive derivatives of nucleoside 5'-mono-, di- and triphosphates — affinity reagents, *Bioorg. Khim. (USSR)*, 9, 1157, 1983.

111. **Knorre, D. G., Kurbatov, V. A., and Samukov, V. V.**, General method for the synthesis of ATP gamma-derivatives, *FEBS Lett.*, 70, 105, 1976.

112. **Grachev, M. A., Knorre, D. G., and Lavrik, O. I.**, γ-Derivatives of nucleoside-5'-triphosphates as probes for enzyme and protein factor active sites, in *Soviet Scientific Reviews, Section D*, (Biology Reviews), Vol. 2, Skulachev, V. P., Ed., Harwood Academic, New York, 1981, 107.

113. **Godovikova, T. S., Zarytova, V. F., and Khalimskaya, L. M.**, Reactive phosphoamidates of mono- and dinucleotides, *Bioorg. Khim. (USSR)*, 12, 475, 1986.

114. **Zarytova, V. F., Graifer, D. M., Ivanova, E. M., Knorre, D. G., Lebedev, A. V., and Rezvukhin, A. I.**, ^{31}P and ^1H NMR investigation of the structure of the phosphorylating intermediates in the phosphodiester approach to the oligonucleotide synthesis, *Nucl. Acids Res.*, Spec. Publ. No. 4, s209, 1978.

115. **Knorre, D. G., Buneva, V. N., Baram, G. I., Godovikova, T. S., and Zarytova, V. F.**, Dynamic aspects of affinity labelling as revealed by alkylation and phosphorylation of pancreatic ribonuclease with reactive deoxyribodinucleotide derivatives, *FEBS Lett.*, 194, 64, 1986.

116. **Cohen, J. A., Oosterbaan, R. A., and Berends, F.**, Organophosphorus compounds, *Enzyme Structure*, (Methods in Enzymology Series, Vol. 11), Hirs, C. H. W., Ed., Academic Press, New York, 1967, 686.

117. **Salamone, S. J. and Jordan, F.**, Synthesis of 9-(3,4-dioxopentyl)hypoxantine, the first arginine-directed purine derivative: an irreversible inactivator for purine nucleoside phosphorylase, *Biochemistry*, 21, 6383, 1982.

118. **Rayford, R., Anthony, D. D., Jr., O'Neill, R. E., Jr., and Merrick, W. C.**, Reductive alkylation with oxidized nucleotides, use in affinity labeling or affinity chromatography, *J. Biol. Chem.*, 260, 15708, 1985.

119. **Borchardt, R. T. and Huber, J. A.**, Catechol O-methyl-transferase. II. Inactivation by 5-hydroxy-3-mercapto-4-methoxybenzoic acid, *J. Med. Chem.*, 25, 321, 1982.

120. **Birchmeier, W., Kohler, C. E., and Schatz, G.**, Interaction of integral and peripheral membrane proteins: affinity labeling of yeast cytochrome oxidase by modified yeast cytochrome C, *Proc. Natl. Acad. Sci. U.S.A.*, 73, 4334, 1976.

121. **Collier, G. E. and Nishimura, J. S.**, Affinity labeling of succinyl-CoA synthetase from porcine heart and *Escherichia coli* with oxidized coenzyme A disulfide, *J. Biol. Chem.*, 253, 4938, 1978.

122. **Lwowski, W.**, Nitrenes in photoaffinity labeling: speculations of an organic chemist, *Ann. N.Y. Acad. Sci.*, 346, 491, 1980.

123. **Chowdhry, V. and Westheimer, F. H.**, Photoaffinity labeling of biological systems, *Annu. Rev. Biochem.*, 48, 293, 1979.

124. **Turro, N. J.**, Structure and dynamics of important reactive intermediates involved in photobiological systems, *Ann. N.Y. Acad. Sci.*, 346, 1, 1980.

125. **Lwowski, W., Ed.**, *Nitrenes*, John Wiley & Sons, New York, 1970.

126. **Knowles, J. R.**, Photogenerated reagents for biological receptor-site labeling, *Acc. Chem. Res.*, 5, 155, 1972.

127. **Weber, H., Junge, W., Hoppe, J., and Sebald, W.**, Laser-activated carbene labels the same residues in the proteolipid subunit of the ATP synthase in energized and nonenergized chloroplasts and mitochondria, *FEBS Lett.*, 202, 23, 1986.

128. **Nielsen, P. E. and Buchardt, O.**, Arylazides as photoaffinity labels. A. Photochemical study of some 4-substituted aryl azides, *Photochem. Photobiol.*, 35, 317, 1982.

129. **Badashkeyeva, A. G., Gall, T. S., Efimova, E. V., Knorre, D. G., Lebedev, A. V., and Mysina, S. D.**, Reactive derivative of adenosine-5'-triphosphate formed by irradiation of ATP γ-p-azidoanilide, *FEBS Lett.*, 155, 263, 1983.

130. **Brunswick, D. J. and Cooperman, B. S.**, Synthesis and characterization of photoaffinity labels for adenosine 3':5'-cyclic monophosphate and adenosine 5'-monophosphate, *Biochemistry*, 12, 4074, 1973.

131. **Chowdhry, V., Vaughan, R., and Westheimer, F. H.**, 2-diazo-3,3,3-trifluoropropionyl chloride: reagent for photoaffinity labeling, *Proc. Natl. Acad. Sci. U.S.A.*, 73, 1406, 1976.

132. **Brunner, J. and Richards, F. M.**, Analysis of membranes photolabeled with lipid analogues: reaction of phospholipids containing a disulfide group and a nitrene or carbene precursor with lipids and with gramicidin A, *J. Biol. Chem.*, 255, 3319, 1980.

133. **Cavalla, D. and Neff, N. H.**, Chemical mechanisms for photoaffinity labeling of receptors, *Biochem. Pharmacol.*, 34, 2821, 1985.

134. **Campbell, P. and Gioannini, T. L.**, The use of benzophenone as a photoaffinity label. Labeling in p-benzoyl-phenylacetyl chymotrypsin at unit efficiency, *Photochem. Photobiol.*, 29, 883, 1979.

135. **Ogez, J. R., Tivol, W. F., and Benisek, W. F.**, A novel chemical modification of Δ⁵-3-ketosteroid isomerase occuring during its 3-oxo-4-estren-17β-yl acetate dependent photoinactivation, *J. Biol. Chem.*, 262, 6151, 1977.

136. **Möhler, H., Battersby, M. K., and Richards, J. G.**, Benzodiazepine receptor protein identified and visualized in brain tissue by a photoaffinity label, *Proc. Natl. Acad. Sci. U.S.A.*, 77, 1666, 1980.

137. **Jelenc, P. C., Cantor, C. R., and Simon, S. R.**, High yield photoreagents for protein cross-linking and affinity labeling, *Proc. Natl. Acad. Sci. U.S.A.*, 75, 3564, 1978.

138. **Castello, A., Marquet, J., Moreno-Manas, M., and Sirera, X.**, The nucleophilic aromatic photosubstitution in photoaffinity labeling. A model study of a cycloheximide derivative, *Tetrahedron Lett.*, 26, 2489, 1985.

139. **Cimino, G. D., Gamper, H. B., Isaacs, S. T., and Hearst, J. E.**, Psoralens as photoactive probes of nucleic acid structure and function: organic chemistry, photochemistry and biochemistry, *Annu. Rev. Biochem.*, 54, 1151, 1985.

140. **Fairall, L., Finch, J. T., Hui, C. -F., Cantor, C. R., and Butler, P. J. G.**, Studies of tobacco mosaic virus reassembly with an RNA tail blocked by a hybridized and cross-linked probe, *Eur. J. Biochem.*, 156, 459, 1986.

141. **Schwartz, D. C., Saffran, W., Welsh, J., Haas, R., Goldenberg, M., and Cantor, C. R.,** New techniques for purifying large DNA's and studying their properties and packaging, *Cold Spring Harbor Symp. Quant. Biol.,* 47, 189, 1982.

142. **Stockmann, M. I.,** Nonlinear laser photomodification of macromolecules: possibility and applications, *Phys. Lett.,* 76A, 191, 1980.

143. **Benimetskaya, L. Z., Bulychev, N. V., Kozionov, A. L., Lebedev, A. V., Nesterikhin, Yu. E., Novozhilov, S. Yu., Rautian, S. G., and Stockman, M. I.,** Two-quantum selective laser scission of polyadenylic acid in the complementary complex with a dansyl derivative of oligothymidilate, *FEBS Lett.,* 163, 144, 1983.

144. **Benimetskaya, L. Z., Bulychev, N. V., Kozionov, A. L., Lebedev, A. V., Novozhylov, S. Yu., and Stockman, M. I.,** Two-quantum selective laser modification of poly- and oligonucleotides in complementary complexes with dansyl derivatives of oligodeoxynucleotides, *Nucleic Acids Res., Symp.* Ser. No. 14, 323, 1984.

145. **Shetlar, M. D., Christensen, J., and Hom, K.,** Photochemical addition of amino acids and peptides to DNA, *Photochem. Photobiol.,* 39, 125, 1984.

146. **Shetlar, M. D., Hom, K., Carbone, J., Moy, D., Steady, E., and Watanabe, M.,** Photochemical addition of amino acids and peptides to homopolyribonucleotides of the major DNA bases, *Photochem. Photobiol.,* 39, 135, 1984.

147. **Shetlar, M. D., Carbone, J., Steady, E., and Hom, K.,** Photochemical addition of amino acids and peptides to polyuridylic acid, *Photochem. Photobiol.,* 39, 141, 1984.

148. **Cadet, J., Voituriez, L., Grand, A., Hruska, F. E., Vigny, P., and Kan, L. -S.,** Recent aspects of the photochemistry of nucleic acids and related model compounds, *Biochimie,* 67, 277, 1985.

149. **Cadet, J., Berger, M., Decarroz, C., Wagner, J. R., Van Lier, J. E., Ginot, Y. M., and Vigny, P.,** Photosensitized reactions of nucleic acids, *Biochimie,* 68, 813, 1986.

150. **Prince, J. B., Taylor, B. H., Thurlow, D. L., Ofengand, J., and Zimmermann, R. A.,** Covalent crosslinking of tRNA$_1^{Val}$ to 16S RNA at the ribosomal P site: identification of crosslinked residues, *Proc. Natl. Acad. Sci. U.S.A.,* 79, 5450, 1982.

151. **Saito, I. and Matsuura, T.,** Chemical aspects of UV-induced cross-linking of proteins to nucleic acids. Photoreactions with lysine and tryptophan, *Acc. Chem. Res.,* 18, 134, 1985.

152. **Carroll, S. F., McCloskey, J. A., Crain, P. F., Oppenheimer, N. J., Marschner, T. M., and Collier, R. J.,** Photoaffinity labeling of diphtheria toxin fragment A with NAD: structure of the photoproduct at position 148, *Proc. Natl. Acad. Sci. U.S.A.,* 82, 7237, 1985.

153. **Jericevic, Z., Kucan, I., and Chambers, R. W.,** Photochemical cleavage of phosphodiester bond in oligoribonucleotides, *Biochemistry,* 21, 6563, 1982.

154. **Budowsky, E. I., Axentyeva, M. S., Abdurashidova, G. G., Simukova, N. A., and Rubin, L. B.,** Induction of polynucleotide-protein cross-linkages by ultraviolet irradiation. Peculiarities of the high-intensity laser pulse irradiation, *Eur. J. Biochem.,* 159, 95, 1986.

155. **Hockensmith, J. W., Kubasek, W. L., Vorachek, W. R., and von Hippel, P. H.,** Laser cross-linking of nucleic acids to proteins. Methodology and first applications to the phage T4 DNA replication system, *J. Biol. Chem.,* 261, 3512, 1986.

156. **Walsh, C. T.,** Suicide substrates: mechanism-based enzyme inactivators, *Tetrahedron,* 38, 871, 1982.

157. **Walsh, C. T.,** Suicide substrates, mechanism-based enzyme inactivators: recent developments, *Annu. Rev. Biochem.,* 53, 493, 1984.

158. **Rando, R. R.,** Mechanism-based enzyme inactivators, *Pharmacol. Rev.,* 36, 111, 1984.

159. **Bloch, K.,** Hydroxydecanoyl thioester dehydrase, in *The Enzymes,* Vol. 5, 3rd ed., Boyer, P., Ed., Academic Press, New York, 1971, 441.

160. **Batzold, F. H. and Robinson, C. H.,** Irreversible inhibition of Δ^5-3-ketosteroid isomerase by 5,10-secosteroids, *J. Am. Chem. Soc.,* 97, 2576, 1975.

161. **Penning, T. M., Heller, D. N., Balasubramanian, T. M., Fenselau, C. C., and Talalay, P.,** Mass spectrometric studies of a modified active-site tetrapeptide from Δ^5-3-ketosteroid isomerase of *Pseudomonas testosteroni, J. Biol. Chem.,* 257, 12589, 1982.

162. **Bey, P.,** *Enzyme-Activated Irreversible Inhibitors,* Elsevier, New York, 1978, 27.

163. **Johnston, M., Raines, R., Walsh, C., and Firestone, R. A.,** Mechanism-based enzyme inactivation using an allyl sulfoxide-allyl sulfenate ester rearrangement, *J. Am. Chem. Soc.,* 102, 4241, 1980.

164. **Ortiz de Montellano, P. R. and Correia, M. A.,** Suicidal destruction of cytochrome P-450 during oxidative drug metabolism, *Annu. Rev. Pharmacol. Toxicol.,* 23, 481, 1983.

165. **Silverman, R. B. and Zieske, P. A.,** Mechanism of inactivation of monoamine oxidase by 1-phenylcyclopropylamine, *Biochemistry,* 24, 2128, 1985.

166. **Corey, E. J. and Park, H.,** Irreversible inhibition of the enzymic oxidation of arachidonic acid to 15-(hydroperoxy) 5,8,11 (Z), 13(E)-eicosatetraenoic acid (15-HPETE) by 14,15-dehydroarachidonic acid, *J. Am. Chem. Soc.,* 104, 1750, 1982.

167. **Daniels, S. B., Cooney, E., Sofia, M. J., Chakravarty, P. K., and Katzenellenbogen, J. A.,** Haloenol lactones. Potent enzyme-activated irreversible inhibitors for α-chymotrypsin, *J. Biol. Chem.,* 258, 15046, 1983.

168. **Donadio, S., Perks, H. M., Tsuchiya, K., and White, E. H.,** Alkylation of amide linkages and cleavage of the C chain in the enzyme-activated-substrate inhibition of α-chymotrypsin with N-nitrosamines, *Biochemistry,* 24, 2447, 1985.

169. **Brenner, D. G. and Knowles, J. R.,** Penicillanic acid sulfone: nature of irreversible inactivation of RTEM β-lactamase from *Escherichia coli, Biochemistry,* 23, 5833, 1984.

170. **Ringe, D., Seaton, B. A., Gelb, M. H., and Abeles, R. H.,** Inactivation of chymotrypsin by 5-benzyl-6-chloro-2-pyrone: ^{13}C NMR and X-ray diffraction analysis of the inactivator-enzyme complex, *Biochemistry,* 24, 64, 1985.

171. **Laura, R., Robison, D. J., and Bing, D. H.,** (p-Amidinophenyl)methanesulfonyl fluoride, an irreversible inhibitor of serine proteases, *Biochemistry,* 19, 4859, 1980.

172. **Santi, D. and McHenry, C.,** 5'-Fluoro-2'-deoxyuridylate: covalent complex with thymidylate synthetase, *Proc. Natl. Acad. Sci. U.S.A.,* 69, 1855, 1972.

173. **Cogoli, M., Lubini, D., and Christen, P.,** Paracatalytic self-inactivation of enzymes, in *Enzyme Inhibitors,* Brodbeck, U., Ed., Verlag Chemie, Weinheim, West Germany, 1980, 27.

174. **Goeldner, M. P. and Hirth, C. G.,** Specific photoaffinity labeling induced by energy transfer: application to irreversible inhibition of acetylcholinesterase, *Proc. Natl. Acad. Sci. U.S.A.,* 77, 6439, 1980.

175. **Eid, P., Goeldner, H. P., Hirth, C. G., and Jost, P.,** Photosuicide inactivation of acetylcholinesterase by nitrosamine derivatives, *Biochemistry,* 20, 2251, 1981.

176. **Kawamura, T. and Tawada, K.,** Dissociation of actomyosin by vanadate plus ADP, and decomposition of the myosin-ADP-vanadate complex by actin, *J. Biochem.,* 91, 1293, 1982.

177. **Goodno, C. C. and Taylor, E. W.,** Inhibition of actomyosin ATPase by vanadate, *Proc. Natl. Acad. Sci. U.S.A.,* 79, 21, 1982.

178. **Ransom, S. C., Colanduoni, J. A., Eads, C. D., Gibbs, E. J., and Villafranca, J. J.,** Affinity labeling of *Escherichia coli* glutamine synthetase by β,γ-Cr(III) (H$_2$O)$_4$ ATP, *Biochem. Biophys. Res. Commun.,* 130, 418, 1985.

179. **Granot, J., Mildvan, A. S., Bramson, H. N., and Kaiser, E. T.,** Magnetic resonance measurements of intersubstrate distances at the active site of protein kinase using substitution-inert cobalt (III) and chromium (III) complexes of adenosine 5'-(β,γ-methylene triphosphate), *Biochemistry,* 19, 3537, 1980.

180. **Howe-Grant, M. E. and Lippard, S. J.,** Aqueous platinum (II) chemistry; binding to biological molecules, *Met. Ions Biol. Syst.,* 11, 63, 1980.

181. **Lippard, S. J. and Hoeschele, J. D.,** Binding of cis- and trans-dichlorodiammineplatinum (II) to nucleosome core, *Proc. Natl. Acad. Sci. U.S.A.,* 76, 6091, 1979.

182. **Fillipski, J., Kohn, K. W., Prather, R., and Bonner, W. M.,** Thiourea reverses cross-links and restores biological activity in DNA treated with dichlorodiammine platinum (II), *Science,* 204, 181, 1979.

183. **Vlassov, V. V., Gall, A. A., Godovikov, A. A., Zarytova, V. F., Kazakov, S. A., Kutiavin, I. V., Shishkin, G. V., and Mamayev, S. V.,** Complementary addressed modification of oligonucleotides and polynucleotides with reactive derivatives of oligonucleotides prepared by derivatization with new heterobifunctional reagents, *Ric. Sci.,* 113, 71, 1984.

184. **Vlassov, V. V., Gorn, V. V., Ivanova, E. M., Kazakov, S. A., and Mamaev, S. V.,** Complementary addressed modification of oligonucleotide d(pGpGpCpGpGpA) with platinum derivative of oligonucleotide d(pTpCpCpGpCpCpTpTpT), *FEBS Lett.,* 162, 286, 1983.

185. **Hecht, S. M., Ed.,** *Bleomycin: Chemical, Biochemical and Biological Aspects,* Springer-Verlag, Heidelberg, West Germany, 1979.

186. **Burger, R. M., Peisach, J., and Horwitz, S. B.,** Mechanism of bleomycin action: in vitro studies, *Life Sci.,* 28, 715, 1981.

187. **Sugiyama, H., Kilkuskie, R. E., Chang, L. -H., Ma, L. -T., and Hecht, S. M.,** DNA strand scission by bleomycin: catalytic cleavage and strand selectivity, *J. Am. Chem. Soc.,* 108, 3852, 1986.

188. **Uesugi, S., Shida, T., Ikehara, M., Kobayashi, Y., and Kyogoku, Y.,** Identification of oligonucleotide fragments produced in a strand scission reaction of the d(C-G-C-G-C-G) duplex by bleomycin, *Nucleic Acids Res.,* 12, 1581, 1984.

189. **Wu, J. C., Kozarich, J. W., and Stubbe, J.,** The mechanism of free base formation from DNA by bleomycin. A proposal based on site specific tritium release from poly(dA.dU), *J. Biol. Chem.,* 258, 4694, 1983.

190. **Halliwell, B. and Gutteridge, J. M. C.,** Oxygen toxicity, oxygen radicals, transition metals and disease, *Biochem. J.,* 219, 1, 1984.

191. **Dervan, P. B.,** Design of sequence-specific DNA-binding molecules, *Science,* 232, 464, 1986.

192. **Sigman, D. S.,** Nuclease activity of 1,10-phenanthroline-copper ion, *Acc. Chem. Res.,* 19, 180, 1981.

193. **Lown, J. W. and Joshua, A. V.,** Bleomycin models. Haemin-acridines which bind to DNA and cause oxygen-dependent scission, *J. Chem. Soc. D,* 1298, 1982.

194. **Boutorin, A. S., Vlassov, V. V., Kazakov, S. A., Kutiavin, I. V., and Podyminogin, M. A.,** Complementary addressed reagents carrying EDTA-Fe(II) groups for direct cleavage of single-stranded nucleic acids, *FEBS Lett., 172,* 43, 1984.

195. **Dreyer, G. B. and Dervan, P. B.,** Sequence-specific cleavage of single-stranded DNA: oligodeoxynucleotide-EDTA · Fe(II), *Proc. Natl. Acad. Sci. U.S.A., 82,* 968, 1985.

196. **Chu, B. C. F. and Orgel, L. E.,** Nonenzymatic sequence-specific cleavage of single-stranded DNA, *Proc. Natl. Acad. Sci. U.S.A., 82,* 963, 1985.

197. **Lavrik, O. I. and Moor, N. A.,** Interaction of aminoacyl-tRNA synthetases with amino acids, *Mol. Biol. (USSR), 18,* 1208, 1984.

198. **Lavrik, O. I., Nevinsky, G. A., Khodyreva, S. N., and Moor, N. A.,** Synthesis of photoreactive analogs of phenylalanine and investigation of their interaction with *E. coli* phenylalanine tRNA synthetase, *Bioorg. Khim. (USSR), 7,* 884, 1981.

199. **Lee, Y. M. and Benisek, W. F.,** Inactivation of phosphorylase b by potassium ferrate. Identification of a tyrosine residue involved in the binding of adenosine 5'-monophosphate, *J. Biol. Chem., 253,* 5460, 1978.

200. **Kirley, J. W. and Day, R. A.,** Irreversible inhibition of carbonic anhydrase by the carbon dioxide analog cyanogen, *Biochem. Biophys. Res. Commun., 126,* 457, 1985.

201. **Blech, D. M. and Borderes, C. L.,** Hydroperoxide anion, HO_2^- is an affinity reagent for the inactivation of yeast Cu, Zn superoxide dismutase: modification of one histidine per subunit, *Arch. Biochem. Biophys., 224,* 579, 1983.

202. **Meyer, J.,** Comparison of carbon monoxide, nitric oxide, and nitrite as inhibitors of the nitrogenase from *Clostridium pasteurianum, Arch. Biochem. Biophys., 210,* 246, 1981.

203. **Lagutina, I. O., Sklyankina, V. A., and Avaeva, S. M.,** Affinity labelling of inorganic pyrophosphatase from yeast by methylphosphate, *Biokhimija (USSR), 45,* 1568, 1980.

204. **Lauquin, G., Pougeois, R., and Vignais, P. V.,** 4-Azido-2-nitrophenyl phosphate, a new photoaffinity derivative of inorganic phosphate binding site of beef heart mitochondrial adenosine triphosphatase, *Biochemistry, 19,* 4620, 1980.

205. **Gibson, D. R., Gracy, R. W., and Hartman, F. C.,** Affinity labeling and characterization of the active site histidine of glucosephosphate isomerase, *J. Biol. Chem., 255,* 9369, 1980.

206. **Colanduoni, J. and Villafranca, J. J.,** Labeling of specific lysine residues at the active site of glutamine synthetase, *J. Biol. Chem., 260,* 15042, 1985.

207. **Palumaa, P., Mähar, A., and Järv, J.,** Kinetic analysis of butyrylcholineesterase inhibition with N,N-dimethyl-2-phenyl-azyridinium ion, *Bioorg. Chem., 11,* 394, 1982.

208. **Zeppezauer, E., Jörnvall, H., and Ohlsson, I.,** Carboxymethylation of horse-liver alcohol dehydrogenase in the crystalline state. The active-site zinc region and general anion-binding site of the enzyme correlated in primary and tertiary structures, *Eur. J. Biochem., 58,* 95, 1975.

209. **Reynolds, C. H. and McKinley-McKee, J. S.,** Anion-binding to liver alcohol dehydrogenase, studied by rate of alkylation, *Eur. J. Biochem., 10,* 474, 1969.

210. **Nakamaye, K. L., Wells, J. A., Bridenbaugh, R. L., Okamoto, Y., and Yount, R. G.,** 2-[(4-azido-2-nitrophenyl) amino] ethyl triphosphate, a novel chromophoric and photoaffinity analogue of ATP. Synthesis, characterization and interaction with myosin subfragment. I, *Biochemistry, 24,* 5226, 1985.

211. **Thill, G. and Lane, R. S.,** Inactivation of histidine decarboxylase by the substrate analog 3-bromoacetylpyridine, *Fed. Proc., Fed. Am. Soc. Exp. Biol., 38,* 566, 1979.

212. **Farley, R. A. and Faller, L. D.,** The amino acid sequence of an active site peptide from the H,K-ATPase of gastric mucosa, *J. Biol. Chem., 260,* 3899, 1985.

213. **Clonis, Y. D. and Lowe, C. R.,** Triazine dyes, a new class of affinity labels for nucleotide-dependent enzymes, *Biochem. J., 191,* 247, 1980.

214. **Batra, S. P. and Nicholson, B. H.,** 9-Azidoacridine, a new photoaffinity label for nucleotide — and aromatic — binding sites in proteins, *Biochem. J., 207,* 101, 1982.

215. **Dahl, K. H., McKinley-McKee, J. S., Beyerman, H. C., and Noordam, A.,** Stereoselective metal-directed affinity labelling. A model inactivation study with alcohol dehydrogenase from liver, *FEBS Lett., 99,* 313, 1979.

216. **Yielding, K. L. and Yielding, L. W.,** Photoaffinity labeling of DNA, *Ann. N.Y. Acad. Sci., 346,* 368, 1980.

217. **Bowler, B. E. and Lippard, S. J.,** Modulation of platinum antitumor drug binding to DNA by linked and free intercalators, *Biochemistry, 25,* 3031, 1986.

218. **Kean, J. M., White, S. A., and Draper, D. E.,** Detection of high-affinity intercalator sites in a ribosomal RNA fragment by the affinity cleavage intercalator methidiumpropyl-EDTA-Iron (II), *Biochemistry, 24,* 5062, 1985.

219. **Vary, C. P. H. and Vournakis, J. N.,** RNA structure analysis using methidiumpropyl-EDTA Fe(II): a base-pair-specific RNA structure probe, *Proc. Natl. Acad. Sci. U.S.A., 81,* 6978, 1984.

220. **Robson, R. J., Radhakrishnan, R., Ross, A. H., Takagaki, Y., and Khorana, H. G.,** Photochemical cross-linking in studies of lipid-protein interactions, in *Lipid-Protein Interactions,* Vol. 2, Jost, P. C. and Griffith, O. P., Eds., Wiley-Interscience, New York, 1982, 149.

221. **Gitler, C. and Bercovici, T.,** Use of lipophilic photoactivatable reagents to identify the lipid-embedded domains of membrane proteins, *Ann. N.Y. Acad. Sci.,* 346, 199, 1980.

222. **Richards, F. M. and Brunner, J.,** General labeling of membrane proteins, *Ann. N.Y. Acad. Sci.,* 346, 144, 1980.

223. **Bayley, H.,** Photoactivated hydrophobic reagents for integral membrane proteins, in *Membranes and Transport,* Vol. 1, Martonosi, A. N., Ed., Plenum Press, New York, 1982, 185.

224. **Sigrist, H. and Zahler, P.,** Hydrophobic labeling and cross-linking of membrane proteins, in *Membranes and Transport,* Vol. 1, Martonosi, A. N., Ed., Plenum Press, New York, 1982, 173.

225. **Burnett, B. K., Robson, R. J., Takagaki, Y., Radhakrishnan, R., and Khorana, H. G.,** Synthesis of phospholipids containing photoactivatable carbene precursors in the headgroups and their cross-linking with membrane proteins, *Biochim. Biophys. Acta,* 815, 57, 1985.

226. **Moonen, P., Haagsman, H. P., Van Deenen, L. L. M., and Wirtz, K. W. A.,** Determination of the hydrophobic binding site of phosphatidylcholine exchange protein with photosensitive phosphatidylcholine, *Eur. J. Biochem.,* 99, 439, 1979.

227. **El Kebbaj, M. S., Berrez, J. -M., Lakhlifi, T., Morpain, C., and Latruffe, N.,** Photolabelling of D-β-hydroxybutyrate apodehydrogenase with azidoaryl phospholipids, *FEBS Lett.,* 182, 176, 1985.

228. **Ganjian, I., Pettei, M. J., Nakamishi, K., and Kaissling, K. E.,** A photoaffinity-labelled insect sex pheromone for the moth *Antheraeae polyphemus, Nature,* 271, 157, 1978.

229. **Coulson, A. F. W. and Knowles, J. R.,** Active-site-directed inhibition of triosephosphate isomerase, *J. Chem. Soc.,* 7, 1970.

230. **Marver, D., Chiu, W. -H., Wolff, M. E., and Edelman, I. S.,** Photoaffinity site-specific covalent labeling of human corticosteroid-binding globulin, *Proc. Natl. Acad. Sci. U.S.A.,* 73, 4462, 1976.

231. **Wolff, M. E., Feldman, D., Catsoulacos, P., Funder, J. W., Hancock, C., Amano, Y., and Edelman, I. S.,** Steroidal 21-diazo ketones: photogenerated corticosteroid receptor labels, *Biochemistry,* 14, 1750, 1975.

232. **Pons, M., Nicolas, J. -C., Boussioux, A. -M., Descomps, B., and Crastes de Paulet, A.,** Affinity labelling of the estradiol-17β dehydrogenase from human placenta with substrate analoges, *Eur. J. Biochem.,* 68, 385, 1976.

233. **Thamm, P., Saunders, D., and Brandenburg, D.,** Photoreactive insulin derivatives: preparation and characterization, in *Insulin: Chemistry, Structure, and Function and Related Hormones,* Proc. Int. Insulin Symposium, Brandenburg, D. and Wollmer, A., Eds., Walter de Gruyter, Berlin, 1980, 309.

234. **Moreland, R. B., Smith, P. K., Fujimoto, E. K., and Dockter, M. E.,** Synthesis and characterization of N-(4-azidophenylthio)phthalimide: a cleavable, photoactivable crosslinking reagent that reacts with sulfhydryl groups, *Anal. Biochem.,* 121, 321, 1982.

235. **Bisson, R., Azzi, A., Gutweniger, H., Colonna, R., Montecucco, C., and Lanotti, A.,** Interaction of cytochrome c with cytochrome c oxidase. Photoaffinity labeling of beef heart cytochrome c oxidase with arylazidocytochrome c, *J. Biol. Chem.,* 253, 1874, 1978.

236. **Golinska, B., Millon, R., Backendorf, C., Olomucki, M., Ebel, J. -P., and Ehresmann, B.,** Identification of a 16S RNA fragment crosslinked to protein S1 within *Escherichia coli* ribosomal 30S subunit by the use of a crosslinking reagent: ethyl 4-azidobenzoylaminoacetimidate, *Eur. J. Biochem.,* 115, 479, 1981.

237. **Gorman, J. J. and Folk, J. E.,** Transglutaminase amine substrates for photochemical labeling and cleavable cross-linking of proteins, *J. Biol. Chem.,* 255, 1175, 1980.

238. **Kurzchalia, T. V., Wiedmann, M., Girshovich, A. S., Bochkareva, E. S., Bielka, H., and Rapoport, T. A.,** The signal sequence of nascent preprolactin interacts with the 54K polypeptide of the signal recognition particle, *Nature,* 320, 634, 1986.

239. **Bulichev, N. A., Lavrik, O. I., and Nevinsky, G. A.,** Comparative analysis of affinity modification of several aminoacyl-tRNA synthetases with γ-(p-azidoanilide)-ATP, *Mol. Biol. (USSR),* 14, 558, 1980.

240. **Buneva, V. N., Dobrikova, E. Y., Knorre, D. G., Pacha, I. O., and Chimitova, T. A.,** Affinity labelling of yeast hexokinase with benzylamide derivatives of adenosine mono- and triphosphates bearing an alkylating group, *FEBS Lett.,* 135, 159, 1981.

241. **Haley, B. E.,** Development and utilization of 8-azidopurine nucleotide photoaffinity probes, *Fed. Proc., Fed. Am. Soc. Exp. Biol.,* 42, 2831, 1983.

242. **Mirsalikhova, N. M., Baranova, L. A., Tunitskaya, V. L., and Gulyaev, N. N.,** Interaction of Na, K-ATPase with modifying ATP analogs and chloromethylphosphonic acid, *Biokhimija (USSR),* 46, 314, 1981.

243. **Boulay, F., Dalbon, P., and Vignais, P. V.,** Photoaffinity labeling of mitochondrial adenosinetriphosphatase by 2-azidoadenosine $5'$-[α-^{32}P]diphosphate, *Biochemistry,* 24, 7372, 1985.

234   *Affinity Modification of Biopolymers*

244. **Hampton, A. and Nomura, A.**, Inosine 5'-phosphate dehydrogenase. Site of inhibition by guanosine 5'-phosphate and of inactivation by 6-chloro- and 6-mercaptopurine ribonucleoside-5'-phosphates, *Biochemistry,* 6, 679, 1967.

245. **Hampton, A. and Slotin, L. A.**, N⁶-o- and p-fluorobenzoyladenosine 5'-triphosphates, in *Methods in Enzymology,* Vol. 46, Jakoby, W. B. and Wilchek, M., Eds., Academic Press, New York, 1977, 295.

246. **Suzuki, K., Eguchi, C., and Imahori, K.**, Affinity labeling of adenine nucleotide-related enzymes with reactive adenine nucleotide analogs. Affinity labeling of phosphoglycerate kinase with a reactive AMP analog, *J. Biochem.,* 81, 1393, 1977.

247. **Nevinsky, G. A., Podust, V. N., Ankilova, V. N., and Lavrik, O. I.**, Phenylalanyl-tRNA synthetase from *E. coli:* enzyme modification by chemically reactive analogs of fluorescent nucleotides, *Bioorg. Khim. (USSR),* 9, 936, 1983.

248. **Brouillette, C. B., Change, C. T. -C., and Mertes, M. P.**, 5(α-bromoacetyl)-2'-deoxyuridine-5'-phosphate: a mechanism based affinity label for thymidylate synthetase, *Biochem. Biophys. Res. Commun.,* 87, 613, 1979.

249. **Armstrong, V., Sternbach, H., and Eckstein, F.**, 5-Formyl-UPT for DNA-dependent RNA polymerase, in *Methods in Enzymology,* Vol. 46, Jakoby, W. B. and Wilchek, M., Eds., Academic Press, New York, 1977, 346.

250. **Evans, R. K., Johnson, J. D., and Haley, B. E.**, 5-Azido-2'-deoxyuridine 5'-triphosphate: a photoaffinity labeling reagent and tool for the enzymatic synthesis of photoactive DNA, *Proc. Natl. Acad. Sci. U.S.A.,* 83, 5382, 1986.

251. **Chan, J. K. and Anderson, B. M.**, A novel diazonium-sulfhydryl reaction in the inactivation of yeast alcohol dehydrogenase by diazotized 3-aminopyridine adenine dinucleotide, *J. Biol. Chem.,* 250, 67, 1975.

252. **Hixson, S. H., Brownie, T. F., Chua, C. C., Crapster, B. B., Satlin L. M., Hixson, S. S., Boyce, C. O'D. L., Ehrich, M., and Novak, E. K.**, Bifunctional aryl azides as probes of the active sites of enzymes, *Ann. N.Y. Acad. Sci.,* 346, 104, 1980.

253. **Standring, D. N. and Knowles, J. R.**, Photoaffinity labeling of lactate dehydrogenase by the carbene derived from the 3-diazirino analogue of nicotinamide adenine dinucleotide, *Biochemistry,* 19, 2811, 1980.

254. **Foucaud, B. and Biellmann, J. -F.**, Inactivation of yeast alcohol dehydrogenase by a reactive coenzyme analogue: 3-chloroacetylpyridine-adenine dinucleotide, *Biochimie,* 64, 941, 1982.

255. **Jörnvall, H., Woenckhaus, C., Schattle, E., and Jeck, R.**, Modification of alcohol dehydrogenase with two NAD⁺-analogues containing substituents on the functional side of the molecule, *FEBS Lett.,* 54, 297, 1975.

256. **Guillory, R. J. and Jeng, S. J.**, Arylazido nucleotide analogs in a photoaffinity approach to receptor site labeling, in *Methods in Enzymology,* Vol. 46, Jakoby, W. B. and Wilchek, M., Eds., Academic Press, New York, 1977, 259.

257. **Knorre, D. G., Vlassov, V. V., Zarytova, V. F., and Karpova, G. G.**, Nucleotide and oligonucleotide derivatives as enzyme and nucleic acid targeted irreversible inhibitors. Chemical aspects, in *Advances in Enzyme Regulation,* Vol. 24, Weber, J., Ed., Pergamon Press, Oxford, 1985, 277.

258. **Galardy, R. E., Craig, L. C., Jamieson, J. D., and Printz, M. P.**, Photoaffinity labeling of peptide hormone binding sites, *J. Biol. Chem.,* 249, 3510, 1974.

259. **Severin, E. S., Gulyaev, N. N., Bulargina, T. V., and Kochetkova, M. N.**, Specific inhibition of cyclic AMP-dependent protein kinase, adenylate cyclase and phosphodiesterase by ATP and cyclic AMP analogs, in *Advances in Enzyme Regulation,* Vol. 17, Weber, G., Ed., Pergamon Press, Oxford, 1979, 251.

260. **Hampton, A., Harper, P. J., Sasaki, T., and Howgate, P.**, Carboxylic-phosphoric anhydrides isosteric with adenine nucleotides, in *Methods in Enzymology,* Vol. 46, Jakoby, W. B. and Wilchek, M., Eds., Academic Press, New York, 1977, 302.

261. **Chladek, S., Quiggle, K., Chinali, G., Kohut, J., III, and Ofengand, J.**, Synthesis and properties of nucleoside 5'-phosphoazidates derived from guanosine and adenosine nucleotides: effect on elongation factors G and Tu dependent reactions, *Biochemistry,* 16, 4312, 1977.

262. **Blaas, D., Patzelt, E., and Kuechler, E.**, Cap-recognizing protein of influenza virus, *Virology,* 116, 339, 1982.

263. **Schäfer, H. -J., Mainka, L., Rathgeber, G., and Zimmer, G.**, Photoaffinity cross-linking of oligomycin-sensitive ATPase from beef heart mitochondria by 3'-arylazido-8-azido ATP, *Biochem. Biophys. Res. Commun.,* 111, 732, 1983.

264. **Pongs, O., Stöffler, G., and Bald, R. W.**, Localization of protein S1 of *Escherichia coli* ribosomes at the A-site of the codon binding site. Affinity labeling studies with a 3'-modified A-U-G analog, *Nucleic Acids Res.,* 3, 1635, 1976.

265. **Pongs, O., Stöffler, G., and Lanka, E.**, The codon binding site of the *Escherichia coli* ribosome as studied with a chemically reactive A-U-G analog, *J. Mol. Biol.,* 99, 301, 1975.

266. **Hountondji, C., Fauat, G., and Blanquet, S.**, Complete inactivation and labeling of methionyl-tRNA synthetase by periodate-treated initiator tRNA in presence of sodium cyanohydridoborate, *Eur. J. Biochem.,* 102, 247, 1979.

267. **Towbin, H. and Elson, D.,** A photoaffinity labelling study of the messenger RNA-binding region of *Escherichia coli* ribosomes, *Nucleic Acids Res.,* 5, 3389, 1978.

268. **Leitner, M., Wilchek, M., and Zamir, A.,** Identification by affinity labelling of potential sites in 23-S rRNA interacting with the 3' end of tRNA, *Eur. J. Biochem.,* 125, 49, 1982.

269. **Hansske, F. H., Seela, F., Watanabe, K., and Cramer, F.,** Modification of the rare nucleoside X in *Escherichia coli* tRNAs with antigenic determining, photolabile, and paramagnetic residues, *Methods in Enzymology,* Vol. 59, Colowick, S., Ed., Academic Press, New York, 1979, 166.

270. **Budker, V. G., Knorre, D. G., Kravchenko, V. V., Lavrik, O. I., Nevinsky, G. A., and Teplova, N. M.,** Photoaffinity reagents for modification of aminoacyl-tRNA synthetases, *FEBS Lett.,* 49, 159, 1974.

271. **Chen, J. -K., Franke, L. A., Hixson, S. S., and Zimmermann, R. A.,** Photochemical cross-linking of tRNA$_1^{Arg}$ to the 30S ribosomal subunit using aryl azide reagents attached to the anticodon loop, *Biochemistry,* 24, 4777, 1985.

272. **Chen, J. -K., Krauss, J. H., Hixson, S. S., and Zimmermann, R. A.,** Covalent cross-linking of tRNA^{Gly} to the ribosomal P site via the dihydrouridine loop, *Biochim. Biophys. Acta,* 825, 161, 1985.

273. **Bochkareva, E. S., Budker, V. G., Girshovich, A. S., Knorre, D. G., and Teplova, N. M.,** An approach to specific labelling of ribosome in the region of peptydyl-transferase center using N-acyl-aminoacyl-tRNA with an active alkylating group, *FEBS Lett.,* 19, 121, 1971.

274. **Pellegrini, M., Oen, H., Eilat, D., and Cantor, C. R.,** Mechanism of covalent reaction of bromoacetylphenylalanyl-transfer RNA with the peptidyl-transfer RNA binding site of the *Escherichia coli* ribosome, *J. Mol. Biol.,* 88, 809, 1974.

275. **Czernilofsky, A. P., Collatz, E. E., Stöffler, G., and Kuechler, E.,** Proteins at the tRNA binding sites of *Escherichia coli* ribosomes, *Proc. Natl. Acad. Sci. U.S.A.,* 71, 230, 1974.

276. **Bispink, L. and Matthaei, H.,** Photoaffinity labeling of 23S rRNA in *Escherichia coli* ribosomes with poly(U)-coded ethyl-2-diazomalonyl-Phe-tRNA, *FEBS Lett.,* 37, 291, 1973.

277. **Hsiung, N., Reines, S. A., and Cantor, C. R.,** Investigation of the ribosomal peptidyl transferase center using a photoaffinity label, *J. Mol. Biol.,* 88, 841, 1974.

278. **Girshovich, A. S., Bochkareva, E. S., Kramarov, V. M., and Ovchinnikov, Yu. A.,** *E. coli* 30S and 50S ribosomal subparticle components in the localization region of the tRNA acceptor terminus, *FEBS Lett.,* 45, 213, 1974.

279. **Barta, A., Steiner, G., Brosius, J., Noller, H. F., and Kuechler, E.,** Identification of a site on 23S ribosomal RNA located at the peptidyl transferase center, *Proc. Natl. Acad. Sci. U.S.A.,* 81, 3607, 1984.

280. **Grachev, M. A. and Oshevsky, S. I.,** Studying on the "addressed" chemical modification of a single-stranded DNA fragment, *Dokl. Akad. Nauk SSSR,* 272, 1259, 1983.

281. **Ebel, J. P., Ehresmann, B., Ehresmann, C., Giege, R., Riehl, N., and Romby, P.,** Macromolecular interactions in protein synthesis: the interaction of tRNA with its biological partners (aminoacyl-tRNA synthetase, elongation factor EF-Tu, ribosome), *Pure Appl. Chem.,* 57, 405, 1985.

282. **Hajnsdorf, E., Favre, A., and Expert-Benzacon, A.,** Multiple crosslinks of proteins S7, S9, S13 to domains 3 and 4 of 16S RNA in the 30S particle, *Nucleic Acids Res.,* 14, 4009, 1986.

283. **Cartwright, I. L. and Hutchinson, D. W.,** Azidopolynucleotides as photoaffinity reagents, *Nucleic Acids Res.,* 8, 1675, 1980.

284. **Salganic, R. I., Dianov, G. L., Ovchinnikova, L. P., Voronina, E. N., Kokoza, E. B., and Mazin, A. V.,** Gene-directed mutagenesis in bacteriophage T7 provided by polyalkylating RNAs complementary to selected DNA sites, *Proc. Natl. Acad. Sci. U.S.A.,* 77, 2769, 1980.

285. **Summerton, J.,** Intracellular inactivation of specific nucleotide sequences: a general approach to the treatment of viral diseases and virally-mediated cancers, *J. Theor. Biol.,* 78, 77, 1979.

286. **Vlassov, V. V., Lavrik, O. I., Khodyreva, S. N., Chiszikov, V. E., Shvalie, A. F., and Mamaev, S. V.,** Chemical modification of the phenylalanyl-tRNA synthetase and ribosomes of *Escherichia coli* with derivatives of tRNA^{Phe} carrying photoreactive groups on guanosine residues, *Mol. Biol. (USSR),* 14, 531, 1980.

287. **Levina, E. S., Bavykin, S. G., Shick, V. V., and Mirzabekov, A. D.,** The method of crosslinking histones to DNA partly depurinated at neutral pH, *Anal. Biochem.,* 110, 93, 1981.

288. **Babkina, G. T., Bausk, E. V., Graifer, D. M., Karpova, G. G., and Matasova, N. B.,** The effect of aminoacyl- or peptidyl-tRNA at the A-site on the arrangement of deacylated tRNA at the ribosomal P-site, *FEBS Lett.,* 170, 290, 1984.

289. **Gimautdinova, O. I., Gorn, V. V., Gorshkova, I. I., Graifer, D. M., Karpova, G. G., Mundus, D. A., and Teplova, N. M.,** Alkylation of tRNA^{Phe} with 4-(N-2-chloroethyl-N-methylamino)benzyl-5'-phosphamide of d(ATTTTCA), *Bioorg. Khim. (USSR),* 12, 490, 1986.

290. **Staros, J. V., Bayley, H., Standring, D. M., and Knowles, J. R.,** Reduction of aryl azides by thiols: implications for the use of photoaffinity reagents, *Biochem. Biophys. Res. Commun.,* 80, 568, 1978.

291. **Takagaki, Y., Gupta, C. M., and Khorana, H. G.,** Thiols and the diazo group in photoaffinity labels, *Biochem. Biophys. Res. Commun.,* 95, 589, 1980.

292. **Westcoff, K. R., Olwin, B. B., and Storm, D. R.,** Inhibition of adenylate cyclase by the 2',3'-dialdehyde of adenosine triphosphate, *J. Biol. Chem.,* 225, 8767, 1980.

293. **Syvertsen, C. and McKinley-McKee, J. S.,** Affinity labelling of liver alcohol dehydrogenase. Effects of pH and buffers on affinity labelling with iodoacetic acid and (R,S)-2-bromo-3-(5-imidazolyl)propionic acid, *Eur. J. Biochem.,* 117, 165, 1981.

294. **CaJacob, C. A. and Ortiz de Montellano, P. R.,** Mechanism-based in vitro inactivation of lauric acid hydroxylases, *Biochemistry,* 25, 4705, 1986.

295. **Holmes, T. J., Vennerstrom, J. L., and John, V.,** Electrolysis-mediated irreversible inactivation of lipoxygenase directed toward electroaffinity labelling, *Biochem. Biophys. Res. Commun.,* 123, 156, 1984.

296. **Borchardt, R. T. and Bhatia, P.,** Catechol O-methyltransferase. XII. Affinity labeling the active site with the oxidation products of 5,6-dihydroxyindole, *J. Med. Chem.,* 25, 263, 1982.

297. **Vandest, P., Labee, J. -P., and Kassab, R.,** Photoaffinity labelling of arginine kinase and creatine kinase with a γ-p-substituted arylazido analogue of ATP, *Eur. J. Biochem.,* 104, 433, 1980.

298. **Wouters, W., Dun, J. V., Leysen, J. E., and Laduron, P. M.,** Photoaffinity probes for serotonin and histamine receptors. Synthesis and characterization of two azide analogues of ketanserin, *J. Biol. Chem.,* 260, 8423, 1985.

299. **Marinetti, T. D., Okamura, M. Y., and Feher, G.,** Localization of the primary quinone binding site in reaction centers from *Rhodopseudomonas sphaeroides* R-26 by photoaffinity labeling, *Biochemistry,* 18, 3126, 1979.

300. **Ferguson, J. J., Jr.,** Irradiation in the frozen state: a technique for direct photoaffinity labeling, *Photochem. Photobiol.,* 32, 137, 1980.

301. **Park, C. S., Hillel, Z., and Wu, C. -W.,** Molecular mechanism of promoter selection in gene transcription. I. Development of a rapid mixing-photocrosslinking technique to study the kinetics of *Escherichia coli* RNA polymerase binding to T7 DNA, *J. Biol. Chem.,* 257, 6944, 1982.

302. **Cooperman, B. S. and Branswick, D. Y.,** On the photoaffinity labeling of rabbit muscle phosphofructokinase with $O^{2'}$-(ethyl-2-diazomalonyl) adenosine 3':5'-cyclic monophosphate, *Biochemistry,* 12, 4079, 1973.

303. **Vlassov, V. V., Grineva, N. I., Zarytova, V. F., and Knorre, D. G.,** The effectivity of the acetals of 4-(N-2-chloroethyl-N-methylamino)-benzaldehyde derivatives of oligonucleotides in the tRNA alkylation, *Mol. Biol. (USSR),* 4, 201, 1970.

304. **Bailey, J. M. and Colman, R. F.,** Affinity labeling of $NADP^+$-specific isocitrate dehydrogenase by a new fluorescent nucleotide analogue, 2-[(4-bromo-2,3-dioxobutyl)thio]-1,N^6-ethenoadenosine 2',5'-biphosphate, *Biochemistry,* 24, 5367, 1985.

305. **Denisov, A. Yu., Nevinsky, G. A., and Lavrik, O. I.,** Affinity modification of creatine kinase from skeletal muscle of rabbit by fluorescent analog of ATP: γ-(p-azidoanilidate)-1,N^6-ethenoadenosine triphosphate, *Biokhimiya,* 47, 184, 1982.

306. **Hart, G. J. and O'Brien, R. D.,** Recording spectrophotometric method for determination of dissociation and phosphorylation constants for the inhibition of acetylcholinesterase by organophosphates in the presence of substrate, *Biochemistry,* 12, 2940, 1973.

307. **Tian, W. -X. and Tsou, C. -L.,** Determination of the rate constant of enzyme modification by measuring the substrate reaction in the presence of the modifier, *Biochemistry,* 21, 1028, 1982.

308. **Walker, B. and Elmore, D. T.,** The irreversible inhibition of urokinase, kidney-cell plasminogen activator, plasmin and β-trypsin by 1-(N-6-amino-n-hexyl) carbamoylimidazole, *Biochem. J.,* 221, 277, 1984.

309. **Tudela, J., Garcia-Canovas, E., Garcia-Carmona, F., Iborra, J. L., and Lozano, J. A.,** Irreversible inhibition of trypsin by TLCK. A continuous method for kinetic study of irreversible enzymatic inhibitors in the presence of substrate, *Int. J. Biochem.,* 18, 285, 1986.

310. **Walker, B., Wikstrom, P., and Shaw, E.,** Evaluation of inhibitor constants and alkylation rates for a series of thrombin affinity labels, *Biochem. J.,* 230, 645, 1985.

311. **Chicheportiche, R., Balerna, M., Lombet, A., Romey, G., and Lazdunski, M.,** Synthesis and mode of action of axonal membranes of photoactivable derivatives of tetrodotoxin, *J. Biol. Chem.,* 254, 1552, 1979.

312. **Nevinsky, G. A., Gazaryants, M. G., and Mkrtchyan, Z. S.,** Affinity modification of creatine kinase from rabbit skeletal muscle by 2',3'-dialdehyde derivatives of ADP and ATP, *Bioorg. Khim. (USSR),* 9, 487, 1983.

313. **Nevinsky, G. A., Lavrik, O. I., Favorova, O. O., and Kisselev, L. L.,** Finding of two types of the nucleotide-binding sites in the tryptophanyl-tRNA-synthetase and their modification, *Bioorg. Khim. (USSR),* 5, 352, 1979.

314. **Birch, P. L., El-Obeid, H. A., and Akhtar, M.,** The preparation of chloromethylketone analogues of amino acids: inhibition of leucine aminopeptidase, *Arch. Biochem. Biophys.,* 148, 447, 1972.

315. **Jones, J. B. and Ship, S.,** Inhibition of the 3-ketosteroid $\Delta^5 \rightarrow \Delta^4$-isomerase of *Pseudomonas testosteroni* by some bromo-3-ketosteroid derivatives, *Biochim. Biophys. Acta,* 258, 800, 1972.

316. **Kim, I. -Y., Zhang, C. -Y., Ginlio, L. C., Montgomery, J. A., and Chiang, P. K.,** Inactivation of S-adenosylhomocysteine hydrolase by nucleosides, *Biochim. Biophys. Acta,* 829, 150, 1985.

317. **Babkina, G. T., Karpova, G. G., and Matassova, N. B.,** Affinity labelling of *Escherichia coli* ribosome near acceptor tRNA-binding site, *Mol. Biol. (USSR),* 5, 1287, 1984.

318. **Howard, C. A. and Traut, R. R.,** Separation and radioautography of microgram quantities of ribosomal proteins by two-dimensional PAAG gel electrophoresis, *FEBS Lett.,* 29, 177, 1973.

319. **Gimautdinova, O. I., Karpova, G. G., Knorre, D. G., and Kobetz, N. D.,** The proteins of the messenger RNA binding site of *Escherichia coli* ribosomes, *Nucleic Acids, Res.,* 9, 3465, 1981.

320. **Ballmer-Hofer, K., Schlup, V., Burn, P., and Burger, M. M.,** Isolation of in situ crosslinked ligand-receptor complexes using anticrosslinker specific antibody, *Anal. Biochem.,* 126, 246, 1982.

321. **Schouten, J. P.,** Hybridization selection of covalent nucleic acid-protein complexes. II. Cross-linking of proteins to specific *Escherichia coli* mRNAs and DNA sequences by formaldehyde treatment of intact cells, *J. Biol. Chem.,* 260, 9929, 1985.

322. **Hajnsdorf, E., Favre, A., and Expert-Bezancon, A.,** Multiple crosslinks of proteins S7, S9, S13 to domains 3 and 4 of 16S RNA in the 30S particle, *Nucleic Acids Res.,* 14, 4009, 1986.

323. **Phillips, A. T.,** Differential labeling: a general technique for selective modification of binding sites, in *Methods in Enzymology,* Vol. 46, Jakoby, W. B. and Wilchek, M., Eds., Academic Press, New York, 1977, 59.

324. **Grachev, M. A., Kolocheva, I. I., Lukhtanov, E. A., and Mustaev, A. A.,** Studies on the functional topography of *Escherichia coli* RNA polymerase. Highly selective affinity labelling by analogues of initiating substrates, *Eur. J. Biochem.,* 163, 113, 1987.

325. **Nevinsky, G. A., Gazaryants, M. G., and Lavrik, O. I.,** Affinity modification of creatine kinase from rabbit skeletal muscle by ATP γ-amide, a derivative of nitrogenous mustard, *Bioorg. Khim. (USSR),* 10, 656, 1984.

326. **Johanson, R. A. and Henkin, J.,** Affinity labeling of dihydrofolate reductase with an antifolate glyoxal, *J. Biol. Chem.,* 260, 1465, 1985.

327. **Murdock, G. L. and Warren, J. C.,** Isolation of histidyl peptides of the steroid-binding site of human placental estradiol 17β-dehydrogenase, *Steroids,* 39, 165, 1982.

328. **Nevinsky, G. A., Lavrik, O. I., Favorova, O. O., and Kisselev, L. L.,** A method for purification of peptides from hydrolysates of proteins modified by chemically active analogs of substrates containing cis diol groups, *Mol. Biol. Rep.,* 4, 181, 1978.

329. **Hearne, M. and Benisek, W. F.,** Use of solid-phase photoaffinity reagent to label a steroid binding site: application to the Δ⁵-3-ketosteroid isomerase of *Pseudomonas testosteroni, Biochemistry,* 24, 7511, 1985.

330. **Grachev, M. A., Mustaev, A. A., and Kolocheva, I. I.,** Identification of the active site histidine residue of *E. coli* RNA polymerase, *Dokl. Akad. Nauk SSSR,* 281, 723, 1985.

331. **Stringer, C. D. and Hartman, F. C.,** Sequences of two active-site peptides from spinach ribulosebis-phosphate carboxylase/oxygenase, *Biochem. Biophys. Res. Commun.,* 80, 1043, 1978.

332. **Garin, J., Boulay, F., Issartel, J. P., Lunardi, J., and Vignais, P. V.,** Identification of amino acid residues photolabeled with 2-azido [α-³²P]adenosine diphosphate in the β subunit of beef heart mitochondrial F₁-ATPase, *Biochemistry,* 25, 4431, 1986.

333. **Okamoto, Y. and Yount, R. G.,** Identification of an active site peptide of skeletal myosin after photoaffinity labeling with N-(4-azido-2-nitrophenyl)-2-aminoethyl diphosphate, *Proc. Natl. Acad. Sci. U.S.A.,* 82, 1575, 1985.

334. **Lee, Y. M. and Preiss, J.,** Covalent modification of substrate-binding sites of *Escherichia coli* ADP-glucose synthetase. Isolation and structural characterization of 8-azido-ADP-glucose-incorporated peptides, *J. Biol. Chem.,* 261, 1058, 1986.

335. **Schoemaker, H. J. P., Budzik, G. P., Giege, R., and Schimmel, P. R.,** Three photo-cross-linked complexes of yeast phenylalanine specific transfer ribonucleic acid with aminoacyl transfer ribonucleic acid synthetases, *J. Biol. Chem.,* 250, 4440, 1975.

336. **Maly, P., Rinke, J., Ulmer, E., Zwieb, C., and Brimacombe, R.,** Precise localization of the site of cross-linking between protein L4 and 23S ribonucleic acid induced by mild ultraviolet irradiation of *Escherichia coli* 50S ribosomal subunits, *Biochemistry,* 19, 4179, 1980.

337. **Wilkins, K. J.,** Site-specific analysis of drug interactions and damage in DNA using sequencing techniques, *Anal. Biochem.,* 147, 267, 1985.

338. **Vlasov, V. V., Zarytova, V. F., Kutiavin, I. V., Mamaev, S. V., and Podyminogin, M. A.,** Complementary addressed modification and cleavage of a single stranded DNA fragment with alkylating oligonucleotide derivatives, *Nucleic Acids Res.,* 14, 4065, 1986.

339. **Barta, A., Steiner, G., Brosius, J., Noller, H. F., and Kuechler, E.,** Identification of a site on 23S ribosomal RNA located at the peptidyl transferase center, *Proc. Natl. Acad. Sci. U.S.A.,* 81, 3607, 1984.

340. **Baker, B. R., Lee, W. W., Tong, E., and Ross, L. O.,** Potential anticancer agents. LXVI. Nonclassical antimetabolites. III. 4-(Iodoacetamido)-salicylic acid, an exoalkylating irreversible inhibitor of glutamic dehydrogenase, *J. Am. Chem. Soc.,* 83, 3713, 1961.

341. **Kitz, R. and Wilson, I. B.,** Esters of methanesulfonic acid as irreversible inhibitors of acetylcholinesterase, *J. Biol. Chem.,* 237, 3245, 1962.

342. **Malcolm, A. D. B. and Radda, G. K.,** The reaction of glutamate dehydrogenase with 4-iodoacetamido salicylic acid, *Eur. J. Biochem.,* 15, 555, 1970.

343. **Meloche, H. P.,** Bromopyruvate inactivation of 2-keto-3-deoxy-6-phosphogluconic aldolase. I. Kinetic evidence for active site specificity, *Biochemistry,* 6, 2273, 1967.

344. **Gorshkova, I. I. and Chimitova, T. A.,** Kinetic features of affinity modification, in *Affinity Modification of Biopolymers,* Knorre, D. G., Ed., Nauka, Novosibirsk, Soviet Union, 1983, 56.

345. **Hollaway, M. R., Clarke, P. H., and Ticho, T.,** Chloroacetone as an active-site-directed inhibitor of the aliphatic amidase from *Pseudomonas aeruginosa, Biochem. J.,* 191, 811, 1980.

346. **Cornish-Bowden, A.,** Validity of a "steady-state" treatment of inactivation kinetics, *Eur. J. Biochem.,* 93, 383, 1979.

347. **Brocklehurst, K.,** The equilibrium assumption is valid for the kinetic treatment of most time-dependent protein modification reactions, *Biochem. J.,* 181, 775, 1979.

348. **Childs, R. E. and Bardsley, W. G.,** Time-dependent inhibition of enzymes by active-site-directed reagents. A theoretical treatment of the kinetics of affinity labelling, *J. Theor. Biol.,* 53, 381, 1975.

349. **Leytus, S. P., Peltz, S. W., and Mangel, W. F.,** Adaptation of acyl-enzyme kinetic theory and an experimental method for evaluating the kinetics of fast acting, irreversible protease inhibitors, *Biochim. Biophys. Acta,* 742, 409, 1983.

350. **Palumaa, P. J., Raidaru, G. I., Järv, J. L., Schevchenko, V. P., and Myasoyedov, N. F.,** Synthesis of tritium labelled N,N-dimethyl-2-phenylaziridinium and its reaction with acetylcholinesterase, *Bioorg. Khim.,* 11, 1348, 1985.

351. **Fritzsch, G. and Koepsell, H.,** An analysis of biphasic time courses: the inactivation of $(Na^+ + K^+)$-ATPase and Ca^{2+}-ATPase by ATP-analogs, *J. Theor. Biol.,* 102, 469, 1983.

352. **Ray, W. J., Jr. and Koshland, D. E., Jr.,** A method for characterizing the type and numbers of groups involved in enzyme action, *J. Biol. Chem.,* 236, 1973, 1961.

353. **Rakitzis, E. T.,** Kinetics of irreversible enzyme inhibition: co-operative effects, *J. Theor. Biol.,* 67, 49, 1977.

354. **Rakitzis, E. T.,** Kinetics of irreversible enzyme inhibition by an unstable inhibitor: co-operative effects, *J. Theor. Biol.,* 75, 239, 1978.

355. **Purdie, J. E. and Heggie, R. M.,** The kinetics of the reaction of N,N-dimethyl-2-phenylaziridinium ion with bovine erythrocyte acetylcholinesterase, *Can. J. Biochem.,* 48, 244, 1970.

356. **Jacobson, M. A. and Colman, R. F.,** Affinity labelling of a guanosine 5'-triphosphate site of glutamate dehydrogenase by a fluorescent nucleotide analogue, 5'-[p-(fluorosulfonyl)benzoyl]-1,N^6-ethenoadenosine, *Biochemistry,* 21, 2177, 1982.

357. **Knorre, D. G. and Chimitova, T. A.,** Equations of the kinetic curves of affinity labelling of biopolymers with the reagents consumed in parallel reactions in solution, *FEBS Lett.,* 131, 249, 1981.

358. **Palumaa, P., Raba, R., and Järv, J.,** Site-specificity of butyrylcholinesterase alkylation with N,N-dimethyl-2-phenylaziridinium ion, *Biochim. Biophys. Acta,* 791, 15, 1984.

359. **Gorshkova, I. I., Kostikova, I. V., Lavrik, O. I., and Mustaev, A. A.,** Investigation of specificity of nucleoside-5'-trimethaphosphates: aminoacyl-tRNA synthetases interaction, *Izv. Sib. Otd. Akad. Nauk SSSR, Ser. Khim. Nauk,* (4), 93, 1984.

360. **Rakitzis, E. T.,** Kinetics of irreversible enzyme inhibition by an unstable inhibitor, *Biochem. J.,* 141, 601, 1974.

361. **Topham, C. M.,** Chemical modification of enzymes: reaction with an unstable inhibitor, *Biochem. J.,* 227, 1025, 1985.

362. **Ashani, Ya., Wins, P., and Wilson, I. B.,** The inhibition of cholinesterase by diethyl phosphorochloridate, *Biochem. Biophys. Acta,* 284, 427, 1972.

363. **Barnett, P. and Rosenberry, T. L.,** Inactivation of *Electrophorus electricus* acetylcholinesterase by benzene-methane sulfonylfluoride, *Arch. Biochem. Biophys.,* 190, 202, 1978.

364. **Waley, S. G.,** Kinetics of suicide substrates. Practical procedures for determining parameters, *Biochem. J.,* 227, 843, 1985.

365. **Tatsunami, S., Yago, N., and Hosoe, M.,** Kinetics of suicide substrates. Steady-state treatments and computer aided exact solutions, *Biochem. Biophys. Acta,* 662, 226, 1981.

366. **Dahl, K. H. and McKinley-McKee, J. S.,** The imidazole-promoted inactivation of horse-liver alcohol dehydrogenase, *Eur. J. Biochem.,* 120, 451, 1981.

367. **Syversten, Ch. and McKinley-McKee, J. S.,** Binding of ligands to the catalytic zinc ion in horse liver alcohol dehydrogenase, *Arch. Biochem. Biophys.,* 228, 159, 1984.

368. **Syversten, Ch. and McKinley-McKee, J. S.,** Ionization-linked cooperative interactions in the active site of horse liver alcohol dehydrogenase, *Arch. Biochem. Biophys.,* 223, 213, 1983.

369. **Western, A., Syversten, Ch., and McKinley-McKee, J. S.,** The binding of sulfonamides to horse liver alcohol dehydrogenase, *Biochem. Pharmacol.,* 33, 731, 1984.

370. **Gorshkova, I. I. and Lavrik, O. I.,** The influence of ATP, amino acid and their analogs on the kinetics of affinity labelling of phenylalanyl-tRNA-synthetase, *FEBS Lett.,* 52, 135, 1975.

371. **Beznedelnaya, N. I., Gorshkova, I. I., and Lavrik, O. I.,** Kinetics of affinity labelling as an approach to elucidate the cooperativity between substrate-recognized centers of dimeric enzymes, *Mol. Biol. (USSR),* 15, 1102, 1981.

372. **Gorshkova, I. I. and Lavrik, O. I.,** Co-operative effects in affinity labelling reveal tRNA recognition centers of phenylalanyl-tRNA synthetase, *Biochim. Biophys. Acta,* 746, 202, 1983.

373. **Knorre, D. G. and Chimitova, T. A.,** The isotherms of the chemical modification of biopolymers for the affinity reagents forming active intermediates, *Mol. Biol. (USSR),* 12, 814, 1978.

374. **Gorshkova, I. I. and Knorre, D. G.,** Initial rate of affinity labelling by the reagents forming active intermediates, *Bioorg. Khim. (USSR),* 6, 230, 1980.

375. **Ankilova, V. N., Knorre, D. G., Kravchenko, V. V., Lavrik, O. I., and Nevinsky, G. A.,** Investigation of the phenylalanyl-tRNA synthetase modification with γ-(p-azidoanilide)-ATP, *FEBS Lett.,* 60, 172, 1975.

376. **Groman, E. V., Schultz, R. M., and Engel, L. L.,** Catalytic competence: a direct criterion for affinity labelling, in *Methods in Enzymology,* Vol. 46, Jakoby, W. B. and Wilchek, M., Eds., Academic Press, New York, 1977, 54.

377. **Sundaram, P. V. and Eckstein, F., Eds.,** *Theory and Practice in Affinity Techniques,* Academic Press, New York, 1978.

378. **Branlant, G., Tritsch, D., and Biellmann, J. -F.,** Evidence for the presence of anion-recognition sites in pig-liver aldehyde reductase. Modification by phenyl glyoxal and p-carboxyphenyl glyoxal of an arginyl residue located close to the substrate-binding site, *Eur. J. Biochem.,* 116, 505, 1981.

379. **Inano, H. and Engel, L. L.,** Photoaffinity labeling of human placental estradiol dehydrogenase with 3-(arylazido-β-alanine) estrone, *J. Biol. Chem.,* 255, 7694, 1980.

380. **Kovar, J., Simek, K., Kucera, I., and Matyska, L.,** Steady-state kinetics of horse-liver alcohol dehydrogenase with a covalently bound coenzyme analogue, *Eur. J. Biochem.,* 139, 585, 1984.

381. **Eresinska, M., Vandercool, J. M., and Wilson, D. F.,** Cytochrome c interactions with membranes, *Arch. Biochem. Biophys.,* 171, 108, 1975.

382. **Borchardt, R. T. and Thakker, D.,** Affinity labeling of catechol-O-methyltransferase with N-iodoacetyl-3,5-dimethoxy-4-hydroxy-phenylethylamine, *Biochem. Biophys. Res. Commun.,* 54, 1233, 1973.

383. **Belikova, A. M., Vakhrusheva, T. E., Vlassov, V. V., Grineva, N. I., Knorre, D. G., and Kurbatov, V. A.,** Interaction of 4-(N-2-chloroethyl-N-methylamino)-benzaldehyde with transfer RNA, *Mol. Biol. (USSR),* 3, 210, 1969.

384. **Knorre, D. G., Kutyavin, I. V., Levina, A. S., Pichko, N. P., Podust, L. M., and Fedorova, O. S.,** Investigation of complementary-addressed alkylation of oligodeoxyribonucleotides with 2',3'-0-{4-[N-(2-chloroethyl)N-methylamino]} benzylidene derivatives having a short oligonucleotide address, *Bioorg. Khim. (USSR),* 12, 230, 1986.

385. **Gimautdinova, O. I., Gorshkova, I. I., Graifer, D. M., Karpova, G. G., and Kutyavin, I. V.,** Selective alkylation of G^{24} residue in *Escherichia coli* tRNA^{Phe}, *Mol. Biol. (USSR),* 18, 1419, 1984.

386. **Vlassov, V. V. and Skobeltsyna, L. M.,** Chemical modification study on the macrostructure of tRNA^{Phe} (E. coli), *Bioorg. Khim. (USSR),* 4, 550, 1978.

387. **Jencks, W. P.,** *Catalysis in Chemistry and Biology,* McGraw-Hill, New York, 1969.

388. **Denisov, A. Yu., Nevinsky, G. A., and Lavrik, O. I.,** Creatine kinase from rabbit skeletal muscle: investigation of active sites using photoreactive analogs of 1,N^6-ethenoadenosine-5'-di- and triphosphates, *Bioorg. Khim. (USSR),* 8, 72, 1982.

389. **Nevinsky, G. A., Podust, V. N., Khodyreva, S. N., Gorshkova, I. I., and Lavrik, O. I.,** Adenosine- and ethenoadenosine-5'-trimethaphosphate: influence of covalent bond formation with the enzyme on the interaction of affinity label with the active site of phenylalanyl-tRNA synthetase, *Mol. Biol. (USSR),* 18, 1311, 1984.

390. **Kuo, D. J. and Jordan, F.,** Direct spectroscopic observation of brewer's yeast pyruvate decarboxylase bound enamine intermediate produced from a suicide substrate, *J. Biol. Chem.,* 258, 13415, 1983.

391. **Doerge, D. R.,** Mechanism-based inhibition of lactoperoxidase by thiocarbamide goitrogens, *Biochemistry,* 25, 4724, 1986.

392. **Wenz, A., Ghisla, S., and Thorpe, C.,** Studies with general acyl-CoA dehydrogenase from pig kidney, *Eur. J. Biochem.,* 147, 553, 1985.

393. **Skibo, E. B.,** Noncompetitive and irreversible inhibition of xanthine oxidase by benzimidazole analogues acting at the functional flavin adenine dinucleotide cofactor, *Biochemistry,* 25, 4189, 1986.

394. **Wong, Y.-H. H. and Frey, P. A.,** Uridine diphosphate galactose 4-epimerase. Alkylation of enzyme-bound diphosphopyridine nucleotide by p-(bromoacetamido)phenyl uridyl pyrophosphate, an active-site-directed irreversible inhibitor, *Biochemistry,* 18, 5337, 1979.

395. **Ondetti, M. A., Condon, M. E., Reid, J., Sabo, E. F., Cheung, H. S., and Cushman, D. W.,** Design of potent and specific inhibitors of carboxypeptidases A and B, *Biochemistry,* 18, 1427, 1979.

396. **Bartmann, P., Hanke, T., Hammer-Raber, B., and Holler, E.,** Selective labelling of the β-subunit of L-phenylalanyl-tRNA synthetase from *E. coli* with N-bromoacetyl-L-phenylalanyl-tRNA^{Phe}, *Biochem. Biophys. Res. Commun.*, 60, 743, 1974.

397. **Khodyreva, S. N., Moor, N. A., Ankilova, V. N., and Lavrik, O. I.,** Phenylalanyl-tRNA synthetase from *E. coli* MRE-600: analysis of the active site distribution on the enzyme subunits by affinity labeling, *Biochim. Biophys. Acta*, 830, 206, 1985.

398. **Weber, J., Lücken, U., and Schäfer, G.,** Total number and differentiation of nucleotide binding sites on mitochondrial F$_1$-ATP-ase. An approach by photolabeling and equilibrium binding studies, *Eur. J. Biochem.*, 148, 41, 1985.

399. **Sloothaak, J. B., Berden, J. A., Herweijer, M. A., and Kemp, A.,** The use of 8-azido-ATP and 8-azido-ADP as photoaffinity labels of the ATP synthase in submitochondrial particles: evidence for a mechanism of ATP hydrolysis involving two independent catalytic sites?, *Biochim. Biophys. Acta*, 809, 27, 1985.

400. **Lübben, M., Lücken, U., Weber, J., and Schäfer, G.,** Azidonaphthoyl-ADP: a specific photolabel for the high-affinity nucleotide-binding sites of F$_1$-ATPase, *Eur. J. Biochem.*, 143, 483, 1984.

401. **van Dongen, M. B. M. and Berden, J. A.,** Demonstration of two exchangeable noncatalytic and cooperative catalytic sites in isolated bovine heart mitochondrial F$_1$, using photoaffinity labels [2-^{3}H]8-azido-ATP and [2-^{3}H]8-azido-ADP, *Biochim. Biophys. Acta*, 850, 121, 1986.

402. **Schäfer, H.-J., Rathgeber, G., Dose, K., Masafumi, Y., and Kagawa, Y.,** Photoaffinity cross-linking of F$_1$ ATPase from the thermophilic bacterium PS3 by 3'-arylazido-β-alanyl-8-azido ATP, *FEBS Lett.*, 186, 275, 1985.

403. **Schäfer, H.-J. and Dose, K.,** Photoaffinity cross-linking of the coupling factor 1 from *Micrococcus luteus* by 3'-arylazido-8-azido-ATP, *J. Biol. Chem.*, 259, 15301, 1984.

404. **Strickler, R. C., Covey, D. F., and Tobias, B.,** Study of 3α, 20β-hydroxysteroid dehydrogenase with an enzyme-generated affinity alkylator: dual enzyme activity at a single active site, *Biochemistry*, 19, 4950, 1980.

405. **Sweet, F. and Samant, B. R.,** Bifunctional enzyme activity at the same active site: study of 3α and 20β activity of affinity alkylation of 3α, 20β-hydroxysteroid dehydrogenase with 17-(bromoacetoxy)steroids, *Biochemistry*, 19, 978, 1980.

406. **Tobias, B., Covey, D. F., and Strickler, R. C.,** Inactivation of human placental 17β-estradiol dehydrogenase and 20α-hydroxysteroid dehydrogenase with active site-directed 17β-propynyl-substituted progestin analogs, *J. Biol. Chem.*, 257, 2783, 1982.

407. **Strickler, R. C., Tobias, B., and Covey, D. F.,** Human placental 17β-estradiol dehydrogenase and 20α-hydroxysteroid dehydrogenase. Two activities at a single enzyme active site, *J. Biol. Chem.*, 256, 316, 1981.

408. **Tobias, B. and Strickler, R. C.,** Affinity labeling of human placental 17β-estradiol dehydrogenase and 20α-hydroxysteroid dehydrogenase with 5'-[p-(fluorosulfonyl)benzoyl]adenosine, *Biochemistry*, 20, 5546, 1981.

409. **Buneva, V. N., Gorshkova, I. I., Knorre, D. G., and Pacha, I. O.,** Modification of catalytic center of *Escherichia coli* ATP(CTP): tRNA-nucleotidyltransferase with reactive derivatives of adenosine and cytidinetriphosphates with reactive group in the triphosphate fragment, *Dokl. Akad. Nauk SSSR*, 267, 1492, 1982.

410. **Gillard, B. K., White, R. C., Zingaro, R. A., and Nelson, T. E.,** Amylo-1,6-glucosidase/4-α-glucanotransferase. Reaction of rabbit muscle debranching enzyme with an active site-directed irreversible inhibitor, 1-s-dimethyl-arsino-1-thio-β-D-glucopyranoside, *J. Biol. Chem.*, 255, 8451, 1980.

411. **Baba, T., Muth, J., and Allen, C. M.,** Photoaffinity labeling of undecaprenyl pyrophosphate synthetase with a farnesyl pyrophosphate analogue, *J. Biol. Chem.*, 260, 10467, 1985.

412. **Eriksson, S.,** Direct photoaffinity labelling of ribonucleotide reductase from *Escherichia coli*. Evidence for enhanced binding of the allosteric effector dTTP by the presence of substrates, *J. Biol. Chem.*, 258, 5674, 1983.

413. **Mackenzie, C. W., III, Tse, J., and Donnelly, T. E., Jr.,** Inactivation of cyclic GMP-dependent protein kinase by N^α-tosyl-L-lysine chloromethylketone, *Life Sci.*, 29, 1235, 1981.

414. **Levitzki, A., Stallcup, W. B., and Koshland, D. E.,** Half-of-the-site reactivity and conformational states of cytidine triphosphate synthetase, *Biochemistry*, 10, 3371, 1971.

415. **Lazdunski, M.,** "Half-of-the-site" reactivity and the role of subunit interactions in enzyme catalysis, *Prog. Bioorg. Chem.*, 3, 81, 1974.

416. **Scheurich, P., Schäfer, H.-J., and Dose, K.,** 8-Azidoadenosine 5'-triphosphate as a photoaffinity label for bacterial F$_1$ ATPase, *Eur. J. Biochem.*, 88, 253, 1978.

417. **Chang, G.-G., Chang, T.-C., and Huang, T.-M.,** Involvement of lysine residue in the nucleotide binding of pigeon liver malic enzyme: modification with affinity label periodate-oxidized NADP, *Int. J. Biochem.*, 14, 621, 1982.

418. **Akhverdyan, V. Z., Kisselev, L. L., Knorre, D. G., Lavrik, O. I., and Nevinski, G. A.,** Affinity labelling of tryptophanyl-transfer RNA synthetase, *J. Mol. Biol.,* 113, 475, 1975.

419. **Zinoviev, V. V., Rubtsova, N. G., Lavrik, O. I., Malygin, E. G., Akhverdyan, V. Z., Favorova, O. O., and Kisselev, L. L.,** Comparison of the ATP-[^{32}P]pyrophosphate exchange reactions catalyzed by native (two-site) and chemically modified (one-site) tryptophanyl-tRNA synthetase, *FEBS Lett.,* 82, 130, 1977.

420. **Lavrik, O. I. and Nevinski, G. A.,** Phenylalanyl-tRNA synthetase from *Escherichia coli* MRE-600. Activation by nucleotides and affinity modification of the effector binding sites, *FEBS Lett.,* 109, 13, 1980.

421. **Marsour, T. L. and Colman, R. F.,** Affinity labelling of AMP-ADP sites in heart phosphofructokinase by 5'-p-fluorosulfonyl adenosine, *Biochem. Biophys. Res. Commun.,* 81, 1370, 1978.

422. **Ogilvie, J. W.,** Determination of the free-energy coupling between ATP and an affinity label attached to rabbit muscle phosphofructokinase, *Biochemistry,* 24, 317, 1985.

423. **Pal, P. K. and Colman, R. F.,** Affinity labeling of allosteric GTP site of bovine liver glutamate dehydrogenase by 5'-p-fluorosulfonylbenzoylguanosine, *Biochemistry,* 18, 838, 1979.

424. **Colman, R. F., Huang, Y.-C., King, M. M., and Erb, M.,** 6-[(4-Bromo-2,3-dioxybutyl)thio]-6-deaminoadenosine 5'-monophosphate and 5'-diphosphate: new affinity labels for purine nucleotide sites in proteins, *Biochemistry,* 23, 3281, 1984.

425. **Philip, G., Gringel, G., and Palm, D.,** Rabbit muscle phosphorylase derivatives with oligosaccharides covalently bound to the glycogen storage site, *Biochemistry,* 21, 3043, 1982.

426. **Striniste, G. F. and Smith, D. A.,** Induction of stable linkage between the DNA-dependent RNA polymerase and d(A-T)-$_n$·d(A-T)$_n$ by ultraviolet light, *Biochemistry,* 13, 485, 1974.

427. **Okada, M., Vergne, J., and Brahms, J.,** Subunit topography of RNA polymerase (*E. coli*) in the complex with DNA, *Nucleic Acids Res.,* 5, 1845, 1978.

428. **Frischauf, A. M. and Scheit, K. H.,** Affinity labeling of *E. coli* RNA polymerase with substrate and template analogues, *Biochem. Biophys. Res. Commun.,* 53, 1227, 1973.

429. **Ovchinnikov, Yu. A., Efimov, V. A., Chakhmakhcheva, O. G., Skiba, N. P., Lipkin, V. M., and Modyanov, N. N.,** Covalent cross-linking of *E. coli* RNA polymerase and photosensitive analogs of decathymidylic acid, *Bioorg. Khim. (USSR),* 5, 1410, 1979.

430. **Sverdlov, E. D., Tsarev, S. A., and Levitan, T. L.,** Interaction of *E. coli* RNA polymerase with substrates during initiation of RNA synthesis at different promoters, in *Macromolecules in Functioning Cell,* Salvatore, F. and Marino, G. N. Y., Eds., Plenum Press, New York, 1979, 149.

431. **Skiba, N. P., Efimov, V. A., Lipkin, V. M., Chakhmakhcheva, O. G., and Ovchinnikov, Yu. A.,** Localization of binding sites on *E. coli* DNA-dependent RNA polymerase for photosensitive template analogues, *Bioorg. Khim. (USSR),* 12, 1023, 1986.

432. **Hillel, Z. and Wu, Ch.,** Photochemical cross-linking studies on the interaction of *E. coli* RNA polymerase with T7 DNA, *Biochemistry,* 17, 2954, 1978.

433. **Park, C. S., Hillel, Z., and Wu, C.-W.,** DNA strand specificity in promoter recognition by RNA polymerase, *Nucleic Acids Res.,* 8, 5895, 1980.

434. **Simpson, R. B.,** The molecular topography of RNA polymerase-promoter interaction, *Cell,* 18, 277, 1979.

435. **Ovchinnikov, Yu. A., Efimov, V. A., Chakhmakhcheva, O. G., Skiba, N. P., and Lipkin, V. M.,** Interaction of DNA dependent RNA polymerase *E. coli* with analogs of the promoter region of bacteriophage fd DNA containing 5-bromouracil residues in the template strand, *Bioorg. Khim. (USSR),* 7, 485, 1981.

436. **Chenchik, A. S., Beabealashvilli, R. S., and Mirzabekov, A. D.,** Topography of interaction of *E. coli* RNA polymerase subunit with Lac-UV5 promoter, *FEBS Lett.,* 128, 46, 1981.

437. **Park, C. S., Wu, F. Y.-H., and Wu, C.-W.,** Molecular mechanism of promoter selection in gene transcription. II. Kinetic evidence for promoter search by a one-dimensional diffusion of RNA polymerase molecule along the DNA template, *J. Biol. Chem.,* 257, 6950, 1982.

438. **Grachev, M. A. and Zaychikov, E. F.,** ATP γ-anilidate: a substrate of RNA polymerase, *FEBS Lett.,* 49, 163, 1974.

439. **Sverdlov, E. D., Tiaryov, S. A., Modyanov, N. N., Lipkin, V. M., Grachev, M. A., Zaychikov, E. F., and Pletnyov, A. G.,** Photoaffinity labeling of *E. coli* RNA polymerase with substrate analogs in a transcribing complex, *Bioorg. Khim. (USSR),* 4, 1278, 1978.

440. **Grachev, M. A. and Zaychikov, E. F.,** Photo-modification studies of the contacts of the 5'-terminus of growing RNA with the substunits of RNA-polymerase, *FEBS Lett.,* 130, 23, 1981.

441. **Hanna, M. M. and Meares, C. F.,** Topography of transcription: path of the leading end of nascent RNA through the *Escherichia coli* transcription complex, *Proc. Natl. Acad. Sci. U.S.A.,* 80, 4238, 1983.

442. **Grachev, M. A., Zaychikov, E. F., Ivanova, E. M., Komarova, N. I., Kutyavin, I. V., Sidelnikova, N. P., and Frolova, I. P.,** Oligonucleotides complementary to a promotor over the region −8 . . . +2 as transcription primers for *E. coli* RNA polymerase, *Nucl. Acids Res.,* 12, 9509, 1984.

443. **Grachev, M. A., Zaychikov, E. F., Kutyavin, I. V., and Tsarev, I. G.,** Affinity modification of *E. coli* RNA polymerase in a complex with a promoter by phosphorylating derivatives of primer oligonucleotides, *Bioorg. Khim. (USSR),* 12, 1678, 1986.

444. **Smirnov, Yu. V., Lipkin, V. M., Ovchinnikov, Yu. A., Grachev, M. A., and Mustaev, A. A.,** Affinity labeling of the binding site for initiation substrate in *E. coli* DNA-dependent RNA polymerase by adenosine 5'-trimetaphosphate, *Bioorg. Khim. (USSR)*, 7, 1113, 1981.

445. **Grachev, M. A., Hartmann, G. R., Maximova, T. G., Mustaev, A. A., Schäffner, A. R., Sieber, H., and Zaychikov, E. F.,** Highly selective affinity labelling of RNA polymerase B (II) from wheat germ, *FEBS Lett.,* 200, 287, 1986.

446. **Freund, E. and McGuire, P. M.,** Identification of a nucleoside triphosphate binding site on calf thymus RNA polymerase II, *Biochemistry,* 25, 276, 1986.

447. **Freund, E. and McGuire, P. M.,** Characterization of RNA polymerase type II from human term placenta, *J. Cell. Physiol.,* 127, 432, 1986.

448. **Nevinsky, G. A., Podust, V. N., Levina, A. S., Khalabuda, O. V., and Lavrik, O. I.,** Human placenta DNA polymerase. Efficiency of the interaction between oligothymidylates of varying length and template specific site, *Bioorg. Khim. (USSR)*, 12, 357, 1986.

449. **Nevinsky, G. A., Doronin, S. V., and Lavrik, O. I.,** *E. coli* DNA polymerase I: primer-template-dependent enzyme inactivation by imidazolides of deoxynucleoside-5'-triphosphates, *Biopolym. Kletka,* 1, 247, 1985.

450. **Hardesty, B. and Kramer, G., Eds.,** *Structure, Function and Genetics of Ribosomes,* Springer-Verlag, New York, 1986.

451. **Pellegrini, M., Oen, H., and Cantor, C. R.,** Covalent attachment of a peptidyl-transfer RNA to the 50S subunit of *E. coli* ribosomes, *Proc. Natl. Acad. Sci. U.S.A.,* 69, 837, 1972.

452. **Czernilofsky, A. P. and Kuechler, E.,** Affinity label for the tRNA binding site on the *E. coli* ribosome, *Biochim. Biophys. Acta,* 272, 667, 1972.

453. **Budker, V. G., Girshovich, A. S., and Skobeltsina, L. M.,** N-acylaminoacyl-tRNA covalently linked to ribosome by its N-acyl group in the region of peptidyltransferase center is able to initiate polypeptide synthesis, *Dokl. Akad. Nauk. SSSR,* 207, 215, 1972.

454. **Pellegrini, M. and Cantor, C. R.,** Affinity labelling of ribosomes, in *Molecular Mechanisms of Protein Biosynthesis,* Weissbach, H. and Pestka, S., Eds., Academic Press, New York, 1977, 203.

455. **Cooperman, B. S.,** Functional sites on the *E. coli* ribosome as defined by affinity labelling, in *Ribosomes: Structure, Function and Genetics,* Chambliss, G., Craven, G. R., Davies, J., Davies, K., Kahan, L., and Nomura, M., Eds., University Park Press, Baltimore, 1980, 531.

456. **Johnson, A. E. and Cantor, C. R.,** Factor-dependent affinity labelling of the *E. coli* ribosomes, *J. Mol. Biol.,* 138, 273, 1980.

457. **Eilat, D., Pellegrini, M., and Oen, H.,** A chemical mapping technique for exploring the location of proteins along the ribosome-bound peptide chain, *J. Mol. Biol.,* 88, 831, 1974.

458. **Vladimirov, S. N., Graifer, D. M., Karpova, G. G., Semenkov, Yu. P., Makhno, V. I., and Kirillov, S. V.,** The effect of GTP hydrolisis and transpeptidation on the arrangement of aminoacyl-tRNA at the A site of *Escherichia coli* 70S ribosomes, *FEBS Lett.,* 181, 367, 1985.

459. **Hsu, L., Lin, F.-L., Nurse, K., and Ofengand, J.,** Covalent cross-linking of *Escherichia coli* Phe-tRNA and Val-tRNA to the ribosomal A site via photoaffinity probes attached to 4-thiouridine residues, *J. Mol. Biol.,* 1972, 57, 1984.

460. **Lin, F.-L., Kahan, L., and Ofengand, J.,** Cross-linking of Phe-tRNA to the ribosomal A site via photoaffinity probe attached to the 4-thiouridine residue is exclusively to ribosomal protein S19, *J. Mol. Biol.,* 172, 77, 1984.

461. **Gornicki, P., Ciesiolka, J., and Ofengand, J.,** Cross-linking of the anticodon of P and A site bound tRNAs to the ribosome via aromatic azides of variable length: involvement of 16S rRNA at the A site, *Biochemistry,* 24, 4924, 1985.

462. **Ciesiolka, J., Gornicki, P., and Ofengand, J.,** Identification of the site of cross-linking in 16S rRNA of an aromatic azide photoaffinity probe attached to the 5'-anticodon base of A site bound tRNA, *Biochemistry,* 24, 4931, 1985.

463. **Abdurashidova, G. G., Baskayeva, I. O., Chernyi, A. A., Kaminir, L. B., and Budowsky, E. I.,** Structural characteristics and classification of some tRNA-binding sites of elongating *Escherichia coli* ribosomes, *Eur. J. Biochem.,* 159, 103, 1986.

464. **Budker, V. G., Girshovich, A. S., Grineva, N. I., Karpova, G. G., Knorre, D. G., and Kobets, N. D.,** Specific chemical modification of ribosomes in the region of mRNA binding center, *Dokl. Acad. Nauk SSSR,* 211, 725, 1973.

465. **Luhrmann, R., Schwarz, U., and Gassen, H. G.,** Covalent binding of uridine-oligonucleotides to 70S *E. coli* ribosomes, *FEBS Lett.,* 32, 55, 1973.

466. **Gimautdinova, O. I., Karpova, G. G., and Kozyreva, N. A.,** Affinity labeling of ribosomes from *Escherichia coli* with 4-(N-2-chloroethyl-N-methylamino)-benzyl-5'-phosphamides of oligouridylates of different length, *Mol. Biol.,* 16, 752, 1982.

467. **Gimautdinova, O. I., Karpova, G. G., Knorre, D. G., and Frolova, S. B.,** Direct cross-linking of heptauridylate to *E. coli* ribosomes by water-soluble carbodiimide in the complex stabilized by codon-anticodon interaction at both A and P sites, *FEBS Lett.,* 185, 221, 1985.

468. **Babkina, G. T., Veniaminova, A. G., Vladimirov, S. N., Karpova, G. G., Yamkovoy, V. I., Bersin, V. A., Gren, E. I., and Cielens, I. E.,** Affinity labelling of *Escherichia coli* ribosomes with a benzylidene derivative of AUGU$_6$ within initiation and pretranslocational complexes, *FEBS Lett.,* 202, 340, 1986.

469. **Stahl, J. and Kobetz, N. D.,** Affinity labelling of proteins at the mRNA binding site of rat liver ribosomes by an analogue of octauridylate containing an alkylating group attached to the 3'-end, *FEBS Lett.,* 123, 269, 1981.

470. **Stahl, J. and Kobetz, N. D.,** Affinity labelling of rat liver ribosomal protein S26 by heptauridylate containing a 5'-terminal alkylating group, *Mol. Biol. Rep.,* 9, 219, 1983.

471. **Stahl, J. and Karpova, G. G.,** Investigation on the messenger RNA binding site of eukaryotic ribosomes by using reactive oligo(U) derivatives, *Biomed. Biochem. Acta,* 44, 1057, 1985.

472. **Cooperman, B. S.,** Photolabile antibiotics as probes of ribosomal structure and function, *Ann. N.Y. Acad. Sci.,* 346, 302, 1980.

473. **Olson, H. M., Grant, P. G., Gleitz, D. G., and Cooperman, B. S.,** Immunoelectron microscopic localization of the site of photo-induced affinity labelling of the small ribosomal subunit with puromycin, *Proc. Natl. Acad. Sci. U.S.A.,* 77, 890, 1980.

474. **Olson, H. M., Nicholson, A. W., Cooperman, B. S., and Glitz, D. G.,** Localization of sites of photoaffinity labelling of the large subunit of *Escherichia coli* ribosomes by arylazide derivative of puromycin, *J. Biol. Chem.,* 260, 10326, 1985.

475. **Ehresmann, Ch., Moine, H., Mougl, M., Dondon, J., Grunberg-Manago, M., Ebel, J.-P., and Ehresmann, B.,** Cross-linking of initiation factor IF-3 to *Escherichia coli* 30S ribosomal subunit by transdiamminedichloroplatinum (II): characterization of two cross-linking sites in 16S rRNA; a possible way of functioning for IF 3, *Nucleic Acids Res.,* 14, 4803, 1986.

476. **Girshovich, A. S., Pozdnyakov, V. A., and Ovchinnikov, Yu. A.,** Localization of the GTP-binding site in the ribosome·elongation factor-G·GTP complex, *Eur. J. Biochem.,* 69, 321, 1976.

477. **Maassen, J. A. and Möller, W.,** Elongation factor G dependent binding of a photoreactive GTP analogue to *E. coli* ribosome results in labeling of protein L11, *J. Biol. Chem.,* 253, 2777, 1978.

478. **Vlassov, V. V., Godovikov, A. A., Kobetz, N. D., Ryte, A. S., Yurchenko, L. V., and Bukrinskaya, A. G.,** Nucleotide and oligonucleotide derivatives as enzyme and nucleic acid targeted irreversible inhibitors. Biochemical aspects, in *Advances in Enzyme Regulation,* Weber, G., Ed., Pergamon Press, Oxford, 1985, 301.

479. **Grineva, N. I., Karpova, G. G., Kuznetsova, L. M., Venkstern, T. V., and Bayev, A. A.,** Complementary addressed modification of yeast tRNA$_1^{Val}$ with alkylating derivative of d(pC-G)rA. The positions of the alkylated nucleotides and the course of the alkylation in the complex, *Nucleic Acids Res.,* 4, 1609, 1977.

480. **Vlassov, V. V., Gaidamakov, S. A., Gorn, V. V., and Grachev, S. A.,** Sequence-specific chemical modification of a 365-nucleotide-long DNA fragment with an alkylating oligonucleotide derivative, *FEBS Lett.,* 182, 415, 1985.

481. **Grineva, N. I.,** Complementary addressed DNA fragmentation and its possible applications in genetic engineering, *Vestn. Akad. Med. Nauk SSSR,* 2, 83, 1981.

482. **Gorshkova, I. I., Zenkova, M. A., Karpova, G. G., Levina, A. S., and Solov'eyv, V. V.,** Complementary addressed modification of 16S rRNA from *Escherichia coli* with 2',3'-O-[4-N-(2-chloroethyl)-N-methylamino]benzylidene derivatives of deoxyribooligonucleotides. I. Kinetic parameters of alkylation, *Mol. Biol. (USSR),* 20, 1084, 1986.

483. **Letsinger, R. L. and Schott, M. E.,** Selectivity in binding a phenanthridinium-dinucleotide derivative to homopolynucleotides, *J. Am. Chem. Soc.,* 103, 7394, 1981.

484. **Asseline, U., Delarue, M., Lancelot, G., Toulme, F., Thuong, N. T., Montenay-Garestier, T., and Helene, C.,** Nucleic acid-binding molecules with high affinity and base sequence specificity: intercalating agents covalently linked to oligodeoxynucleotides, *Proc. Natl. Acad. Sci. U.S.A.,* 81, 3297, 1984.

485. **Zarytova, V. F., Kytyavin, I. V., Silnikov, V. N., and Shishkin, G. V.,** Modification of nucleic acids in stabilized complementary complexes. I. Synthesis of alkylating derivatives of oligodeoxyribonucleotides having a 5'-terminal N-(2-oxyethylphenazinium) residue, *Bioorg. Khim. (USSR),* 12, 911, 1986.

486. **Boidot-Forget, M., Thuong, N. T., Chassignol, M., and Helene, C.,** Artificial nucleases: sequence-specific cleavage of single-stranded nucleic acids by oligodeoxynucleotides covalently linked to EDTA and to an intercalating agent, *C. R. Acad. Sc. Paris,* 302, 75, 1986.

487. **Belyaev, N. D., Budker, V. G., Dubrovskaya, V. A., and Matveeva, N. M.,** Structure of DNA in metaphase chromosomes of mouse fibroblasts, *FEBS Lett.,* 154, 285, 1983.

488. **Belyaev, N. D., Vlassov, V. V., Kobetz, N. D., Ivanova, E. M., and Yakubov, L. A.,** Complementary addressed modification of DNA in metaphase chromosomes and interphase chromatin, *Dokl. Akad. Nauk SSSR,* 291, 234, 1986.

489. **Webb, T. R. and Matteucci, M. D.,** Hybridization triggered cross-linking of deoxyoligonucleotides, *Nucleic Acids Res.,* 14, 7661, 1986.

490. **Baker, B. F. and Dervan, P. B.,** Sequence-specific cleavage of double-helical DNA N-bromoacetyldistamycin, *J. Am. Chem. Soc.,* 107, 8266, 1985.

491. **Hazum, E.,** Photoaffinity labeling of peptide hormone receptors, *Endocr. Rev.,* 4, 352, 1983.

492. **Gronemeyer, H.,** Photoaffinity labelling of steroid hormone binding sites, *Trends Biochem. Sci.,* 10, 264, 1985.

493. **Simons, S. S., Jr. and Thompson, E. B.,** Affinity labeling of glucocorticoid receptors: new methods in affinity labeling, in *Biochemical Actions of Hormones,* Vol. 9, Academic Press, New York, 1982, 221.

494. **Fedan, J. S., Hogaboom, G. K., and O'Donnell, J. P.,** Photoaffinity labels as pharmacological tools, *Biochem. Pharmacol.,* 33, 1167, 1984.

495. **Prestwich, G. D., Koeppe, J. K., Kovalick, G. E., Brown, J. J., Chang, E. S., and Sing, A. K.,** Experimental techniques for photoaffinity labeling of juvenile hormone binding proteins of insects with epoxytarnesyl diazoautate, in *Methods in Enzymology,* Vol. 111, Academic Press, New York, 1985, 509.

496. **Eberle, A. N. and Graan, P. N. E.,** General principles for photoaffinity labeling of peptide hormone receptors, in *Methods in Enzymology,* Vol. 109, Academic Press, New York, 1985, 129.

497. **Linsley, P. S., Das, M., and Fox, C. F.,** Affinity labeling of hormone receptors and other ligand binding proteins, in *Membrane Receptors: Methods for Purification and Characterization, Receptors and Recognition,* (Series B), Vol. 11, Jacobs, S. and Cuatrecasas, P., Eds., Chapman and Hall, London, 1981, 87.

498. **Katzenellenbogen, J. A., Kilbourn, H. R., and Carlson, K. E.,** Photosensitive steroids as probes of estrogen receptor sites, *Ann. N. Y. Acad. Sci.,* 346, 18, 1980.

499. **Witzemann, V. and Blanchard, S. G.,** The use of photochemical probes for studies of structure and function of purified acetylcholine receptor preparations, *Ann. N. Y. Acad. Sci.,* 346, 458, 1980.

500. **Sutoh, K., Yamamoto, K., and Wakabayashi, T.,** Electron microscopic visualization of the ATPase site of myosin by photoaffinity labeling with a biotinylated photoreactive ADP analog, *Proc. Natl. Acad. Sci. U.S.A.,* 83, 212, 1986.

501. **Gronemeyer, H. and Pongs, O.,** Localization of ecdysterone on polytene chromosomes of *Drosophila melanogaster, Proc. Natl. Acad. Sci. U.S.A.,* 77, 2108, 1980.

502. **Satoh, S. and Fujii, T.,** Detection and evaluation of serine enzymes by [^{3}H]-DFP affinity labeling in spinach plants, *Plant Cell Physiol.,* 23, 1383, 1982.

503. **Seely, J. E., Pösö, H., and Pegg, A. E.,** Measurement of the number of ornithine decarboxylase molecules in rat and mouse tissues under various physiological conditions by binding of radiolabelled α-difluoromethylornithine, *Biochem. J.,* 206, 311, 1982.

504. **Zagon, I. S.,** Autoradiographic identification of ornithine decarboxylase in mouse kidney by means of α-[5-^{14}C] difluoromethylornithine, *Science,* 217, 68, 1982.

505. **Rando, R. R.,** The chemical labeling of glutamate decarboxylase in vivo, *J. Biol. Chem.,* 256, 1111, 1981.

506. **Barrett, A. J., Kembhavi, A. A., Brown, M. A., Kirschke, H., Knight, C. G., Tamai, M., and Hanada, K.,** L-trans-epoxy-succinyl-leucylamido (4-guanidino)butane (E-64) and its analogues as inhibitors of cysteine proteinases including cathepsins B, H and L, *Biochem. J.,* 201, 189, 1982.

507. **Kettner, C. and Shaw, E.,** Inactivation of trypsin-like enzymes with peptides of arginine chloromethyl ketone, in *Methods in Enzymology,* Vol. 80, Academic Press, New York, 1981, 826.

508. **Docherty, K., Carroll, R. J., and Steiner, D. F.,** Conversion of proinsulin to insulin: involvement of a 31,500 molecular weight thiol protease, *Proc. Natl. Acad. Sci. U.S.A.,* 79, 4613, 1982.

509. **Shaw, E. and Dean, R. T.,** The inhibition of macrophage protein turnover by a selective inhibitor of thiol proteinases, *Biochem. J.,* 186, 385, 1980.

510. **Cornwell, M. M., Safa, A. R., Felsted, R. L., Gottesman, M. M., and Pastan, I.,** Membrane vesicles from multidrug-resistant human cancer cells contain a specific 150- to 170-kDa protein detected by photoaffinity labeling, *Proc. Natl. Acad. Sci. U.S.A.,* 83, 3847, 1986.

511. **Haring, R., Kloog, Y., and Sokolovsky, M.,** Identification of polypeptides of the phencyclidine receptor of rat hippocampus by photoaffinity labeling with [^{3}H]-azidophencyclidine, *Biochemistry,* 25, 612, 1986.

512. **Haring, R., Kloog, Y., and Sokolovsky, M.,** Regional heterogeneity of rat brain phencyclidine (PCP) receptors revealed by photoaffinity labeling with [^{3}H]-azidophencyclidine, *Biochem. Biophys. Res. Commun.,* 131, 1117, 1985.

513. **Möhler, H., Richards, J. G., and Wu, J.-Y.,** Autoradiographic localization of benzodiazepine receptors in immunocytochemically identified γ-aminobutyrergic synapses, *Proc. Natl. Acad. Sci. U.S.A.,* 78, 1935, 1981.

514. **Möhler, H., Battersby, M. K., and Richards, J. G.,** Benzodiazepine receptor protein identified and visualized in brain tissue by a photoaffinity label, *Proc. Natl. Acad. Sci. U.S.A.,* 77, 1666, 1980.

515. **Carpentier, J.-L., Fehlmann, M., Van Obberghen, E., Gorden, P., and Orci, L.,** Insulin receptor internalization and recycling: mechanism and significance, *Biochimie,* 67, 1143, 1985.

516. **Huecksteadt, T., Olefsky, J. M., Brandenburg, D., and Heidenreich, K. A.,** Recycling of photoaffinity-labeled insulin receptors in rat adipocytes. Dissociation of insulin-receptor complexes is not required for receptor recycling, *J. Biol. Chem.,* 261, 8655, 1986.

517. **Fehlmann, M., Carpentier, J.-L., Van Obberghen, E., Freychet, P., Thamm, P., Saunders, D., Brandenburg, D., and Orci, L.,** Internalized insulin receptors are recycled to the cell surface in rat hepatocytes, *Proc. Natl. Acad. Sci. U.S.A.,* 79, 5921, 1982.

518. **Marburg, S., Tolman, R. L., and Callahan, L. T., III,** Pseudomonas exotoxin A: toxoid preparation by photoaffinity inactivation, *Proc. Natl. Acad. Sci. U.S.A.,* 80, 2870, 1983.

519. **Mazin, A. V., Dianov, G. L., Ovchinnikova, L. P., and Salganic, R. I.,** Induction of specific mutations in tetracycline resistance gene in plasmid pBR 322 using complementary DNA fragments carrying alkylating groups, *Dokl. Akad. Nauk SSSR,* 268, 979, 1983.

520. **Chatterjee, P. K. and Cantor, C. R.,** Preparation of psoralen-cross-linked R-loops and generation of large deletions by their repair in vivo, *J. Biol. Chem.,* 257, 9173, 1982.

521. **Zoller, M. J. and Smith, M.,** Oligonucleotide-directed mutagenesis using M13-derived vectors: an efficient and general procedure for the production of point mutations in any fragment of DNA, *Nucleic Acids Res.,* 10, 6487, 1982.

522. **Albert, A.,** *Selective Toxicity. The Physico-Chemical Basis of Therapy,* 6th ed., Chapman and Hall, London, 1979.

523. **Himmelweit, F.,** *The Collected Papers of Paul Ehrlich,* Vol. 3, Pergamon Press, Elmsford, N.Y., 1960.

524. **Morrison, J. F.,** The slow-binding and slow, tight-binding inhibition of enzyme-catalysed reactions, *Trends Biochem. Sci.,* 7, 102, 1982.

525. **Wilchek, M.,** Affinity therapy and polymer bound drugs, *Makromol. Chem. Suppl.,* 2, 207, 1979.

526. **Gros, L., Ringsdorf, H., and Schupp, H.,** Polymeric antitumor agents on a molecular and on a cellular level?, *Angew. Chem. Int. Ed. Engl.,* 20, 305, 1981.

527. **Vogl, O.,** Functional polymers, *Makromol. Chem. Suppl.,* 7, 1, 1984.

528. **Lloyd, J. B., Duncan, R., and Kopecek, J.,** Synthetic polymers as targetable carriers for drugs, *Pure Appl. Chem.,* 56, 1301, 1984.

529. **Freeman, A. I. and Mayhew, E.,** Targeted drug delivery, *Cancer,* 58, 573, 1986.

530. **Seiler, F. R., Gronski, P., Kuvrle, R., Lüben, G., Harthus, H.-P., Ax, W., Bosslet, K., and Schwick, H.-G.,** Monoclonal antibodies: their chemistry, functions, and possible uses, *Angew. Chem. Int. Ed. Engl.,* 24, 139, 1985.

531. **Vitetta, E. S., Krolick, K. A., Miyama-Inaba, M., Cushley, W., and Uhr, J. W.,** Immunotoxins: a new approach to cancer therapy, *Science,* 219, 644, 1983.

532. **Uhr, J. W.,** Immunotoxins: harnessing nature's poisons, *J. Immunol.,* 133, i, 1984.

533. **Vitetta, E. S. and Uhr, J. W.,** Immunotoxins: redirecting nature's poisons, *Cell,* 41, 653, 1985.

534. **De Clercq, E.,** Specific targets for antiviral drugs, *Biochem. J.,* 205, 1, 1982.

535. **Edvards, D. C.,** Targeting potential of antibody conjugates, *Pharmacol. Ther.,* 23, 147, 1983.

536. **Gregoriadis, G.,** *Liposome Technology, Vol. 3,* CRC Press, Boca Raton, Fla., 1984.

537. **Matthay, K. K., Heath, T. D., Badger, C. C., Bernstein, I. D., and Papahadjopoulos, D.,** Antibody-directed liposomes: comparison of various ligands for association, endocytosis and drug delivery, *Cancer Res.,* 46, 4904, 1986.

538. **Singhal, A. and Gupta, C. M.,** Antibody-mediated targeting of liposomes to red cells in vivo, *FEBS Lett.,* 201, 321, 1986.

539. **Ho, R. J. Y., Rouse, B. T., and Huang, L.,** Target-sensitive immunoliposomes: preparation and characterization, *Biochemistry,* 25, 5500, 1986.

540. **Weber, G., Ed.,** *Enzyme Pattern-Targeted Chemotherapy,* (Advances in Enzyme Regulation Series, Vol. 24), Pergamon Press, Oxford, 1985.

541. **Seiler, N., Jung, M. J., and Koch-Weser, J., Eds.,** *Enzyme-Activated Irreversible Inhibitors,* Elsevier/North Holland, Amsterdam, 1978.

542. **Ghisla, S., Wenz, A., and Thorpe, C.,** in *Enzyme Inhibitors,* Brodbeck, U., Ed., Verlag Chemie, Weinheim, West Germany, 1980, 43.

543. **Massey, V., Komai, H., Palmer, G., and Elion, G.,** On the mechanism of inactivation of xanthine oxidase by allopurinol and other pyrazolo [3,4-d] pyrimidines, *J. Biol. Chem.,* 245, 2837, 1970.

544. **Osawa, Y., Yarborough, C., and Osawa, Y.,** Norethisterone, a major ingredient of contraceptive pills is a suicide inhibitor of estrogen biosynthesis, *Science,* 215, 1249, 1982.

545. **Youdim, M. B. H. and Salach, J. I.,** The active site of monoamine oxidase: binding studies with [^{14}C]-acetylenic and non-acetylenic inhibitors, in *Enzyme-Activated Irreversible Inhibitors,* Seiler, N., Jung, M. J., and Koch-Weser, J., Eds., Elsevier/North Holland, Amsterdam, 1978, 232.

546. **Sandler, M., Glover, V., Elsworth, J. D., Lewinsohn, R., and Reveley, M. A.,** Monoamine oxidase inhibition: some chemical dimensions, in *Enzyme Inhibitors as Drugs,* Sandler, M., Ed., Macmillan, London, 1980, 173.

547. **Knoll, J.**, Monoamine oxidase inhibitors: chemistry and pharmacology, in *Enzyme Inhibitors as Drugs*, Sandler, M., Ed., Macmillan, London, 1980, 151.

548. **Malt, R. A., Kingsnorth, A. N., LaMuraglia, G. M., Lacaine, F., and Ross, J. S.**, Chemoprevention and chemotherapy by inhibition of ornithine decarboxylase activity and polyamine synthesis: colonic, pancreatic, mammary, and renal carcinomas, in *Advances in Enzyme Regulation*, Vol. 24, Weber, G., Ed., Pergamon Press, Oxford, 1985, 93.

549. **Heby, O.**, Ornithine decarboxylase as target of chemotherapy, in *Advances in Enzyme Regulation*, Vol. 24, Weber, G., Ed., Pergamon Press, Oxford, 1985, 103.

550. **Bitonti, A. J., McCann, P. P., and Sjoerdsma, A.**, Restriction of bacterial growth by inhibition of polyamine biosynthesis by using monofluoromethylornithine, difluoromethylarginine and dicyclohexylammonium sulphate, *Biochem. J.*, 208, 435, 1982.

551. **Fischer, J. F. and Knowles, J. R.**, The inactivation of β-lactamase by mechanism-based reagents, in *Enzyme Inhibitors as Drugs*, Sandler, M., Ed., Macmillan, London, 1980, 209.

552. **Frere, J. M., Duez, C., Dusart, J., Coyette, J., Leyh-Bouille, M., Ghuysen, J. M., Dibederg, O., and Knox, J.**, Mode of action of β-lactam antibiotics at the molecular level, in *Enzyme Inhibitors as Drugs*, Sandler, M., Ed., Macmillan, London, 1980, 183.

553. **Neidle, S. and Waring, M. J., Eds.**, *Molecular Aspects of Anti-Cancer Drug Action*, Verlag Chemie, Weinheim, West Germany, 1983.

554. **Cornelissen, A. W. C. A., Verspieren, M. P., Toulme, J.-J., Swinkels, B. W., and Borst, P.**, The common 5'-terminal sequence on trypanosome mRNAs: a target for anti-messenger oligodeoxynucleotides, *Nucleic Acids Res.*, 14, 5605, 1986.

555. **Cazenave, C., Lorean, N., Toulme, J.-J., and Helene, C.**, Anti-messenger oligodeoxynucleotides: specific inhibition of rabbit β-globin synthesis in wheat germ extracts and *Xenopus* oocytes, *Biochimie*, 68, 1063, 1986.

556. **Toulme, J. J., Krisch, H. M., Loreau, N., Thuong, N. T., and Helene, C.**, Specific inhibition of mRNA translation by complementary oligonucleotides covalently linked to intercalating agents, *Proc. Natl. Acad. Sci. U.S.A.*, 83, 1227, 1986.

557. **Blake, K. R., Murakami, A., and Miller, P. S.**, Inhibition of rabbit globin mRNA translation by sequence-specific oligodeoxyribonucleotides, *Biochemistry*, 24, 6132, 1985.

558. **Zamecnik, P. C. and Stephenson, M. L.**, Inhibition of Rous sarcoma virus replication and cell transformation by a specific oligodeoxynucleotide, *Proc. Natl. Acad. Sci. U.S.A.*, 75, 280, 1978.

559. **Stephenson, M. L. and Zamecnik, P. C.**, Inhibition of Rous sarcoma viral RNA translation by a specific oligodeoxyribonucleotides, *Proc. Natl. Acad. Sci. U.S.A.*, 75, 285, 1978.

560. **Zamecnik, P. C., Goodchild, J., Taguchi, J., and Sarin, P. S.**, Inhibition of replication and expression of human T-cell lymphotropic virus type III in cultured cells by exogenous synthetic oligonucleotides complementary to viral RNA, *Proc. Natl. Acad. Sci. U.S.A.*, 83, 4143, 1986.

561. **Blake, K. R., Murakami, A., Spitz, S. A., Glave, S. A., Reddy, M. P., Ts'o, P. O. P., and Miller, P. S.**, Hybridization arrest of globin synthesis in rabbit reticulocyte lysates and cells by oligodeoxyribonucleoside methylphosphonates, *Biochemistry*, 24, 6139, 1985.

562. **Miller, P. S., Agris, C. H., Aurelian, L., Blake, K. R., Murakami, A., Reddy, M. P., Spitz, S. A., and Ts'o, P. O. P.**, Control of ribonucleic acid function by oligonucleoside methylphosphonates, *Biochimie*, 67, 769, 1985.

563. **Smith, C. C., Aurelian, L., Reddy, M. P., Miller, P. S., and Ts'o, P. O. P.**, Antiviral effect of an oligo(nucleosidemethylphosphonate) complementary to the splice junction of herpes simplex virus type 1 immediate early pre-mRNAs 4 and 5, *Proc. Natl. Acad. Sci. U.S.A.*, 83, 2787, 1986.

564. **Geen, P. J., Pines, O., and Inouye, M.**, The role of antisense RNA in gene regulation, *Annu. Rev. Biochem.*, 55, 569, 1986.

565. **Karpova, G. G., Knorre, D. G., Ryte, A. S., and Stephanovich, L. S.**, Selective alkylation of RNA inside the cell with derivative ethyl ether of oligothymidilate bearing 2-chloroethylamino group, *FEBS Lett.*, 122, 21, 1980.

566. **Dupuis, G. and Radwan, F.**, Synthesis of two photoreactive heterobifunctional reagents derived from hexanoic acid, *Can. J. Biochem. Cell Biol.*, 61, 99, 1983.

567. **Olomucki, M., Jerram, M., Parfait, R., Bollen, A., and Gros, F.**, Synthesis and properties of a new cleavable nucleic acid-protein cross-linking reagent, *Bioorg. Chem.*, 10, 455, 1981.

568. **Wollenweber, H.-W. and Morrison, D. C.**, Synthesis and biochemical characterization of a photoactivatable, iodinatable, cleavable bacterial lipopolysaccharide derivative, *J. Biol. Chem.*, 260, 15068, 1985.

569. **Rinke, J., Meinke, M., Brimacombe, R., Fink, G., Rommel, W., and Fasold, H.**, The use of azidoaryl imidoesters in RNA-protein cross-linking studies with *Escherichia coli* ribosomes, *J. Mol. Biol.*, 137, 301, 1980.

570. **Schwartz, M. A.**, A [125I]-radiolabel transfer crosslinking reagent with a novel cleavable group, *Anal. Biochem.*, 149, 142, 1985.

571. **Sigrist, H., Haldemann, A., Sigrist-Nelson, K., and Wolf, M.,** Heterobifunctional cross-linking of bacteriorhodopsin by 4-azido-azobenzene-4′-isothiocyanate, a site-directed cleavable photoreagent, *Experientia,* 40, 612, 1984.

572. **Oste, C., Parfait, R., Bollen, A., and Crichton, R. R.,** A new nucleic acid-protein cross-linking reagent, *Mol. Gen. Genet.,* 152, 253, 1977.

573. **Escher, E. H. F., Robert, H., and Guillemette, G.,** 4-Azidoaniline, a versatile protein and peptide modifying agent for photoaffinity labeling, *Helv. Chim. Acta,* 62, 1217, 1979.

574. **Bose, K. and Bothner-By, A. A.,** Two new bifunctional protein modification reagents and their application to the study of parvalbumin, *Biochemistry,* 22, 1342, 1983.

575. **Moreland, R. B. and Dockter, M. E.,** Interaction of the "back" of yeast iso-1-cytochrome c with yeast cytochrome c oxidase, *Biochem. Biophys. Res. Commun.,* 99, 339, 1981.

576. **Seela, F.,** Synthesis of 5-azido-3-nitro-ω-bromoacetophenone — a photochemically active bifunctional reagent for the cross-linking of biopolymers, *Z. Naturforsch.,* 31c, 389, 1976.

577. **Millon, R., Ebel, J.-P., Le Coffic, F., and Ehresmann, B.,** Ribonucleic acid-protein cross-linking in *Escherichia coli* ribosomal 30S subunits by the use of two new heterobifunctional reagents: 4-azido-2,3,5,6-tetrafluoropyridine and 4-azido-3,5-dichloro-2,6-difluoropyridine, *Biochem. Biophys. Res. Commun.,* 101, 784, 1981.

578. **Hixson, S. H., Burroughs, S. F., Caputo, T. M., Crapster, B. B., Daly, M. V., Lowrie, A. W., and Wasko, M. L.,** Reaction of p-azidophenacyl iodoacetate, a photolabile reagent, with yeast alcohol dehydrogenase, *Arch. Biochem. Biophys.,* 192, 296, 1979.

579. **Klausner, Y. S., Feigenbaum, A. M., de Groot, N., and Hochberg, A. A.,** N-Bromoacetyl-p-azido-L-phenylalanine methyl ester: a bifunctional chemically and photoactivated cross-linking reagent, *Arch. Biochem. Biophys.,* 185, 151, 1978.

580. **Baba, T. and Allen, Ch. M.,** Inactivation of undecaprenylpyrophosphate synthetase with a photolabile analogue of farnesyl pyrophosphate, *Biochemistry,* 23, 1312, 1984.

581. **Fasold, H., Bäumert, H. G., and Meyer, C.,** Cross-linking reagents, in *Modern Methods in Protein Chemistry,* Vol. 2, Tschesche, H., Ed., Walter de Gruyter, Berlin, 1985, 261.

582. **Cheng, S., Merlino, G. T., and Pastan, I. H.,** A versatile method for the coupling of protein to DNA: synthesis of α_2-macroglobulin-DNA conjugates, *Nucleic Acids Res.,* 11, 659, 1983.

583. **Hultin, J.,** Selective high-efficiency cross-linking of mammalian ribosomal proteins with cleavable thiol-directed heterobifunctional reagents: identification and binding directions of major protein complexes, *Biochim. Biophys. Acta,* 872, 226, 1986.

584. **Senter, P. D., Tansey, M. J., Lambert, J. M., and Blättler, W. A.,** Novel photocleavable protein crosslinking reagents and their use in the preparation of antibody-toxin conjugates, *Photochem. Photobiol.,* 42, 231, 1985.

585. **Kitagawa, J., Shimozono, J., Aikawa, J., Yoshida, J., and Nishimura, H.,** Preparation and characterization of heterobifunctional crosslinking reagents for protein modifications, *Chem. Pharmacol. Bull.,* 29, 1130, 1981.

586. **Bernatowicz, M. S. and Matsueda, G. R.,** The N-hydroxysuccinimide ester of B_{oc}-[S-(3-nitro-2-pyridinesulfenyl]-cysteine: a heterobifunctional cross-linking agent, *Biochem. Biophys. Res. Commun.,* 132, 1046, 1985.

587. **Weltman, J. K., Johnson, S. A., Langevin, J., and Riester, E. F.,** N-succinimidyl(4-iodoacetyl)aminobenzoate: a new heterobifunctional crosslinker, *BioTechniques,* 1, 148, 1983.

588. **Tao, T., Lamkin, M., and Scheiner, C. J.,** The conformation of the C-terminal region of actin: a site-specific photocrosslinking study using benzophenone-4-maleimide, *Arch. Biochem. Biophys.,* 240, 627, 1985.

589. **Blättler, W. A., Kuenzi, B. S., Lambert, J. M., and Senter, P. D.,** New heterobifunctional protein cross-linking reagent that forms an acid-labile link, *Biochemistry,* 24, 1517, 1985.

590. **Oste, C. and Brimacombe, R.,** The use of sym-triazine trichloride in RNA-protein cross-linking studies with *Escherichia coli* ribosomal subunits, *Mol. Gen. Genet.,* 168, 81, 1979.

591. **Ohnishi, M., Sugimoto, H., Yamada, H., Imoto, T., Zaitsu, K., and Ohkura, Y.,** Heterobifunctional reagents for cross-linking of sugar with protein, *Chem. Pharm. Bull.,* 33, 674, 1985.

592. **Ulmer, E., Meinke, H., Ross, A., Fink, G., and Brimacombe, R.,** Chemical cross-linking of protein to RNA within intact ribosomal subunits from *Escherichia coli, Mol. Gen. Genet.,* 160, 183, 1978.

593. **Bechet, J. J., Dupaix, A., Yon, J., Wakselman, M., Robert, J. C., and Vilkas, M.,** Inactivation of α-chymotrypsin by a bifunctional reagent, 3,4-dihydro-3,4-dibromo-6-(bromomethyl) coumarin, *Eur. J. Biochem.,* 35, 527, 1973.

594. **Fries, R. W., Bohlken, D. P., Murch, B. P., Leidal, K. G., and Plapp, B. V.,** ω-Haloalkyl esters of 5′-adenosine monophosphate as potential active-site-directed reagents for dehydrogenases, *Archiv. Biochem. Biophys.,* 225, 110, 1983.

595. **Alston, T. A., Mela, L., and Bright, H. J.,** Inactivation of alcohol dehydrogenase by 3-butyn-1-ol, *Arch. Biochem. Biophys.,* 197, 516, 1979.

596. **Dahl, K. H., Eklund, H., and McKinley-McKee, J. S.,** Enantioselective affinity labelling of horse liver alcohol dehydrogenase. Correlation of inactivation kinetics with the three-dimensional structure of the enzyme, *Biochem. J.,* 211, 391, 1983.

597. **Dahl, K. H. and McKinley-McKee, J. S.,** Affinity labelling of alcohol dehydrogenases chemical modification of the horse liver and the yeast enzyme with α-bromo-β(5-imidazolyl)-propionic acid and 1,3-dibromoacetone, *Eur. J. Biochem.,* 81, 223, 1977.

598. **Dahl, K. H. and McKinley-McKee, J. S.,** Enzymatic catalysis in the affinity labelling of liver alcohol dehydrogenase with haloacids, *Eur. J. Biochem.,* 118, 507, 1981.

599. **Borson, W. F., Yin, S.-J., Dwulet, F. E., and Li, T.-K.,** Selective carboxymethylation of cysteine-174 of the β₂β₂ and β₁β₁ human liver alcohol dehydrogenase isoenzymes by iodoacetate, *Biochemistry,* 25, 1876, 1986.

600. **Sogin, D. C. and Plapp, B. V.,** Inactivation of horse liver alcohol dehydrogenase by modification of cysteine residue 174 with diazonium-1H-tetrazole, *Biochemistry,* 15, 1087, 1976.

601. **Woenckhaus, C., Jeck, R., and Jörnvall, H.,** Affinity labeling of yeast and liver alcohol dehydrogenases with the NAD analogue 4-(3-bromoacetylpyridinio)butyldiphosphoadenosine, *Eur. J. Biochem.,* 93, 65, 1979.

602. **Tamura, J. K., Rakov, R. D., and Cross, R. L.,** Affinity labeling of nucleotide-binding sites on kinases and dehydrogenases by pyridoxal 5'-diphospho-5'-adenosine, *J. Biol. Chem.,* 261, 4126, 1986.

603. **Biellmann, J. F., Branlant, G., Foucaud, B. Y., and Jung, M. J.,** Preparation of 3-chloroacetylpyridine adenine oligonucleotide: an alkylating analogue of NAD⁺, *FEBS Lett.,* 40, 29, 1974.

604. **Chen, S. and Guillory, R. J.,** Arylazido-β-alanine NAD⁺, an NAD⁺ photoaffinity analogue. Preparation and properties, *J. Biol. Chem.,* 252, 8990, 1977.

605. **Mansson, M.-O., Larsson, P.-O., and Mosbach, K.,** Covalent binding of an NAD analogue to liver alcohol dehydrogenase resulting in a enzyme-coenzyme complex not requiring exogenous coenzyme for activity, *Eur. J. Biochem.,* 86, 455, 1978.

606. **Jörnvall, H., Woenckhaus, C., and Johnscher, G.,** Modificaiton of alcohol dehydrogenases with a reactive coenzyme analogue. Identification of labelled residues in the horse-liver and yeast enzymes after treatment with nicotinamide-5-bromoacetyl-4-methyl-imidazol dinucleotide, *Eur. J. Biochem.,* 53, 71, 1975.

607. **Grubmeyer, C. T. and Gray, W. R.,** A cysteine residue (cysteine-116) in the histidinol binding site of histidinol dehydrogenase, *Biochemistry,* 25, 4778, 1986.

608. **Hensel, R., Mayr, U., and Woenckhaus, C.,** Affinity labelling of the allosteric site of the α-lactate dehydrogenase of *Lactobacillus casei, Eur. J. Biochem.,* 135, 359, 1983.

609. **Olson, S. T., Massey, V., Ghisla, S., and Whitfield, C. D.,** Suicide inactivation of the flavoenzyme D-lactate dehydrogenase by α-hydroxybutynoate, *Biochemistry,* 18, 4724, 1979.

610. **Gould, K. and Engel, P. C.,** Modification of mouse testicular lactate dehydrogenase by pyridoxal 5'-phosphate, *Biochem. J.,* 191, 365, 1980.

611. **Yamaguchi, M., Chen, S., and Hategi, Y.,** Photoaffinity labeling of D-(−)-β-hydroxybutyrate dehydrogenase by (arylazido)-β-alanyl-substituted nicotinamide adenine dinucleotide, *Biochemistry,* 24, 4912, 1985.

612. **Yamaguchi, M., Chen, S., and Hatefi, Y.,** Amino acid sequence of the nucleotide-binding site of D-(−)-β-hydroxybutyrate dehydrogenase labeled with arylazido-[3-³H] alanylnicotinamide adenine dinucleotide, *Biochemistry,* 25, 4864, 1986.

613. **Roy, S. and Colman, R. F.,** Affinity labeling of a lysine residue in the coenzyme binding site of pig heart mitochondrial malate dehydrogenase, *Biochemistry,* 18, 4683, 1979.

614. **Chang, G.-G. and Hsu, R. Y.,** Mechanism of pigeon liver malic enzyme. Kinetics, specificity, and half-site stoichiometry of the alkylation of a cysteinyl residue by the substrate-inhibitor bromopyruvate, *Biochemistry,* 16, 311, 1977.

615. **King, M. M. and Colman, R. F.,** Affinity labeling of nicotinamide adenine dinucleotide dependent isocitrate dehydrogenase by the 2',3'-dialdehyde derivative of adenosine 5'-diphosphate. Evidence of the formation of an unusual reaction product, *Biochemistry,* 22, 1656, 1983.

616. **Bednar, R. A., Hartman, F. C., and Colman, R. F.,** 3-bromo-2-ketoglutarate: a substrate and affinity label for diphosphopyridine nucleotide dependent isocitrate dehydrogenase, *Biochemistry,* 21, 3681, 1982.

617. **Bednar, R. A., Hartman, F. C., and Colman, R. F.,** 3,4-didehydro-2-ketoglutarate: an affinity label for diphosphopyridine nucleotide dependent isocitrate dehydrogenase, *Biochemistry,* 21, 3690, 1982.

618. **Huang, Y.-C. and Colman, R. F.,** Affinity labeling of the allosteric ADP activation site of NAD-dependent isocitrate dehydrogenase by 6-(4-bromo-2,3-dioxobutyl)thioadenosine 5'-diphosphate, *J. Biol. Chem.,* 259, 12481, 1984.

619. **Hartman, F. C.,** Interaction of isocitrate dehydrogenase with (RS)-3-bromo-2-ketoglutarate. A potential affinity label for α-ketoglutarate binding sites, *Biochemistry,* 20, 894, 1981.

620. **Bellini, T., Signorini, M., Dallocchio, F., and Rippa, M.,** Affinity labelling of the NADP⁺-binding site of glucose 6-phosphate dehydrogenase from *Candida utilis, Biochem. J.,* 183, 297, 1979.

621. **Haghighi, B., Flynn, G., and Levy, H. R.,** Glucose-6-phosphate dehydrogenase from *Leuconostoc mesenteroides.* Isolation and sequence of a peptide containing an essential lysine, *Biochemistry,* 21, 6415, 1982.

622. **Battais, E., Terouanne, B., Nicolas, J. C., Descomps, B., and Crastes de Paulet, A.,** Characterization of an associate 17-β-hydroxysteroid dehydrogenase activity and affinity labelling of the 3-α-hydroxysteroid dehydrogenase of *Pseudomonas testosteroni, Biochimie,* 59, 909, 1977.

623. **Strickler, R. C., Sweet, F., and Warren, J. C.,** Affinity labeling of steroid binding sites. Study of the active site of 20β-hydrosteroid dehydrogenase with 2α-bromoacetoxyprogesterone and 11α-bromoacetoxyprogesterone, *J. Biol. Chem.,* 250, 7656, 1975.

624. **Samant, B. R. and Sweet, F.,** 5'-Bromoacetamido-5'-deoxyadenosine. A novel reagent for labeling adenine nucleotide sites in proteins, *J. Biol. Chem.,* 258, 12779, 1983.

625. **Murdock, G. L., Chin, C.-C., and Warren, J. C.,** Human placental estradiol 17β-dehydrogenase: sequence of a histidine-bearing peptide in the catalytic region, *Biochemistry,* 25, 641, 1986.

626. **Groman, E. V., Schulz, R. M., and Engel, L. L.,** Catalytic competence, a new criterion for affinity labelling. Demonstration on the reversible enzymatic interconversion of estrone and estradiol-17β covalently bound to human placental estradiol-17β dehydrogenase, *J. Biol. Chem.,* 250, 5450, 1975.

627. **Thomas, J. L., LaRochelle, M. C., Covey, D. F., and Strickler, R. C.,** Inactivation of human placental 17β, 20α-hydroxysteroid dehydrogenase by 16-methylene estrone, an affinity alkylator enzymatically generated from 16-methylene estradiol-17β, *J. Biol. Chem.,* 258, 11500, 1983.

628. **Bhatnagar, Y. M., Chin, C.-C., and Warren, J. C.,** Synthesis of 4-bromoacetamidoesterone methyl ether and study of the steroid binding site of human placental estradiol 17β-dehydrogenase, *J. Biol. Chem.,* 253, 811, 1978.

629. **Thomas, J. L. and Strickler, R. C.,** Human placental 17β-estradiol dehydrogenase and 20α-hydroxysteroid dehydrogenase. Studies with β-bromoacetoxyprogesterone, *J. Biol. Chem.,* 258, 1587, 1983.

630. **Chin, C.-C., Murdock, G. L., and Warren, J. C.,** Identification of two histidyl residues in the active site of human placental estradiol 17β-dehydrogenase, *Biochemistry,* 21, 3322, 1982.

631. **Pons, M., Nicolas, J. C., Boussioux, A. M., Descomps, B., and Crastes de Paulet, A.,** Affinity labelling of an histidine of the active site of the human placental 17β-oestradiol dehydrogenase, *FEBS Lett.,* 36, 23, 1973.

632. **Biellmann, J.-F., Goulas, P. R., Nicolas, J.-C., Descomps, B., and Crastes de Paulet, A.,** Alkylation of estradiol 17β-dehydrogenase from human placenta with 3-chloroacetyl-pyridine-adenine dinucleotide phosphate, *Eur. J. Biochem.,* 99, 81, 1979.

633. **Inano, H. and Tamaoki, B.,** Affinity labeling of the cofactor-binding site of estradiol 17β-dehydrogenase of human placenta by 5'-p-fluorosulfonylbenzoyl adenosine, *J. Steroid Biochem.,* 22, 681, 1985.

634. **Nichols, C. S. and Cromartie, T. H.,** Irreversible inactivation of the flavoenzyme alkohol oxidase with acetylenic alcohols, *Biochem. Biophys. Res. Commun.,* 97, 216, 1980.

635. **Biellmann, J.-F., Eid, P., Hirth, C., and Jörnvall, H.,** Aspartate-β-semialdehyde dehydrogenase from *Escherichia coli.* Affinity labeling with the substrate analogue α-2-amino-4-oxo-5-chloropentanoic acids: an example of half-site reactivity, *Eur. J. Biochem.,* 104, 59, 1980.

636. **Biellmann, J.-F., Eid, P., and Hirth, C.,** Affinity labeling of the *Escherichia coli* aspartate-β-semialdehyde dehydrogenase with an alkylating coenzyme analogue half-site reactivity and competition with the substrate alkylating analogue, *Eur. J. Biochem.,* 104, 65, 1980.

637. **Bayne, S., Sund, H., and Guillory, R. J.,** Photodependent incorporation of arylazido-β-alanyl-NAD$^+$ into the coenzyme binding site of yeast glyceraldehyde-3-phosphate dehydrogenase, *Biochimie,* 63, 569, 1981.

638. **Branlant, G., Eiler, B., Wallen, L., and Biellmann, J.-F.,** Affinity labeling of glyceraldehyde-3-phosphate dehydrogenase from sturgeon and *Bacillus stearothermophilus* by 3-chloroacetyl-pyridine-adenine dinucleotide. Kinetic studies, *Eur. J. Biochem.,* 127, 519, 1982.

639. **Dietz, G., Woenckhaus, C., Jaenicke, R., and Schuster, I.,** Modification of glyceraldehyde-3-phosphate dehydrogenase from rabbit skeletal muscle by 3-(3-bromacetylpyridinio)-propyl-adenosine pyrophosphate, *Z. Naturforsch.* 32c, 85, 1977.

640. **Gilbert, H. J. and Drabble, W. T.,** Active-site modification of native and mutant forms of inosine 5'-monophosphate dehydrogenase from *Escherichia coli* K12, *Biochem. J.,* 191, 533, 1980.

641. **Gomes, B., Fendrich, G., and Abeles, R. H.,** Mechanism of action of glutaryl-CoA and butyryl-CoA dehydrogenases. Purification of glutaryl-CoA dehydrogenase, *Biochemistry,* 20, 1481, 1981.

642. **Fendrich, G. and Abeles, R. H.,** Mechanism of action of butyryl-CoA dehydrogenase: reaction with acetylenic, olefinic and fluorinated substrate analogues, *Biochemistry,* 21, 6685, 1982.

643. **Wenz, A., Thorpe, C., and Ghisla, S.,** Inactivation of general acyl-CoA dehydrogenase from pig kidney by a metabolite of hypoglycin A., *J. Biol. Chem.,* 256, 9809, 1981.

644. **Frerman, F. E., Miziorko, H. M., and Beckmann, J. D.,** Enzyme-activated inhibitors, alternate substrates, and a dead end inhibitors of the general acyl-CoA dehydrogenase, *J. Biol. Chem.,* 255, 11192, 1980.

250 Affinity Modification of Biopolymers

645. **Shaw, L. and Engel, P. C.**, The suicide inactivation of ox liver short-chain acyl-CoA dehydrogenase by propionyl-CoA, *Biochem. J.*, 230, 723, 1985.
646. **Liang, T., Cheung, A. H., Reynolds, G. F., and Rasmusson, G. H.**, Photoaffinity labeling of steroid 5α-reductase of rat liver and prostate microsomes, *J. Biol. Chem.*, 260, 4890, 1985.
647. **Jacobson, M. A. and Colman, R. F.**, Isolation and identification of a tyrosyl peptide labeled by 5'-[p-(fluorosulfonyl)benzoyl]-1,N⁶-ethenoadenosine at a GTP site of glutamate dehydrogenase, *Biochemistry*, 23, 6377, 1984.
648. **Koberstein, R., Cobianchi, L., and Sund, H.**, Interaction of the photoaffinity labels 8-azido-ADP with glutamate dehydrogenase, *FEBS Lett.*, 64, 176, 1976.
649. **Michel, F., Pons, M., Descomps, B., and Crastes de Paulet, A.**, Affinity labelling of the estrogen binding site of glutamate dehydrogenase with iodoacetyldiethylstilbestrol, *Eur. J. Biochem.*, 84, 267, 1978.
650. **Batra, S. P. and Colman, R. F.**, Isolation and identification of cysteinyl peptide labeled by 6-[(4-bromo-2,3-dioxobutyl)thio]-6-deaminoadenosine 5'-diphosphate in the reduced diphosphopyridine nucleotide inhibitory site of glutamate dehydrogenase, *Biochemistry*, 25, 3508, 1986.
651. **Mäntsälä, P. and Zalkin, H.**, Glutamate synthase. Properties of the glutamine-dependent activity., *J. Biol. Chem.*, 251, 3294, 1976.
652. **Ronchi, S., Galliano, M., Minchiotti, L., Curti, B., Rudie, N. G., Porter, D. J., and Bright, H. J.**, An active site tyrosine-containing heptapeptide from D-amino acid oxidase, *J. Biol. Chem.*, 255, 6044, 1980.
653. **Paech, C., Salach, J. I., and Thomas, P. S.**, Suicide inactivation of monoamine oxidase by trans-phenylcyclopropylamine, *J. Biol. Chem.*, 255, 2700, 1980.
654. **Silverman, R. B. and Zieske, P. A.**, 1-Phenylcyclobutylamine, the first of a new class of monoamine oxidase inactivators. Further evidence for a radical intermediate, *Biochemistry*, 25, 341, 1986.
655. **Silverman, R. B. and Hoffman, S. J.**, Mechanism of inactivation of mitochondrial monoamine oxidase by N-cyclopropyl-N-arylalkyl amines, *J. Am. Chem. Soc.*, 102, 884, 1980.
656. **Silverman, R. B. and Hoffman, S. J.**, N-(1-methyl)cyclopropylbenzylamine: a novel inactivator of mitochondrial monoamine oxidase, *Biochem. Biophys. Res. Commun.*, 101, 1396, 1981.
657. **Tipton, K. F., Fowler, C. J., McCrodden, J. M., and Strolin Benedetti, M.**, The enzyme-activated irreversible inhibition of type B monoamine oxidase by 3-{4-[(3-chlorophenyl)methoxy]phenyl}-5-[(methylamino) methyl]-2-oxazolidinone methanesulphonate (compound MD 780236) and the enzyme-catalysed oxidation of this compound as competing reactions, *Biochem. J.*, 209, 235, 1983.
658. **Yu, P. H.**, Studies on the pargyline-binding site of different types of monoamine oxidase, *Can. J. Biochem.*, 59, 5234, 1981.
659. **Edwards, D. J. and Pak, K. Y.**, Selective radiochemical labeling of types A and B active sites of rat liver monoamine oxidase, *Biochem. Biophys. Res. Commun.*, 86, 350, 1979.
660. **Hellerman, L. and Erwin, V. G.**, Mitochondrial monoamine oxidase. Action of various inhibitors for the bovine kidney enzyme catalytic mechanism, *J. Biol. Chem.*, 243, 5234, 1968.
661. **Hevey, R. C., Babson, J., Maycock, A. L., and Abeles, R. H.**, Highly specific enzyme inhibitors. Inhibition of plasma amine oxidase, *J. Am. Chem. Soc.*, 95, 6125, 1973.
662. **Kumar, A. A., Mangum, J. H., Blankenship, D. T., and Freisheim, J. H.**, Affinity labeling of chicken liver dihydrofolate reductase by a substituted 4,6-diaminodihydrotriazine bearing a terminal sulfonyl fluoride, *J. Biol. Chem.*, 256, 8970, 1981.
663. **Sumner, J., Jencks, D. A., Khani, S., and Matthews, R. G.**, Photoaffinity labeling of methylenetetrahydrofolate reductase with 8-azido-S-adenosylmethionine, *J. Biol. Chem.*, 261, 7697, 1986.
664. **Thome, F., Pho, D. B., and Olomucki, A.**, Bromopyruvate, a potential affinity label for octopine dehydrogenase, *Biochimie*, 67, 249, 1985.
665. **Phelps, D. C. and Hatefi, Y.**, Mitochondrial nicotinamide nucleotide transhydrogenase: active site modification by 5'-p-(fluorosulfonyl)benzoyl adenosine, *Biochemistry*, 24, 3503, 1985.
666. **Almeda, S., Bing, D. H., Laura, R., and Friedman, P. A.**, Photoaffinity inhibition of rat liver NAD(P)H dehydrogenase by 3-(α-acetonyl-p-azidobenzyl)-4-hydroxycoumarin, *Biochemistry*, 20, 3731, 1981.
667. **Earley, F. G. P. and Ragan, C. I.**, Photoaffinity labelling of mitochondrial NADH dehydrogenase with arylazidoamorphigenin, an analogue of rotenone, *Biochem. J.*, 224, 525, 1984.
668. **Chen, S. and Guillori, R. J.**, Studies on the interaction of arylazido-β-alanyl NAD⁺ with the mitochondrial NADH dehydrogenase, *J. Biol. Chem.*, 256, 8318, 1981.
669. **Guillori, R. J., Jeng, S. J., and Chen, S.**, Current application of the photoaffinity technique to the study of the structure of complex I, *Ann. N.Y. Acad. Sci.*, 346, 244, 1980.
670. **Bisson, R., Steffens, G. C. M., Capaldi, R. A., and Buse, G.**, Mapping of the cytochrome c binding site on cytochrome c oxydase, *FEBS Lett.*, 144, 359, 1982.
671. **Erecinska, M., Oshino, R., and Wilson, D. F.**, Binding of cytochrome c to cytochrome c-oxidase in intact mitochondria. A study with radioactive photoaffinity-labeled cytochrome c, *Biochem. Biophys. Res. Commun.*, 92, 743, 1980.

672. **Montecucco, C., Schiavo, G., and Bisson, R.,** ATP binding to bovine heart cytochrome c oxidase, *Biochem. J.,* 234, 241, 1986.

673. **Yu, L. and Yu, C.-A.,** The interaction of arylazido ubiquinone derivative with mitochondrial ubiquinol-cytochrome c reductase, *J. Biol. Chem.,* 257, 10215, 1982.

674. **Ho, S. H. K., Das Gupta, U., and Rieske, J. S.,** Detection of antimycin-binding subunits of complex III by photoaffinity-labeling with an azido derivative of antimycin, *J. Bioenerg. Biomembr.,* 17, 269, 1985.

675. **Bisson, R. and Capaldi, R. A.,** Binding of arylazido cytochrome c to yeast cytochrome c peroxydase, *J. Biol. Chem.,* 256, 4362, 1981.

676. **Mol, J. A., Docter, R., Kaptein, E., Jansen, G., Hennemann, G., and Visser, T. J.,** Inactivation and affinity-labeling of rat liver iodothyronine deiodinase with N-bromoacetyl-3,3',5'-triiodothyronine, *Biochem. Biophys. Res. Commun.,* 124, 475, 1984.

677. **Downing, D. T., Ahern, D. G., and Bachta, M.,** Enzyme inhibition by acetylenic compounds, *Biochim. Biophys. Res. Commun.,* 40, 218, 1970.

678. **Schonbrunn, A., Abeles, R. H., Walsh, C. T., Ghisla, S., Ogata, H., and Massey, V.,** The structure of the covalent flavin adduct formed between lactate oxidase and suicide substrate 2-hydroxy-3-butynoate, *Biochemistry,* 15, 1798, 1976.

679. **de Waal, A., de Jong, L., Hartog, A. F., and Kemp, A.,** Photoaffinity labeling of peptide binding sites of prolyl 4-hydroxylase with N-(4-azido-2-nitrophenyl) glycyl-(Pro-Pro-Gly)₅, *Biochemistry,* 24, 6493, 1985.

680. **Gan, L.-S. L., Lu, J.-Y. L., Hershkowitz, D. M., and Alworth, W. L.,** Effects of acetylenic and olefinic pyrenes upon cytochrome P-450 dependent benzo[a]pyrene hydroxylase activity in liver microsomes, *Biochem. Biophys. Res. Commun.,* 129, 591, 1985.

681. **Fitzpatrick, P. F., Flory, D. R., Jr., and Villafranca, J. J.,** 3-Phenylpropenes as mechanism-based inhibitors of dopamine β-hydroxylase: evidence for a radical mechanism, *Biochemistry,* 24, 2108, 1985.

682. **May, S. W., Mueller, P. W., Padgette, S. R., Herman, H. H., and Phillips, R. S.,** Dopamine-B-hydroxylase: suicide inhibition by the novel olefinic substrate, 1-phenyl-1-aminomethylethene, *Biochem. Biophys. Res. Commun.,* 110, 161, 1983.

683. **Tsai, P. K.,** The inactivation and covalent modification of ribonucleotide reductase by the 2',3'-dialdehyde derivatives of ADP and CDP, *Fed. Proc. Fed. Am. Soc. Exp. Biol.,* 40, 1867, 1981.

684. **Eriksson, S. and Sjöberg, B.-M.,** A photoaffinity-labeled allosteric site in *Escherichia coli* ribonucleotide reductase, *J. Biol. Chem.,* 261, 1878, 1986.

685. **Caras, I. W. and Martin, D. W., Jr.,** Direct photoaffinity labeling of an allosteric site of subunit protein M1 of mouse ribonucleotide reductase by dATP, *J. Biol. Chem.,* 257, 9508, 1982.

686. **Harris, G., Ator, M., and Stubbe, J.,** Mechanism of inactivation of *Escherichia coli* and *Lactobacillus leichmannii* ribonucleotide reductase by 2'-chloro-2'-deoxynucleotides: evidence for generation of 2-methylene-3(2H)-furanone, *Biochemistry,* 23, 5214, 1984.

687. **Thelander, L. and Larsson, B.,** Active site of ribonucleoside diphosphate reductase from *Escherichia coli, J. Biol. Chem.,* 251, 1398, 1976.

688. **Chan, R. L. and Carrillo, N.,** Affinity labeling of spinach ferredoxin-NADP⁺ oxidoreductase with periodate oxidized NADP⁺, *Arch Biochem. Biophys.,* 229, 340, 1984.

689. **Hyman, M. R. and Wood, P. M.,** Suicidal inactivation and labelling of ammonia mono-oxygenase by acetylene, *Biochem. J.,* 227, 719, 1985.

690. **Ortiz de Montellano, P. R., Mico, B. A., Mathews, J. M., Kunze, K. L., Miwa, G. T., and Lu, A. Y. H.,** Selective inactivation of cytochrome P-450 isozymes by suicide substrates, *Arch. Biochem. Biophys.,* 210, 717, 1981.

691. **Halpert, J.,** Covalent modification of lysine during the suicide inactivation of rat liver cytochrome P-450 by chloramphenicol, *Biochem. Pharmacol.,* 30, 875, 1981.

692. **Miziorko, H. M., Ahmad, P. M., Ahmad, F., and Behnke, C. E.,** Active site directed inactivation of rat mammary gland fatty acid synthase by 3-chloropropionyl Coenzyme A, *Biochemistry,* 25, 468, 1986.

693. **Tu, S.-C. and Henkin, J.,** Characterization of the aldehyde binding site of bacterial luciferase by photoaffinity labeling, *Biochemistry,* 22, 519, 1983.

694. **Lee, Y., Esch, F. S., and DeLuca, M. A.,** Identification of lysine residue at a nucleotide binding site in the firefly luciferase with p-fluorosulfonyl [¹⁴C] benzoyl-5'-adenosine, *Biochemistry,* 20, 1253, 1981.

695. **Tang, S.-S., Simpson, D. E., and Kagan, H. M.,** β-Substituted ethylamine derivatives as suicide inhibitors of lysyl oxidase, *J. Biol. Chem.,* 259, 975, 1984.

696. **Stepp, L. R. and Reed, L. J.,** Active-site modification of mammalian pyruvate dehydrogenase by pyridoxal 5'-phosphate, *Biochemistry,* 24, 7187, 1985.

697. **Blohm, T. R., Metcalf, B. W., Laudhlin, M. E., Sjoerdsma, A., and Schatzman, G. L.,** Inhibition of testosterone 5α-reductase by a proposed enzyme-activated, active site-directed inhibitor, *Biochem. Biophys. Res. Commun.,* 95, 273, 1980.

698. **Van Zyl, J. M. and Van der Walt, B. J.,** Solubilization of peroxidase from porcine thyroids and characterisation by photoaffinity labelling, *Biochim. Biophys. Acta,* 787, 174, 1984.

699. **Covey, D. F., Hood, W. F., and Parikh, V. D.,** 10β-Propynyl-substituted steroids. Mechanism-based enzyme-activated irreversible inhibitors of estrogen biosynthesis, *J. Biol. Chem., 256,* 1076, 1981.
700. **Marcotte, P. A. and Robinson, C. H.,** Design of mechanism-based inactivators of human placental aromatase, *Cancer Res. Suppl., 42,* 3322s, 1982.
701. **Marcotte, P. A. and Robinson, C. H.,** Inhibition and inactivation of estrogen synthetase (aromatase) by fluorinated substrate analogues, *Biochemistry, 21,* 2773, 1982.
702. **Tan, L. and Petit, A.,** Inactivation of human placental aromatase by 6α- and 6β-hydroperoxyandroste-nedione, *Biochem. Biophys. Res. Commun., 128,* 613, 1985.
703. **Metcalf, B. W., Wright, C. L., Burkhart, J. P. and Johnston, J. O.,** Substrate-induced inactivation of aromatase by allenic and acetylenic steroids, *J. Am. Chem. Soc., 103,* 3221, 1981.
704. **Numazawa, M. and Osawa, Y.,** Synthesis and some reactions of 6-bromoandrogens: potential affinity ligand and inactivator of estrogen synthetase, *Steroids, 34,* 347, 1979.
705. **Osawa, Y., Yarborough, C., and Osawa, Y.,** Norethisterone, a major ingredient of contraceptive pills, is a suicide inhibitor of estrogen biosynthesis, *Science, 215,* 1249, 1982.
706. **Wang, E. A., Kallen, R., and Walsh, C.,** Mechanism-based inactivation of serine transhydroxymethylases by D-fluoroalanine and related amino acids, *J. Biol. Chem., 256,* 6917, 1981.
707. **Kaiser, I. I., Kladianos, D. M., Van Kirk, E. A., and Haley, B. E.,** Photoaffinity labeling of catechol O-methyltransferase with 8-azido-S-adenosyl-methionine, *J. Biol. Chem., 258,* 1747, 1983.
708. **Borchardt, R. T.,** Catechol O-methyltransferase. VII. Affinity labeling with the oxidation products of 6-aminodopamine, *J. Med. Chem., 19,* 30, 1976.
709. **Borchardt, R. T.,** Affinity labeling of catechol O-methyltransferase by the oxidation products of 6-hydroxydopamine, *Mol. Pharmacol., 11,* 436, 1985.
710. **Borchardt, R. T. and Thakker, D. R.,** Catechol O-methyltransferase. VI. Affinity labeling with N-haloacetyl-3,5-dimethoxy-4-hydroxyphenylalkylamines, *J. Med. Chem., 18,* 152, 1975.
711. **Borchardt, R. T., Wu, Y. S., and Wu, B. S.,** Affinity labeling of histamine N-methyltransferase by 2′,3′-dialdehyde derivatives of S-adenosylhomocysteine and S-adenosylmethionine. Kinetics of inactivation, *Biochemistry, 17,* 4145, 1978.
712. **Fujioka, M. and Ishiguro, Y.,** Reaction of rat liver glycine methyltransferase with 5′-p-fluorosulfonyl-benzoyladenosine, *J. Biol. Chem., 261,* 6346, 1986.
713. **Hurst, J. H., Billingsley, M. L., and Lovenberg, W.,** Photoaffinity labelling of methyltransferase enzymes with S-adenosylmethionine: effects of methyl acceptor substrates, *Biochem. Biophys. Res. Commun., 122,* 499, 1984.
714. **Maggiora, L., Chang, C. C. T.-C., Torrence, P. F., and Mertes, M. P.,** 5-Nitro-2′-deoxyuridine 5′-phosphate: a mechanism-based inhibitor of thymidylate synthetase, *J. Am. Chem. Soc., 103,* 3192, 1981.
715. **Wataya, Y., Matsuda, A., and Santi, D. V.,** Interaction of thymidylate synthetase with 5-nitro-2′-deoxyuridylate, *J. Biol. Chem., 255,* 5538, 1980.
716. **Kalman, T. I. and Goldman, D.,** Inactivation of thymidylate synthetase by a novel mechanism-based enzyme inhibitor: 1-(β-D-2′-deoxyribofuranosyl)8-azapurin-2-one-5′-monophosphate, *Biochem. Biophys. Res. Commun., 102,* 682, 1981.
717. **DeClercq, E., Balzarini, J., Chang, C. T.-C., Bigge, C. F., Kalaritis, P., and Mertes, M. P.,** 5(E)(3-azidostyryl)-2′-deoxyuridine 5′-phosphate is a photoactivated inhibitor of thymidylate synthetase, *Biochem. Biophys. Res. Commun., 97,* 1068, 1980.
718. **Sani, B. P., Vaid, A., Cory, J. G., Brockman, R. W., Elliott, R. D., and Montgomery, J. A.,** 5′-Haloacetamido-5′-deoxythymidines novel inhibitors of thymidylate synthase, *Biochim. Biophys. Acta, 881,* 175, 1986.
719. **Poto, E. M. and Wood, H. G.,** Photoaffinity labeling and stoichiometry of the coenzyme A ester sites of transcarboxylase, *J. Biol. Chem., 253,* 2979, 1978.
720. **Lauritzen, A. M. and Lipscomb, W. N.,** Modification of three active site lysine residues in the catalytic subunit of aspartate transcarbamylase by D- and L-bromosuccinate, *J. Biol. Chem., 257,* 1312, 1982.
721. **Mauro, J. M., Lewis, R. V., and Barden, R. E.,** Photoaffinity labelling of carnitine acetyltransferase with S-(p-azidophenacyl)-thiocarnitine, *Biochem. J., 237,* 533, 1986.
722. **Holland, P. C., Clark, M. G., and Bloxham., D. P.,** Inactivation of pig heart thiolase by 3-butynoyl coenzyme A, 3-pentynoyl coenzyme A, and 4-bromocrotonyl coenzyme A, *Biochemistry, 12,* 3309, 1973.
723. **Owens, M. S. and Barden, R. E.,** S-(4-bromo-2,3-dioxobutyl)-CoA: an affinity label for certain enzymes that bind acetyl-CoA, *Arch. Biochem. Biophys., 187,* 299, 1978.
724. **Lau, E. P., Haley, B. E., and Barden, R. E.,** Photoaffinity labelling of acyl-coenzyme A: glycine N-acyltransferase with p-azidobenzoyl-coenzyme A, *Biochemistry, 16,* 2581, 1977.
725. **Lau, E. P., Haley, B. E., and Barden, R. E.,** The 8-azidoadenine analog of S-benzoyl-(3′-dephospho) coenzyme A — a photoaffinity label for acyl CoA:glycine N-acyltransferase, *Biochem. Biophys. Res. Commun., 76,* 843, 1977.
726. **Kleanthous, C., Cullis, P. M., and Shaw, W. V.,** 3-(Bromoacetyl)chloramphenicol, an active site directed inhibitor for chloramphenicol acetyltransferase, *Biochemistry, 24,* 5307, 1985.

727. **Inoue, M., Horiuchi, S., and Morino, Y.,** Affinity labeling of rat-kidney β-glutamyl transpeptidase, *Eur. J. Biochem.,* 73, 335, 1977.

728. **Tate, S. S. and Meister, A.,** Affinity labeling of γ-glutamyl transpeptidase and location of the γ-glutamyl binding site on the light subunit, *Proc. Natl. Acad. Sci. U.S.A.,* 74, 931, 1977.

729. **Takahashi, S., Zukin, R. S., and Steinman, H. M.,** γ-Glutamyl transpeptidase from WI-38 fibroblasts: purification and active site modification studies, *Arch. Biochem. Biophys.,* 207, 87, 1981.

730. **Inoue, M., Horiuchi, S., and Morino, Y.,** Inactivation of γ-glutamyl transpeptidase by phenylmethanesulfonyl flouride, a specific inactivator of serine enzymes, *Biochem. Biophys. Res. Commun.,* 82, 1183, 1978.

731. **Heagy, W., Danner, J., Lenhoff, H., Cobb, M. H., and Marshall, G.,** Araserine affinity labelling of γ-glutamyl transferase of *Hydra attenuata* without inactivation of the glutathione receptor, *J. Exp. Biol.,* 101, 287, 1982.

732. **Shimomura, S., Nakano, K., and Fukui, T.,** Affinity labeling of the cofactor site in glycogen phosphorylase b with a pyridoxal 5′-phosphate analog, *Biochem. Biophys. Res. Commun.,* 82, 462, 1978.

733. **Seery, V. L.,** Interaction of glycogen phosphorylase with 8-azidoadenosine 5′-monophosphate, a photoaffinity analog of AMP, *Biochim. Biophys. Acta,* 612, 195, 1980.

734. **Anderson, R. A., Parrish, R. F., and Graves, D. J.,** Chemistry of the adenosine monophosphate site of rabbit muscle glycogen phosphorylase. II. Properties of 8-[m-(m-fluorosulfonylbenzamido) benzylthio] adenine-modified phosphorylase, *Biochemistry,* 12, 1901, 1973.

735. **Anderson, R. A. and Graves, D. J.,** Chemistry of the adenosine monophosphate site of rabbit muscle glycogen phosphorylase. Hydrophobic nature and affinity labeling of the allosteric site, *Biochemistry,* 12, 1895, 1973.

736. **Tagaya, M., Nakano, K., and Fukui, T.,** A new affinity labelling reagent for the active site of glycogen synthase, *J. Biol. Chem.,* 260, 6670, 1985.

737. **Lee, T. K., Wong, L.-J. C., and Wong, S. S.,** Photoaffinity labeling of lactose synthase with a UDP-galactose analogue, *J. Biol. Chem.,* 258, 13166, 1983.

738. **Burkhardt, A. E., Russo, S. O., Rinehardt, C. G., and Loudon, G. M.,** The interaction of N-acetylglucosamine and an affinity-label analogue with α-lactalbumine and lactose synthetase, *Biochemistry,* 14, 5465, 1975.

739. **Powell, J. T. and Brew, K.,** Affinity labeling of bovine colostrum galactosyltransferase with a uridine 5′-diphosphate derivative, *Biochemistry,* 15, 3499, 1976.

740. **Tso, J. Y., Hermodson, M. A., and Zalkin, H.,** Glutamine phosphoribosylpyrophosphate amidotransferase from cloned *Escherichia coli* pur F, *J. Biol. Chem.,* 257, 3532, 1982.

741. **Bause, E.,** Epoxyalkyl peptide derivatives as active site directed inhibitors of asparagine-N-glycosyltransferases, *Hoppe Seyler's Z. Physiol. Chem.,* 363, 1024, 1982.

742. **Rilling, H. C.,** Photoaffinity substrate analogs for eukaryotic prenyltransferase, *Methods Enzymol.,* 110, 125, 1985.

743. **Brems, D. N., Bruenger, E., and Rilling, H. C.,** Isolation and characterization of a photoaffinity-labeled peptide from the catalytic site of prenyltransferase, *Biochemistry,* 20, 3711, 1981.

744. **Allen, C. M. and Baba, T.,** Photolabile analog of the allylic pyrophosphate substrate of prenyltransferases, *Methods, Enzymol.,* 110, 117, 1985.

745. **Seddon, A. P., Bunni, M., and Douglas, K. T.,** Photoaffinity labelling by S-(p-azidophenacyl)-glutathione of glyoxalase II and glutathione S-transferase, *Biochem. Biophys. Res. Commun.,* 95, 446, 1980.

746. **Tanase, S. and Morino, Y.,** Irreversible inactivation of aspartate aminotransferases during transamination with L-propargylglycine, *Biochem. Biophys. Res. Commun.,* 68, 1301, 1976.

747. **Gehring, H., Rando, R. R., and Christen, P.,** Active-site labeling of aspartate aminotransferases by the β,γ-unsaturated amino acid vinylglycine, *Biochemistry,* 16, 4832, 1977.

748. **Mattingly, J. R., Jr., Farach, H. A., Jr., and Martinez-Carrion, M.,** Properties of the active site lysyl residue of mitochondrial aspartate aminotransferase in solution affinity labeling with a coenzyme analog, *J. Biol. Chem.,* 258, 6243, 1983.

749. **Okamoto, M. and Morino, Y.,** Selective modification of the mitochondrial isozyme of aspartate aminotransferase by β-bromopropionate. Inactivation process and properties of inactivated enzyme, *Biochemistry,* 11, 3188, 1972.

750. **Riva, F., Carotti, D. B., Giartosio, A., and Turano, C.,** Different reactivity of mitochondrial and cytoplasmic aspartate aminotransferases toward an affinity labeling reagent analog of the coenzyme, *J. Biol. Chem.,* 255, 9230, 1980.

751. **Morino, Y., Kojima, H., and Tanase, S.,** Affinity labeling of alanine aminotransferase by 3-chloro-L-alanine, *J. Biol. Chem.,* 254, 279, 1979.

752. **Burnett, G., Marcotte, P., and Walsh,** Mechanism-based inactivation of pig heart L-alamine transaminase by L-propargylglycine. Half-site reactivity, *J. Biol. Chem.,* 255, 3487, 1980.

753. **Alston, T. A., Porter, D. J. T., Mela, L., and Bright, H. J.,** Inactivation of alanine aminotransferase by the neurotoxin β-cyano-L-alanine, *Biochem. Biophys. Res. Commun.,* 92, 299, 1980.

754. **Jung, M. J. and Seiler, N.**, Enzyme-activated irreversible inhibitors of L-ornithine:2-oxoacid aminotransferase. Demonstration of mechanistic features of the inhibition of ornithine aminotransferase by 4-aminohex-5-ynoic acid and gabaculine and correlation with in vivo activity, *J. Biol. Chem.*, 253, 7431, 1978.

755. **Metcalf, B. W. and Jung, M. J.**, Molecular basis for the irreversible inhibition of 4-aminobutyric acid: 2-oxoglutarate and L-ornithine:2-oxoacid aminotransferases by 3-amino-1,5-cyclohexadienyl carboxylic acid (isogabaculine), *Mol. Pharmacol.*, 16, 539, 1979.

756. **Silverman, R. B. and Levy, M. A.**, Irreversible inactivation of pig brain γ-aminobutyric acid-α-ketoglutarate transaminase by 4-amino-5-halopentanoic acids, *Biochem. Biophys. Res. Commun.*, 95, 250, 1980.

757. **Lippert, B., Metcalf, B. W., and Resvick, R. J.**, Enzyme-activated irreversible inhibition of rat and mouse brain 4-aminobutyric acid-α-ketoglutarate transaminase by 5-fluoro-4-oxo-pentanoic acid, *Biochem. Biophys. Res. Commun.*, 108, 146, 1982.

758. **Lippert, B., Metcalf, B. W., Jung, M. J., and Casara, P.**, 4-Amino-hex-5-enoic acid, a selective catalytic inhibitor, of 4-aminobutyric acid aminotransferase in mammalian brain, *Eur. J. Biochem.*, 74, 441, 1977.

759. **Subbarao, B. and Kenkare, U. W.**, Inactivation of brain hexokinase by an adenosine 5'-triphosphate analog, *Arch. Biochem. Biophys.*, 181, 19, 1977.

760. **Weng, L., Heinrikson, R. L., and Mansour, T. E.**, Amino acid sequence at the allosteric site of sheep heart phosphofructokinase, *J. Biol. Chem.*, 255, 1492, 1980.

761. **Craig, D. W. and Hammes, G. G.**, Structural mapping of rabbit muscle phosphofructokinase. Distance between the adenosine cyclic 3',5'-phosphate binding site and a reactive sulfhydryl group, *Biochemistry*, 19, 330, 1980.

762. **Ferguson, J. J., Jr. and MacInnes, M.**, Direct photoaffinity labeling of phosphofructokinase using adenosine derivatives, *Ann. N.Y. Acad. Sci.*, 346, 31, 1980.

763. **Brunswick, D. J. and Cooperman, B. S.**, Photo-affinity labels for adenosine 3':5'-cyclic monophosphate, *Proc. Natl. Acad. Sci. U.S.A.*, 68, 1801, 1971.

764. **Chen, M. S., Chang, P. K., and Prusoff, W. H.**, Photochemical studies and ultraviolet sensitization of *Escherichia coli* thymidylat kinase by various halogenated substrate analog, *J. Biol. Chem.*, 251, 6555, 1976.

765. **Hoyer, P. B., Owens, J. R., and Haley, B. E.**, The use of photoaffinity probes to elucidate molecular mechanisms of nucleotide-regulated phenomena, *Ann. N.Y. Acad. Sci.*, 346, 280, 1980.

766. **Liu, A. Y.-C.**, Differentiation-specific increase of cAMP-dependent protein kinase in the 3T3-L1 cells, *J. Biol. Chem.*, 257, 298, 1982.

767. **Pomerantz, A. H., Rudolph, S. A., Haley, B. E., and Greengard, P.**, Photoaffinity labeling of a protein kinase from bovine brain with 8-azidoadenosine 3',5'-monophosphate, *Biochemistry*, 14, 3858, 1975.

768. **Kerlavage, A. R. and Taylor, S. S.**, Covalent modification of an adenosine 3':5'-monophosphate binding site of the regulatory subunit of cAMP-dependent protein kinase II with 8-azidoadenosine 3':5'-monophosphate. Identification of single modified tyrosine residue, *J. Biol. Chem.*, 255, 8483, 1980.

769. **Uno, I. and Ishikawa, T.**, Adenosine 3',5'-monophosphate-receptor protein and protein kinase in *Corpinus macrorhizus*, *J. Biochem.*, 89, 1275, 1981.

770. **Walter, U. and Greengard, P.**, Photoaffinity labeling of the regulatory subunit of cAMP-dependent protein kinase, *Method Enzymol.*, 99, 154, 1983.

771. **Theurkauf, W. E. and Vallee, R. B.**, Molecular characterization of the cAMP-dependent protein kinase bound to microtubule-associated protein. II, *J. Biol. Chem.*, 257, 3284, 1982.

772. **Geahlen, R.L., Haley, B. E., and Krebs, E. G.**, Synthesis and use of 8-azidoguanosine 3',5'-cyclic monophosphate as a photoaffinity label for cyclic GMP-dependent protein kinase, *Proc. Natl. Acad. Sci. U.S.A.*, 76, 2213, 1979.

773. **Casnellie, J. E., Schlichter, D. J., Walter, U., and Greengard, P.**, Photoaffinity labeling of a guanosine 3':5'-monophosphate-dependent protein kinase from vascular smooth muscle, *J. Biol. Chem.*, 253, 4771, 1978.

774. **Hixson, C. S. and Krebs, E. G.**, Affinity labeling of catalytic subunit of bovine heart muscle cyclic AMP-dependent protein kinase by 5'-p-fluorosulfonylbenzoyladenosine, *J. Biol. Chem.*, 254, 7509, 1979.

775. **Hathaway, G. M., Zoller, M. J., and Traugh, J. A.**, Identification of catalytic subunit of casein kinase II by affinity labeling with 5'-p-fluorosulfonylbenzoyl adenosine, *J. Biol. Chem.*, 256, 11442, 1981.

776. **Zoller, M. J., Nelson, N. C., and Taylor, S. S.**, Affinity labeling of cAMP-dependent protein kinase with p-fluorosulfonylbenzoyl adenosine. Covalent modification of lysine-71., *J. Biol. Chem.*, 256, 10837, 1981.

777. **Hixson, C. S. and Krebs, E. G.**, Affinity labeling of the ATP binding site of bovine lung cyclic GMP-dependent protein kinase with 5'-p-fluorosulfonylbenzoyladenosine, *J. Biol. Chem.*, 256, 1122, 1981.

778. **Hashimoto, E., Takio, K., and Krebs, E. G.**, Amino acid sequence at the ATP-binding site of cGMP-dependent protein kinase, *J. Biol. Chem.*, 257, 727, 1982.

779. **Buhrow, S. A., Cohen, S., and Staros, J. V.**, Affinity labeling of the protein kinase associated with the epidermal growth factor receptor in membrane vesicles from A 431 cells, *J. Biol. Chem.*, 257, 4019, 1982.

780. **Puri, R. N., Bhatnagar, D., and Roskoski, R., Jr.,** Adenosine cyclic 3',5'-monophosphate dependent protein kinase: fluorescent affinity labeling of the catalytic subunit from bovine skeletal muscle with O-phtalaldehyde, *Biochemistry*, 24, 6499, 1985.

781. **Witt, J. J. and Roskoski, R., Jr.,** Adenosine cyclic 3',5'-monophosphate dependent protein kinase: active site directed inhibition by cibacron blue F3GA, *Biochemistry*, 19, 143, 1980.

782. **Hoppe, J. and Freist, W.,** Localization of the high-affinity ATP site in adenosine-3':5'-monophosphate-dependent protein kinase type I. Photoaffinity labelling studies with 8-azidoadenosine 5'-triphosphate, *Eur. J. Biochem.*, 93, 141, 1978.

783. **King, M. M. and Carlson, G. M.,** Affinity labeling of rabbit skeletal muscle phosphorylase kinase by 5'-(p-fluorosulfonylbenzoyl)adenosine, *FEBS Lett.*, 140, 131, 1982.

784. **King, M. M., Carlson, G. M., and Haley, B. E.,** Photoaffinity labeling of the β-subunit of phosphorylase kinase by 8-azidoadenosine 5'-triphosphate and its 2',3'-dialdehyde derivative, *J. Biol. Chem.*, 257, 14058, 1982.

785. **Yun, S. and Suelter, C. H.,** Modification of yeast pyruvate kinase by an active site-directed reagent, bromopyruvate, *J. Biol. Chem.*, 254, 1811, 1979.

786. **Likos, J. J., Hess, B., and Colman, R. F.,** Affinity labelling of the active site of yeast pyruvate kinase by 5'-p-fluorosulfonylbenzoyl adenosine, *J. Biol. Chem.*, 255, 9388, 1980.

787. **Annamalai, A. E., Tomich, J. M., Mas, M. T., and Colman, R. F.,** Evidence for distinguishable site of attack in the reactions of 5'-p-fluorosulfonylbenzoyl adenosine and 5'-p-fluorosulfonylbenzoyl guanosine with rabbit muscle pyruvate kinase, *Arch. Biochem. Biophys.*, 219, 47, 1982.

788. **Hinrichs, M. V. and Eyzaguirre, J.,** Affinity labeling of rabbit muscle pyruvate kinase with dialdehyde-ADP, *Biochim. Biophys. Acta*, 704, 177, 1982.

789. **Kapetanovic, E., Bailey, J. M., and Colman, R. F.,** 2-[(4-bromo-2,3-dioxobutyl)thio] adenosine 5'-monophosphate, a new nucleotide analogue that acts as an affinity label of pyruvate kinase, *Biochemistry*, 24, 7586, 1985.

790. **Kochetkov, S. N., Bulargina, T. V., Sashchenko, L. P., and Severin, E. S.,** Studies on the mechanism of action of histone kinase dependent on adenosine 3':5'-monophosphate. Evidence for involvement of histidine and lysine residues in the phosphotransferase reaction, *Eur. J. Biochem.*, 81, 111, 1977.

791. **Marletta, M. A. and Kenyon, G. L.,** Affinity labeling of creatine kinase by N-(2,3-epoxypropyl)-N-amidinoglycine, *J. Biol. Chem.*, 254, 1879, 1979.

792. **Akopyan, Zh. I., Gazaryants, M. G., Mkrtchyan, Z. S., Nersesova, L. S., Lavrik, O. I., and Popov, R. A.,** Affinity modification of creatine kinase EC.2.7.3.2. from rabbit skeletal muscles using gamma-p-azido anilide ATP, *Biokhimiya*, 46, 262, 1981; *Biochemistry (U.S.S.R.)*, 46, 215, 1981.

793. **Mkrtchyan, Z. S., Nersesova, L. S., Akopyan, G. I., Babkina, G. T., Buneva, V. N., and Knorre, D. G.,** Interaction of rabbit muscle creatine kinase with a reactive ATP derivative-ATP (γ-4-(N-2-chloroethyl-N-methylamino)-benzylamidate, *Biokhimija (U.S.S.R)*, 45, 806, 1980.

794. **Kuznetsov, A. V. and Saks, V. A.,** Affinity modification of creatine kinase and ATP-ADP translocase in heart mitochondria: determination of their molar stoichiometry, *Biochem. Biophys. Res. Commun.*, 134, 359, 1986.

795. **Crivellone, M. D., Hermodson, M., and Axelrod, B.,** Inactivation of muscle adenylate kinase by site-specific destruction of tyrosine 95 using potassium ferrate, *J. Biol. Chem.*, 260, 2657, 1985.

796. **Andreasen, T. J., Keller, C. H., LaPorte, D. C., Edelman, A. M., and Storm, D. R.,** Preparation of azidocalmodulin: a photoaffinity label for calmodulin-binding proteins, *Proc. Natl. Acad. Sci. U.S.A.*, 78, 2782, 1981.

797. **Slepneva, I. A.,** Detection of nucleoside triphosphate binding sites of two types in *E. coli* RNA polymerase by affinity labelling, *Mol. Biol. Rep.*, 6, 31, 1980.

798. **Malcolm, A. D. B. and Moffatt, J. R.,** Intrasubunit nucleotide binding in RNA polymerase, *Biochem. J.*, 175, 189, 1978.

799. **Kimball, A. P.,** DNA-dependent RNA polymerase, *Method Enzymol.*, 46, 353, 1977.

800. **Wu, F. and Wu, C.-W.,** Fluorescent affinity labeling of initiation site on RNA polymerase of *E. coli*, *Biochemistry*, 13, 2562, 1974.

801. **Miller, J. A., Serio, G. F., and Bear, J. L.,** Affinity labeling of a cystein at or near the catalytic center of *E. coli* DNA-dependent RNA polymerase, *Biochim. Biophys. Acta*, 612, 286, 1980.

802. **Woody, A. M., Vader, C. R., Woody, R. W., and Haley, B. E.,** Photoaffinity labeling of DNA-dependent RNA polymearase from *E. coli* with 8-azidoadenosine 5'-triphosphate, *Biochemistry*, 23, 2843, 1984.

803. **Ruetsch, N. and Dennis, D.,** RNA polymerase. Direct evidence for a unique topographical site for initiation, *J. Biol. Chem.*, 260, 16310, 1985.

804. **Panka, D. and Dennis, D.,** RNA polymerase. Direct evidence for two active sites involved in transcription, *J. Biol. Chem.*, 260, 1427, 1985.

805. **De Riemer, L. H. and Meares, C. F.,** Synthesis of mono- and dinucleotide photoaffinity probes of ribonucleic acid polymerase, *Biochemistry*, 20, 1606, 1981.

806. **Morgan, E. M. and Kingsbury, D. W.,** Pyridoxal phosphate as a probe of reovirus transcriptase, *Biochemistry,* 19, 484, 1980.

807. **Hilton, M. D. and Whiteley, H. R.,** UV cross-linking of the *B. subtilus* RNA polymerase to DNA in promoter and nonpromoter complexes, *J. Biol. Chem.,* 260, 8121, 1985.

808. **Gundelfinger, E. D.,** Interaction of nucleic acids with the DNA-dependent RNA polymerases of *Drosophila, FEBS Lett.,* 157, 133, 1983.

809. **Abboud, M. M., Sim, W. J., Loeb, L. A., and Mildvan, A. S.,** Apparent suicidal inactivation of DNA polymerase by adenosine 2′3′-riboepoxide 5′-triphosphate, *J. Biol. Chem.,* 253, 3415, 1978.

810. **Salvo, R. A., Serio, G. F., Evans, J. E., and Kimball, A. P.,** The affinity labelling of amino acids in or about the active site of DNA-dependent DNA polymerase, *Biochemistry,* 15, 493, 1976.

811. **Buneva, V. N., Demidova, I. V., Knorre, D. G., Kudryashova, N. V., Romashchenko, A. G., and Starobrazova, M. G.,** Affinity modification of *E. coli* DNA polymerase I with dTTP and dATP γ-4-(N-2-chloroethyl-N-methylamino)-benzylamidates, *Mol. Biol. (USSR),* 14, 1080, 1980.

812. **Abracham, K. I. and Modak, M. I.,** Affinity labelling of *E. coli* DNA polymerase I with thymidine-5′-triphosphate and 8-azidoadenosine-5′-triphosphate: conditions for optimum labelling, specifity and identification of labelling site, *Biochemistry,* 23, 1176, 1984.

813. **Bisman, S. B. and Kornberg, A.,** Nucleotide triphosphate binding to DNA polymerase III holoenzyme of *E. coli.* A direct photoaffinity labelling study, *J. Biol. Chem.,* 259, 7990, 1984.

814. **Hazra, A. K., Detera-Wadleigh, S., and Wilson, S. H.,** Site-specific modification of *E. coli* DNA polymerase I large fragment with pyridoxal 5′-phosphate, *Biochemistry,* 23, 2073, 1984.

815. **Modak, M. J.,** Pyridoxal-5′-phosphate: a selective inhibitor of oncoviral DNA polymerases, *Biochem. Biophys. Res. Commun.,* 71, 180, 1976.

816. **Markovitz, A.,** Ultraviolet light-induced stable complexes of DNA and DNA polymerase, *Biochim. Biophys. Acta,* 281, 522, 1972.

817. **Larsen, C. E. and Preiss, I.,** Covalent modification of the inhibitor binding site(s) of *Escherichia coli* ADP-glucose synthetase: specific incorporation of the photoaffinity analogue 8-azidoadenosine 5′-monophosphate, *Biochemisty,* 25, 4371, 1986.

818. **Abraham, K. I., Haley, B., and Modak, M. J.,** Biochemistry of terminal deoxynucleotidyltransferase: characterization and properties of photoaffinity labelling with 8-azidoadenosine 5′-triphosphate, *Biochemistry,* 22, 4197, 1983.

819. **Phillips, N. F. B., Goss, N. H., and Wood, H. G.,** Modification of pyruvate, phosphate dikinase with pyridoxal 5′-phosphate: evidence for a catalytically critical lysine residue, *Biochemistry,* 22, 2518, 1983.

820. **Borchardt, R. T., Wu, S. E., and Schasteen, C. S.,** ADP dialdehyde an affinity labeling reagent for phenol sulfotransferase EC 2.8.2.1, *Biochem. Biophys. Res. Commun.,* 81, 841, 1978.

821. **Huang, K.-S. and Law, J. H.,** Photoaffinity labeling of *Crotalus atrox* phospholipase A_2 by a substrate analogue, *Biochemistry,* 20, 181, 1981.

822. **O'Brien, R. D.,** Binding sites of cholnesterases; alkylation by an aziridinium derivative, *Biochem. J.,* 113, 713, 1969.

823. **Gordon, M. A., Carpenter, D. E., Barrett, H. W., and Wilson, I. B.,** Determination of the normality of cholinesterase solutions, *Anal. Biochem.,* 85, 519, 1978.

824. **Chang, G.-G., Wang, S.-C., and Pan, F.,** Periodate-oxidized AMP as a substrate an inhibitor and an affinity label of human placental alkaline phosphatase, *Biochem. J.,* 199, 281, 1981.

825. **Rajababu, C. and Axelrod, B.,** Site-specific inactivation of phosphatases by ferrate ion, *Arch. Biochem. Biophys.,* 188, 31, 1978.

826. **Maccioni, R. B., Hubert, E., and Slebe, J. C.,** Selective modification of fructose 1,6-bisphosphatase by periodate-oxidized AMP, *FEBS Lett.,* 102, 29, 1979.

827. **Riquelme, P. T. and Czarnecki, J. J.,** Conformational and allosteric changes in fructose 1,6-bisphosphatase EC3.1.3.11 upon photoaffinity labeling with 2-azido AMP, *J. Biol. Chem.,* 258, 8240, 1983.

828. **Marcus, F. and Haley, B. E.,** Inhibition of fructose-1,6-bisphosphatase by the photoaffinity AMP analog, 8-azidoadenosine 5′-monophosphate, *J. Biol. Chem.,* 254, 259, 1979.

829. **Pares, X., Llorens, R., Arus, C., and Cuchillo, C. M.,** The reaction of bovine pancreatic ribonuclease A with 6-chloropurineriboside 5′-monophosphate. Evidence on the existence of a phosphate-binding subsite, *Eur. J. Biochem.,* 105, 571, 1980.

830. **Steczko, I., Walker, D. E., Hermodson, M., and Axelrod, B.,** Identification of histidine-119 as the target in the site-specific inactivation of ribonuclease A by ferrate ion, *J. Biol. Chem.,* 254, 3254, 1979.

831. **Alonso, J., Nogues, M. V., and Cuchillo, C. M.,** Modification of bovine pancreatic ribonuclease A with 6-chloropurine riboside, *Arch. Biochem. Biophys.,* 246, 681, 1986.

832. **Sinnott, M. L. and Smith, P. J.,** Active-site-directed irreversible inhibition of *E. coli* β-galactosidase by the "hot" carbonium ion precursor. β-D-Galactopyranosylmethyl-p-nitrophenyltriazene, *J. Chem. Soc. Chem. Commun.,* 6, 223, 1976.

833. **Aiyar, V. N. and Hershfied, M. S.**, Covalent labelling of ligand binding sites of human placental S-adenosylhomocysteine hydrolase with 8-azido derivatives of adenosine and cyclic AMP, *Biochem. J.*, 232, 643, 1985.

834. **Abeles, R. H., Tashjian, A. H., Jr., and Fish, S.**, The mechanism of inactivation of S-adenosylhomocysteinase by 2'-deoxyadenosine, *Biochem. Biophys. Res. Commun.*, 95, 612, 1980.

835. **Ferro, A. J., Vandenbark, A. A., and MacDonald, M. R.**, Inactivation of S-adenosylhomocysteine hydrolase by 5'-deoxy-5'-methylthioadenosine, *Biochem. Biophys. Res. Commun.*, 100, 523, 1981.

836. **Chiang, P. K., Guranowski, A., and Segall, J. E.**, Irreversible inhibition of S-adenosylhomocysteine hydrolase by nucleoside analog, *Arch. Biochem. Biophys.*, 207, 175, 1981.

837. **Gratecos, D., Varesi, L., Knibiehler, M., and Semeriva, M.**, Photoaffinity labeling of membrane-bound porcine aminopeptidase N., *Biochim. Biophys. Acta*, 705, 218, 1982.

838. **Fujiwara, K., Matsumoto, E., Katagawa, T., and Tsuru, D.**, N-α-Carbobenzoxypyroglutamyl diazomethyl ketone as active-site-directed inhibitor for pyroglutamyl peptidase, *Biochim. Biophys. Acta*, 702, 149, 1980.

839. **Fujiwara, K., Kitagawa, T., and Tsuru, D.**, Inactivation of pyroglutamyl aminopeptidase by L-pyroglutamyl chloromethyl ketone, *Biochem. Biophys. Acta*, 655, 10, 1981.

840. **Fujiwara, K., Matsumoto, E., Kitagawa, T., and Tsuru, D.**, Inactivation of pyroglutamyl aminopeptidase by N^α-carbobenzoxy-L-pyroglutamyl chloromethyl ketone, *J. Biochem.*, 90, 433, 1981.

841. **Green, G. D. I. and Shaw, E.**, Peptidyl diazomethyl ketones are specific inactivators of thiol proteinases, *J. Biol. Chem.*, 256, 1923, 1981.

842. **Harris, R. and Wilson, I. B.**, Irreversible inhibition of bovine lung angiotensin I-converting enzyme with P-[N,N-bis(chloroethyl)amino] phenylbutyric acid (chlorambucyl) and chlorambucyl-L-proline with evidence that an active site carboxyl group is labelled, *J. Biol. Chem.*, 257, 811, 1982.

843. **Hass, G. M. and Neurath, H.**, Affinity labelling of bovine carboxypeptidase A_γ^Leu by N-bromoacetyl-N-methyl-L-phenylalanine. I. Kinetics of inactivation, *Biochemistry*, 10, 3535, 1971.

844. **Shaw, E.**, Site-specific reagents for chymotrypsin and trypsin, in *Methods in Enzymology*, Vol. 11, Hirs, C. H. W., Ed., Academic Press, New York, 1967, 677.

845. **Lawson, W. B. and Rao, G. J. S.**, Specificity in the alkylation of methionine at the active site of α-chymotrypsin by aromatic α-bromo amides, *Biochemistry*, 19, 2133, 1980.

846. **Lawson, W. B.**, Specificity in the alkylation of serine at the active site of α-chymotrypsin by aromatic α-bromo amides, *Biochemistry*, 19, 2140, 1980.

847. **Brown, W. and Wold, F.**, Alkyl isocyanates as active-site-specific reagents for serine proteases. Identification of the active-site serine as the site of reaction, *Biochemistry*, 12, 835, 1973.

848. **Escher, E. and Schwyzer, R.**, Photoaffinity labelling of chymotrypsin. Synthesis of photolysable ligands, *Helv. Chim. Acta*, 58, 1465, 1975.

849. **Penny, G. S. and Dyckes, D. F.**, Interaction of dansylated peptidyl chloromethanes with trypsin, chymotrypsin, elastase, and thrombin, *Biochemistry*, 19, 2888, 1980.

850. **Westkaemper, R. B. and Abeles, R. H.**, Novel inactivators of serine proteases based on 6-chloro-2-pyrone, *Biochemistry*, 22, 3256, 1983.

851. **Imperiali, B. and Abeles, R. H.**, Inhibition of serine proteases by peptidyl fluoromethyl ketones, *Biochemistry*, 25, 3760, 1986.

852. **Kurachi, K., Powers, J., and Wilcox, P. E.**, Kinetics of the reaction of chymotrypsin A_α with peptide chloromethyl ketones in relation to its subsite specificity, *Biochemistry*, 12, 771, 1973.

853. **White, E. H., Roswell, D. F., Politzer, I. R., and Branchini, B. R.**, Active site directed inhibition of enzymes utilizing deaminatively produced carbonium ions. Application to chymotrypsin, *J. Am. Chem. Soc.*, 97, 2290, 1975.

854. **Scofield, R. E., Werner, R. P., and Wold, F.**, *p*-Nitrophenyl carbamates as active-site-specific reagents for serine proteases, *Biochemistry*, 16, 2492, 1977.

855. **Collen, D., Lijnen, H. R., DeCock, F., Durieux, I. P., and Loffet, A.**, Kinetic properties of tripeptide lysyl chloromethyl ketone and lysyl *p*-nitroanilide derivatives towards trypsin-like serine proteinases, *Biochim. Biophys. Acta*, 165, 158, 1980.

856. **Romachandran, J., Smirnoff, P., and Birk, Y.**, Photoreactive derivative of Kunitz's soybean trypsin inhibitor. Preparation by selective modification of tryptophan residue and formation of a covalent complex of the modified inhibitor with trypsin, *Int. J. Peptide Protein Res.*, 23, 72, 1984.

857. **Penny, G. S., Dyckes, D. F., and Burleigh, B. D.**, Synthesis, characterization and fluorescence studies on N-α-dansylalanyllysyl trypsino (His-46)-methane, *Int. J. Peptide Protein Res.*, 13, 95, 1979.

858. **Coggins, J. R., Kray, W., and Shaw, E.**, Affinity labeling of proteinases with tryptic specificity by peptides with c-terminal lysine chloromethyl ketone, *Biochem. J.*, 138, 579, 1974.

859. **Walsmann, P., Richter, M., and Markwardt, F.**, Inactivation of trypsin and thrombin by 4-amidinobenzolsulfofluoride and 4-(2-aminoethyl)-benzolsulfofluoride, *Acta Biol. Med. Ger.*, 28, 577, 1972.

860. **Geratz, J. D.**, Kinetic aspects of the irreversible inhibition of trypsin and related enzymes by p-[m(m-fluorosulfonylphenylureido)phenoxyethoxy] benzamidine, *FEBS Lett.*, 20, 294, 1972.

861. **Kettner, C. and Shaw, E.,** Inactivation of trypsin-like enzymes with peptides of arginine chloromethyl ketone, *Method Enzymol.,* 80, 826, 1981.

862. **Griffith, M. J.,** Covalent modification of human α-thrombin with pyridoxal 5′-phosphate. Effect of phosphopyridoxylation on the interaction of thrombin with heparin, *J. Biol. Chem.,* 254, 3401, 1979.

863. **Valenty, V. B., Wos, J. D., Lobo, A. P., and Lawson, W. B.,** 5-Acyloxyoxazoles as serine protease inhibitors. Rapid inactivation of thrombin in contrast to plasmin, *Biochem. Biophys. Res. Commun.,* 88, 1375, 1979.

864. **Bing, D. H., Cory, M., and Fenton, J. W.,** Exo-site affinity labeling of human thrombins. Similar labeling on the A chain and B chain/fragments of clotting α-and nonclotting γ/β-thrombins. II., *J. Biol. Chem.,* 252, 8027, 1977.

865. **Glover, G. and Shaw, E.,** The purification of thrombin and isolation of a peptide containing the active center histidine, *J. Biol. Chem.,* 246, 4594, 1971.

866. **Ryan, T. J.,** Plasmin: photoaffinity labeling of a lysine-binding site which regulates clot lysis, *Biochem. Biophys. Res. Commun.,* 98, 1108, 1981.

867. **Powers, J. C. and Tudy, P.,** Active-site specific inhibitors of elastase, *Biochemistry,* 12, 4767, 1973.

868. **Nayak, P. L. and Bender, M. L.,** Organophosphorus compounds as active site-directed inhibitors of elastase, *Biochem. Biophys. Res. Commun.,* 83, 1178, 1978.

869. **Largman, C., Delmar, E. G., Brodrick, J. W., Fassett, M., and Geokas, M. C.,** Inhibition of human pancreatic elastase 2 by peptide chloromethyl ketones, *Biochim. Biophys. Acta,* 614, 113, 1980.

870. **Visser, L., Sigman, D. S., and Blout, E. R.,** Elastase. I. A new inhibitor, 1-bromo-4-(2,4-dinitrophenyl)butan-2-one, *Biochemistry,* 10, 735, 1971.

871. **Markland, F. S., Kettner, C., Schiffman, S., Shaw, E., Bajwa, S. S., Reddy, K. N. N., Kirakossian, H., Patkos, G. B., Theodor, I., and Pirkle, H.,** Kallikrein-like activity of crotalase, a snake venom enzyme that clots fibrinogen, *Proc. Natl. Acad. Sci. U.S.A.,* 79, 1688, 1982.

872. **Miwa, N., Sawada, T., and Suzuki, A.,** A specific disulfide bond associated with the activity of human urokinase. Its topological identification and reductive cleavage followed by kinetic changes in enzymatic reaction and affinity labeling, *Eur. J. Biochem.,* 140, 539, 1984.

873. **Leary, R. and Shaw, E.,** Inactivation of cathepsin B_1 by diazomethyl ketones, *Biochem. Biophys. Res. Commun.,* 79, 926, 1977.

874. **Evans, B. and Shaw, E.,** Inactivation of cathepsin B by active site-directed exchange. Application in covalent affinity chromatography, *J. Biol. Chem.,* 258, 10227, 1983.

875. **Barrett, A., Kembhavi, A. A., Brown, M. A., Kirschke, H., Knight, C. G., Tamai, M., and Hanada, K.,** L-trans-Epoxysuccinyl-leucylamido(4-guanidino)butane (E-64) and its analogues as inhibitors of cysteine proteinases including cathepsin B, H and L, *Biochem. J.,* 201, 189, 1982.

876. **Watanabe, H., Green, G. D. J., and Shaw, E.,** A comparison of the behavior of chymotrypsin and cathepsin B towards peptidyl diazomethyl ketones, *Biochem. Biophys. Res. Commun.,* 89, 1354, 1979.

877. **Leary, R., Larsen, D., Watanabe, H., and Shaw, E.,** Diazomethyl ketone substrate derivatives as active site-directed inhibitors of thiol proteases. Papain, *Biochemistry,* 16, 5857, 1977.

878. **Stein, M. J. and Liener, I. E.,** Inhibition of ficin by the chloromethyl ketone derivatives of N-tosyl-L-lysine and N-tosyl-L-phenylalanine, *Biochem. Biophys. Res. Commun.,* 26, 376, 1967.

879. **Murachi, T. and Kato, K.,** Inhibition of stem bromelain by chloromethyl ketone derivatives of N-tosyl-L-phenylalanine and N-tosyl-L-lysine, *J. Biochemistry,* 62, 627, 1967.

880. **Porter, W. H., Cunningham, L. W., and Mitchell, W. M.,** Studies of the active site of clostripain. The specific inactivation by the chloromethyl ketone derived from α-N-tosyl-L-lysine, *J. Biol. Chem.,* 246, 7675, 1971.

881. **Kirschke, H. and Shaw, E.,** Rapid inactivation of cathepsin L by Z-Phe-Phe-CHN$_2$ and Z-Phe-Ala-CHN$_2$, *Biochem. Biophys. Res. Commun.,* 101, 454, 1981.

882. **Rao, S. P. and Dunn, B. M.,** Preparation of photoaffinity labels of pepsin with p-nitro, p-azido and p-diazophenyl ligands and a study of the effects of irradiation on pepsin, *Biochim. Biophys. Acta,* 706, 86, 1982.

883. **Hixson, S. H., Hurwitz, J. L., Langridge, K. J., Nichols, D. C., Provost, K. M., and Wolff, A. M.,** Photoaffinity reagents for use with pepsin and other carboxyl proteases, *Biochem. Biophys. Res. Commun.,* 111, 630, 1983.

884. **Ong, E. B. and Perlmann, G. E.,** Specific inactivation of pepsin by benzyloxycarbonyl-L-phenylalanyl-diazomethane, *Nature,* 215, 1492, 1967.

885. **Fry, K. T., Kim, O.-K., Spona, J., and Hamilton, G. A.,** Site of reaction of a specific diazo inactivator of pepsin, *Biochemistry,* 9, 4624, 1970.

886. **Delpierre, G. R. and Fruton, J. S.,** Specific inactivation of pepsin by a diazo ketone, *Proc. Natl. Acad. Sci. U.S.A.,* 56, 1817, 1966.

887. **Charnas, R. L. and Knowles, J. R.,** Inactivation of RTEM β-lactamase from *Escherichia coli* by clavulanic acid and 9-deoxyclavulanic acid, *Biochemistry,* 20, 3214, 1981.

888. **Frere, J.-M., Dormans, C., Lenzini, M., and Duyckaerts,** Interaction of clavulanate with the β-lactomases of *Streptomyces albus* G and *Actinomadura* R 39, *Biochem. J., 207*, 429, 1982.

889. **Pratt, R. F. and Loosemore, M. J.,** 6-β-Bromopenicillanic acid, a potent β-lactamase inhibitor, *Proc. Natl. Acad. Sci. U.S.A., 75*, 4145, 1978.

890. **Knott-Hunziker, V., Waley, S. G., Orlek, B. S., and Sammes, P. G.,** Penicillinase active sites: labeling of serine-44 in β-lactamase I by 6β-bromopenicillanic acid, *FEBS Lett., 99*, 59, 1979.

891. **Heckler, T. G. and Day, R. A.,** *Bacillus cereus* 569/H penicillinase serine-44 acylation by diazotized 6-aminopenicillanic acid, *Biochim. Biophys. Acta, 745*, 292, 1983.

892. **Labia, R., Lelievre, V., and Peduzzi, J.,** Inhibition kinetics of three R-F-factor-mediated β-lactamases by a new β-lactam sulfone (CP 45899), *Biochim. Biophys. Acta, 611*, 351, 1980.

893. **Dmitrienko, G. I., Copeland, C. R., Arnold, L., Savard, M. E., Clarke, A. J., and Viswanatha, T.,** Inhibition of β-lactamase I by 6-β-sulfonamidopenicillanic acid sulfones: evidence for conformational change accompanying the inhibition process, *Bioorg. Chem., 13*, 34, 1985.

894. **Charnas, R. L. and Knowles, J. R.,** Inhibition of the RTEM β-lactamase from *Escherichia coli.* Interaction of enzyme with derivatives of olivanic acid, *Biochemistry, 20*, 2732, 1981.

895. **Weickmann, J. L., Himmel, M. E., Smith, D. W., and Fahrney, D. E.,** Arginine deiminase: demonstration of two active sites and possible half-of-the-sites reactivity, *Biochem. Biophys. Res. Commun., 83*, 107, 1978.

896. **Schaeffer, H. J., Schwartz, M. A., and Odin, E.,** Enzyme inhibitors. XVII. Kinetic studies on the irreversible inhibition of adenosine deaminase, *J. Med. Chem., 10*, 686, 1967.

897. **Ranieri-Raggi, M. and Raggi, A.,** Specific modification of the GTP binding sites of rat 5'-adenylic acid aminohydrolase by periodate-oxidized GTP, *Biochim. Biophys. Acta, 445*, 223, 1976.

898. **Raggi, A. and Ranieri-Raggi, M.,** Negative homotropic cooperativity in rat muscle AMP deaminase. A kinetic study on the inhibition of the enzyme by ATP, *Biochim. Biophys. Acta, 566*, 353, 1979.

899. **Svyato, I. E., Sklyankina, V. A., and Avaeva, S. M.** The active center — a site of binding of affinity inhibitors in baker's yeast inorganic pyrophosphatase, *Biokhimija (USSR), 50*, 1523, 1985.

900. **Avaeva, S. M., Dickov, M. M., Kuznetsov, A. V., and Sklyankina, V. A.,** Specific modification of yeast inorganic pyrophosphatase active site by N-chloroacetylphosphoeethanolamine, *Bioorg. Khim. (USSR), 3*, 943, 1977.

901. **Borschik, I. B., Sklyankina, V. A., and Avaeva, S. M.,** Phosphoric acid monoesters as affinity inhibitors of the *E. coli* inorganic pyrophosphatase, *Bioorg. Khim. (USSR), 11*, 778, 1985.

902. **Lunardi, J., Lauquin, G. J. M., and Vignais, P. V.,** Interaction of azidonitrophenylaminobutyryl-ADP, a photoaffinity ADP analog, with mitochondrial adenosine triphosphatase. Identification of the labeled subunits, *FEBS Lett., 80*, 317, 1977.

903. **Lowe, P. N. and Beechey, R. B.,** Interactions between the mitochondrial adenosinetriphosphatase and periodate-oxidized adenosine 5'-triphosphate, an affinity label for adenosine 5'-triphosphate binding sites, *Biochemistry, 21*, 4073, 1982.

904. **Dupuis, A. and Vignais, P. V.,** Photolabeling of mitochondrial F_1-ATPase by an azido derivative of the oligomycin-sensitivity conferring protein, *Biochem. Biophys. Res. Commun., 129*, 819, 1985.

905. **Cosson, J. J. and Guillory, R. J.,** The use of arylazido-β-alanyl-ATP as a photoaffinity label for the isolated and membrane-bound mitochondrial ATPase complex, *J. Biol. Chem., 254*, 2946, 1979.

906. **Wagenvoord, R. J., Van Der Kraan, I., and Kemp, A.,** Specific photolabelling of beef-heart mitochondrial ATPase by 8-azido-ATP, *Biochim. Biophys. Acta, 460*, 17, 1977.

907. **Hollemans, M., Runswick, M. J., Fearnley, I. M., and Walker, J. E.,** The sites of labeling of the β-subunit of bovine mitochondrial F_1-ATPase with 8-azido-ATP, *J. Biol. Chem., 258*, 9307, 1983.

908. **Van Dongen, M. B. M., DeGeus, J. P., Korver, T., Hartog, A. F., and Berden, J. A.,** Binding and hydrolysis of 2-azido-ATP and 8-azido-ATP by isolated mitochondrial F_1 characterization of high-affinity binding sites, *Biochim. Biophys. Acta, 850*, 359, 1986.

909. **Wakagi, T. and Ohta, T.,** Detection of the conformational change in the catalytic site of adenosine triphosphatase from beef liver mitochondria by affinity labeling with the dialdehyde derivative of ethenoadenosine triphosphate, *J. Biochem., 92*, 1403, 1982.

910. **Fellous, G., Godinot, C., Baubichon, H., DiPietro, A., and Gautheron, D. C.,** Photolabelling on β-subunit of the nucleotide site related to hysteretic inhibition of mitochondrial F_1-ATPase, *Biochemistry, 23*, 5294, 1984.

911. **Esch, F. S. and Allison, W. S.,** Identification of tyrosine residue at a nucleotide binding site in the β-subunit of the mitochondrial ATPase with p-fluorosulfonyl ^{14}C-benzoyl-5'-adenosine, *J. Biol. Chem., 253*, 6100, 1978.

912. **Williams, N. and Coleman, P. S.,** Exploring the adenine nucleotide binding sites on mitochondrial F_1-ATPase with a new photoaffinity probe, 3'-O-(4-benzoyl)benzoyl adenosine 5'-triphosphate, *J. Biol. Chem., 257*, 2834, 1982.

913. **Melese, J. and Boyer, P. D.,** Derivatization of the catalytic subunits of the chloroplast ATPase by 2-azido-ATP and dicyclohexylcarbodiimide. Evidence for catalytically induced interchange of the subunits, *J. Biol. Chem.,* 260, 15398, 1985.

914. **Bullough, D. A., Yoshida, M., and Allison, W. S.,** Sequence of the radioactive tryptic peptide obtained after inactivating the F$_1$-ATPase of the thermophilic bacterium PS3 with 5'-p-fluorosulfonylbenzoyl [³H]adenosine at 65°C, *Arch. Biochem. Biophys.,* 244, 865, 1986.

915. **Bar-Zvi, D., Yoshida, M., and Shavit, N.,** Photoaffinity labeling of the TF$_1$-ATPase from the thermophilic bacterium PS3 with 3'-O-(4-benzoyl)benzoyl ADP, *Biochim. Biophys. Acta,* 807, 293, 1985.

916. **Verheijen, J. H., Postma, P. W., and Van Dam, K.,** Specific labeling of the (Ca^{2+} + Mg^{2+})-ATPase of *E. coli* with 8-azido-ATP and 4-chloro-7-nitrobenzofurazan, *Biochim. Biophys. Acta,* 502, 345, 1978.

917. **Gregory, R., Recktenwald, D., Hess, B., Schäfter, H.-J., Scheurich, P., and Dose, K.,** Photoaffinity labeling of a low-affinity nucleotide binding site on the β-subunit of yeast mitochondrial F$_1$-ATPase, *FEBS Lett.,* 108, 253, 1979.

918. **Bitar, K. G.,** Modification of F$_1$-ATPase from yeast *Saccharomyces cerevisiae* with 5'-p-[³H]fluorosulfonylbenzoyl adenosine, *Biochem. Biophys. Res. Commun.,* 109, 30, 1982.

919. **Pougeois, R., Lauquin, G. J.-M., and Vignais, P. V.,** Evidence that 4-azido-2-nitrophenylphosphate binds to the phosphate site on the β-subunit of *E. coli* BF$_1$-ATPase, *FEBS Lett.,* 153, 65, 1983.

920. **Kumar, G., Kalra, V. K., and Brodie, A. F.,** Affinity labeling of coupling factor-latent ATPase from *Mycobacterium phlei* with 2',3'-dialdehyde derivatives of adenosine 5'-triphosphate and adenosine 5'-diphosphate, *J. Biol. Chem.,* 254, 1964, 1979.

921. **Forbush, B., III, Kaplan, J. H., and Hoffman, J. F.,** Characterization of a new photoaffinity derivative of ouabain: labeling of the large polypeptide and of a proteolipid component of the Na, K-ATPase, *Biochemistry,* 17, 3667, 1978.

922. **Deffo, T., Fullerton, P. S., Kihara, M., McParland, R. H., Becker, R. R., Simat, B. M., From, A. H., Ahmed, K., and Schimerlik, M. I.,** Photoaffinity labeling of the sodium-and potassium-activated adenosinetriphosphatase with a cardiac glycoside containing the photoactive group on the C-17 side chain, *Biochemistry,* 22, 6303, 1983.

923. **Munson, K. B.,** Light-dependent inactivation of (Na$^+$ + K$^+$)-ATPase with a new photoaffinity reagent, chromium arylazido-β-alanyl ATP, *J. Biol. Chem.,* 256, 3223, 1981.

924. **Ponzio, G., Rossi, B., and Lardunski, M.,** Affinity labeling and localization of the ATP binding site in the (Na$^+$, K$^+$)-ATPase, *J. Biol. Chem.,* 258, 8201, 1983.

925. **Ohta, T., Nagano, K., and Yoshida, M.,** The active site structure of Na$^+$/K$^+$-transporting ATPase: location of the 5'-(p-fluorosulfonyl)benzoyladenosine binding site and soluble peptides released by trypsin, *Proc. Natl. Acad. Sci. U.S.A.,* 83, 2071, 1986.

926. **Montecucco, C., Bisson, R., Gache, C., and Johannsson, A.,** Labeling of the hydrophobic domain of the Na$^+$,K$^+$-ATPase, *FEBS Lett.,* 128, 17, 1981.

927. **Goeldner, M., Hirth, C. G., Rossi, B., Ponzio, G., and Lardunski, M.,** Specific photoaffinity labeling of the digitalis binding site of the sodium and potassium ion activated adenosinetriphosphatase induced by energy transfer, *Biochemistry,* 22, 4685, 1983.

928. **Rogers, T. B. and Lardunski, M.,** Photoaffinity labeling of a small protein component of a purified (Na$^+$, K$^+$)-ATPase, *FEBS Lett.,* 98, 373, 1979.

929. **Hall, C. and Ruoho, A.,** Ouabain-binding-site photoaffinity probes that label both subunits of Na$^+$, K$^+$-ATPase, *Proc. Natl. Acad. Sci. U.S.A.,* 77, 4529, 1980.

930. **Ruoho, A. E., Hokin, L. E., Hemingway, R. J., and Kupchan, S. M.,** Hellebrigenin 3-haloacetate: potent site-directed alkylators of transport adenosinetriphosphatase, *Science,* 159, 46, 1968.

931. **Pick, U. and Bassilian, S.,** Modification of the ATP binding site of the Ca^{2+}-ATPase from sarcoplasmic reticulum by fluorescein isothiocyanate, *FEBS Lett.,* 123, 127, 1981.

932. **Carvalho-Alves, P. C., Oliveira, C. R. G., and Verjovski-Almeida, S.,** Stoichiometric photolabeling of two distinct low and high affinity nucleotide sites in sarcoplasmic reticulum ATPase, *J. Biol. Chem.,* 260, 4282, 1985.

933. **Cable, M. B. and Briggs, F. N.,** Labeling the adenine nucleotide binding domain of the sarcoplasmic reticulum Ca, Mg-ATPase with photoaffinity analogs of ATP, *J. Biol. Chem.,* 259, 3612, 1984.

934. **Bragg, P. D., Stan-Lotter, H., and Hou, C.,** Affinity labeling of purified Ca^{2+}, Mg^{2+}-activated ATPase of *E. coli* by the 2',3'-dialdehydes of adenosine 5'-di- and triphosphates, *Arch Biochem. Biophys.,* 207, 290, 1981.

935. **Takemoto, D. J., Haley, B. E., Hansen, J., Pinkett, O., and Takemoto, L. J.,** GTPase from rod outer segments: characterization by photoaffinity labeling and tryptic peptide mapping, *Biochem. Biophys. Res. Commun.,* 102, 341, 1981.

936. **LeBel, D. and Beattie, M.,** Identification of the catalytic subunit of the ATP diphosphohydrolase by photoaffinity labeling of high-affinity ATP-binding sites of pancreatic zymogen granule membranes with 8-azido-[α-³²P]ATP, *Biochem. Cell Biol.,* 64, 13, 1986.

937. **Nashed, N. T., Michaud, D. P., Levin, W., and Jerina, D. M.,** 7-Dehydrocholesterol 5,6-β-oxide as a mechanism-based inhibitor of microsomal cholesterol oxide hydrolase, *J. Biol. Chem.,* 261, 2510, 1986.

938. **Kameji, T. and Hayashi, S.-I.,** Affinity labeling of purified ornitine decarboxylase by α-difluoromethyl-ornithine, *Biochim. Biophys. Acta,* 705, 405, 1982.

939. **Pritchard, M. L., Seely, J. E., Pösö, H., Jefferson, L. S., and Pegg, A. E.,** Binding of radioactive α-difluoromethylornithine to rat liver ornithine decarboxylase, *Biochem. Biophys. Res. Commun.,* 100, 1597, 1981.

940. **Pritchard, M. L., Pegg, A. E., and Jefferson, L. S.,** Ornithine decarboxylase from hepatoma cells and a variant cell line in which the enzyme is more stable, *J. Biol. Chem.,* 257, 5892, 1982.

941. **Kallio, A., McCann, P. P., and Bey, P.,** DL-α-(Difluoromethyl)arginine: a potent enzyme-activated irreversible inhibitor of bacterial arginine decarboxylases, *Biochemistry,* 20, 3163, 1981.

942. **Palfreyman, M. G., Danzin, C., Bey, P., Jung, M. J., Ribereau-Gayon, G., Aubry, M., Vevert, J. P., and Sjoerdsma, A.,** α-Difluoromethyl DOPA, a new enzyme activated irreversible inhibitor of aromatic L-amino acid decarboxylase, *J. Neurochem.,* 31, 927, 1978.

943. **Maycock, A. L., Aster, S. D., and Patchett, A. A.,** Inactivation of 3-(3,4-dihydroxyphenyl)alanine decarboxylase by 2-(fluoromethyl)-3-(3,4-dihydroxyphenyl)alanine, *Biochemistry,* 19, 709, 1980.

944. **Kameshita, I., Tokushige, M., Izui, K., and Katsuki, H.,** Phosphoenolpyruvate carboxylase of *Escherichia coli.* Affinity labeling with bromopyruvate, *J. Biochem.,* 86, 1251, 1979.

945. **Silverstein, R., Rawitch, A. B., and Grainger, D. A.,** Affinity labeling of phosphoenolpyruvate carboxykinase with 1,5-I-AEDANS, *Biochem. Biophys. Res. Commun.,* 87, 911, 1979.

946. **Silverstein, R.,** Phosphoenolpyruvate carboxykinase inactivation by bromopyruvate, *Fed. Proc. Fed. Am. Soc. Exp. Biol.,* 35, 1756, 1976.

947. **Herndon, C. S. and Hartman, F. C.,** 2-(4-Bromoacetamido)-anilino-2-deoxypentitol 1,5-biphosphate, a new affinity label for ribulose biphosphate carboxylase/oxygenase from *Rhodospirillum rubrum, J. Biol. Chem.,* 259, 3102, 1984.

948. **Schloss, J. V., Stringer, C. D., and Hartman, F. C.,** Identification of essential lysyl and cysteinyl residues in spinach ribulosebisphosphate carboxylase/oxygenase modified by the affinity label N-bromo-acetylethanolamine phosphate, *J. Biol. Chem.,* 253, 5707, 1978.

949. **Fraij, B. and Hartman, F. C.,** Isolation and sequencing of an active-site peptide from *Rhodospirillum rubrum* ribulosebisphosphate carboxylase/oxygenase after affinity labelling with 2-[(bromoacetyl)amino]pentitol 1,5-bis-phosphate, *Biochemistry,* 22, 1515, 1983.

950. **Donnelly, M. I. and Hartman, F. C.,** Inactivation of ribulosebisphosphate carboxylase/oxygenase from *Rhodospirillum rubrum* and spinach with the new affinity label 2-bromo-1,5-dihydroxy-3-pentanone 1,5-bisphosphate, *Biochem. Biophys. Res. Commun.,* 103, 161, 1981.

951. **Norton, I. L., Welch, M. H., and Hartman, F. C.,** Evidence for essential lysyl residues in ribulosebisphosphate carboxylase by use of the affinity label 3-bromo-1,4-dihydroxy-2-butanone 1,4-bisphosphate, *J. Biol. Chem.,* 250, 8062, 1975.

952. **Lubini, D. G. E. and Christen, P.,** Paracatalytic modification of aldolase: a side reaction of the catalytic cycle resulting in irreversible blocking of two active-site lysyl residues, *Proc. Natl. Acad. Sci. U.S.A.,* 76, 2527, 1979.

953. **Hartman, F. C. and Norton, I. L.,** Active-site-directed reagents of glycolytic enzymes, *Methods Enzymol.,* 47, 479, 1977.

954. **Magnien, A., Le Clef, B., and Biellmann, J.-F.,** Suicide inactivation of fructose-1,6-bisphosphate aldolase, *Biochemistry,* 23, 6858, 1984.

955. **Meloche, H. P.,** The substrate analog bromopyruvate, as a bridging agent for the active site of 2-keto-3-deoxy-6-phosphogluconic aldolase. Chemical evidence for a carboxylate adjacent to the Schiff's base-forming lysine, *J. Biol. Chem.,* 248, 6945, 1973.

956. **Miziorko, H. M. and Behnke, C. E.,** Active-site-directed inhibition of 3-hydroxy-3-methylglutaryl coenzyme A synthase by 3-chloropropionyl coenzyme A, *Biochemistry,* 24, 3174, 1985.

957. **Basu, A., Subramanian, S., and SivaRaman, C.,** Photoaffinity labeling of *Klebsiella aerogenes* citrate lyase by p-azidobenzoyl coenzyme A, *Biochemistry,* 21, 4434, 1982.

958. **Bower, S. and Zalkin, H.,** Modification of *Serratia marcescens* anthranilate synthase with pyridoxal 5'-phosphate, *Arch. Biochem. Biophys.,* 219, 121, 1982.

959. **Nagano, H., Zalkin, H., and Henderson, E. J.,** The anthranilate synthetase-anthranilate-5-phosphoribosylpyrophosphate phosphoribosyltransferase aggregate, *J. Biol. Chem.,* 245, 3810, 1970.

960. **Tso, J. Y., Bower, S. G., and Zalkin, H.,** Mechanism of inactivation of glutamine amidotransferases by the antitumor drug L-(α-S,5S)-α-amino-3-chloro-4,5-dihydro-5-isoxazoleacetic acid (AT-125), *J. Biol. Chem.,* 255, 6734, 1980.

961. **Nihira, T., Toraya, T., and Fukui, S.,** Pyridoxal-5'-phosphate-sensitized photoinactivation of tryptophanase and evidence for essential histidyl residues in the active sites, *Eur. J. Biochem.,* 101, 341, 1979.

962. **Honda, T. and Tokushige, M.,** Active-site-directed modification of tryptophanase by 3-bromopyruvate, *J. Biochem.,* 97, 851, 1985.

963. **Kirley, J. W., Day, R. A., and Kreishman, G. P.,** Cyanogen-induced γ-glutamyl to imidazol cross-link in carbonic anhydrase, *FEBS Lett.,* 193, 145, 1985.

964. **Cybulsky, D. L., Nagy, A., Kandel, S. I., Kandel, M., and Gornall, A. G.,** Carbonic anhydrase from spinach leaves, Chemical modification and affinity labeling, *J. Biol. Chem.,* 254, 2032, 1979.

965. **Matherly, L. H. and Phillips, A. T.,** Substrate-mediated inactivation of urocanase from *Pseudomonas putida.* Evidence for an essential sulfhydryl group, *Biochemistry,* 19, 5814, 1980.

966. **Johnston, M., Jankowski, D., Marcotte, P., Tahaka, H., Esaki, N., Soda, K., and Walsh, C.,** Suicide inactivation of bacterial cystathionine γ-synthase and methionine γ-lyase during processing of L-propargylglycine, *Biochemistry,* 18, 4691, 1979.

967. **Washtien, W. and Abeles, R. H.,** Mechanism of inactivation of γ-cystathionase by the acetylenic substrate analogue propargylglycine, *Biochemistry,* 16, 2485, 1977.

968. **Fearon, C. W., Rodkey, J. A., and Abeles, R. H.,** Identification of the active-site residue of γ-cystathionase labeled by the suicide inactivator β,β,β-trifluoroalanine, *Biochemistry,* 21, 3790, 1982.

969. **Alston, T. A., Muramatsu, H., Ueda, T., and Bright, H. J.,** Inactivation of γ-cystathionase by γ-fluorinated amino acids, *FEBS Lett.,* 128, 293, 1981.

970. **Seddon, A. P. and Douglas, K. T.,** A photoaffinity label derivative of glutathione and its inhibition of glyoxalase I, *FEBS Lett.,* 110, 262, 1980.

971. **Greenlee, D. V., Andreasen, T. J., and Storm, D. R.,** Calcium-independent stimulation of *Bordetella pertussis* adenylate cyclase by calmodulin, *Biochemistry,* 21, 2759, 1982.

972. **Skurat, A. V., Yurkova, M. S., Khropov, Y. V., Bulargina, T. V., and Severin, E. S.,** 2',3'-Dialdehyde of GTP blocks regulatory functions of adenylate cyclase N_s protein, *FEBS Lett.,* 188, 150, 1985.

973. **Skurat, A. V., Perfilyeva, E. A., Khropov, Yu. V., Bulargina, T. V., and Severin, E. S.,** Irreversible inhibition of adenylate cyclase by oxo-ATP, *Biokhimiya (USSR),* 47, 257, 1982.

974. **Hoyer, P. B., Fletcher, P., and Haley, B. E.,** Synthesis of 2',3'-O-(2,4,6-trinitrocyclohexadienylidine) guanosine 5'-triphosphate and a study of its inhibitory properties with adenylate cyclase, *Arch. Biochem. Biophys.,* 245, 369, 1986.

975. **Morisaki, M. and Bloch, K.,** Inhibition of β-hydroxydecanoyl thioester dehydrase by some allenic acids and their thioesters, *Bioorg. Chem.,* 1, 188, 1971.

976. **Wang, E. A. and Walsh, C.,** Characteristics of β,β-difluoroalanine and β,β,β-trifluoroalanine as a suicide substrates for *E. coli* B alanine racemase, *Biochemistry,* 20, 7539, 1981.

977. **Hartman, F. C.,** Site-specific reagents for triose phosphate isomerase and their potential applicability to aldolase and glycerol phosphate dehydrogenase, *Method Enzymol.,* 25, 661, 1972.

978. **De La Mare, S., Coulson, A. F. W., Knowles, J. R., Priddle, J. D., and Offord, R. E.,** Active-site labeling of triose phosphate isomerase. The reaction of bromohydroxyacetone phosphate with a unique glutamic acid residue and the migration of the label of tyrosine, *Biochem. J.,* 129, 321, 1972.

979. **Steczko, J., Hermodson, M., Axelrod, B., and Dziembor-Kentzer, E.,** Identification of the target amino acids in the site-specific inactivation of triose phosphate isomerase by ferrate anion, *J. Biol. Chem.,* 258, 13148, 1983.

980. **Palmieri, R. H., Gee, D. M., and Noltmann, E. A.,** Isolation and sequence determination of two pyridoxal 5'-phosphate-labeled thermolysin peptides from pig muscle phosphoglucose isomerase, *J. Biol. Chem.,* 257, 7965, 1982.

981. **Chmara, H. and Lähner, H.,** The inactivation of glucosamine synthetase from bacteria by anticapsin, the C-terminal epoxyamino acid of the antibiotic tetaine, *Biochim. Biophys. Acta,* 787, 45, 1984.

982. **Chmara, H., Andruszkiewicz, R., and Borowski, E.,** Inactivation of glucosamine-6-phosphate synthetase from *Salmonella typhimurium* LT2SL 1027 by Nβ-fumarylcarboxyamido-L-2,3-diamino-propionic acid, *Biochem. Biophys. Res. Commun.,* 120, 865, 1984.

983. **Hearne, M. and Benisek, W. F.,** Photoaffinity modification of Δ⁵-3-ketosteroid isomerase by light-activatable steroid ketones covalently coupled to agarose beads, *Biochemistry,* 22, 2537, 1983.

984. **Pollack, R. M., Kayser, R. H., and Bevins, C. L.,** An active-site-directed irreversible inhibitor of Δ⁵-3-ketosteroid isomerase, *Biochem. Biophys. Res. Commun.,* 91, 783, 1979.

985. **Bevins, C. L., Kayser, R. H., Pollack, R. M., Ekiko, D. B., and Sadoff, S.,** Irreversible active-site-directed inhibition of Δ⁵-3-ketosteroid isomerase by steroidal 17-β-oxiranes. Evidence for two modes of binding in steroid-enzyme complexes, *Biochem. Biophys. Res. Commun.,* 95, 1131, 1980.

986. **Kayser, R. H., Bounds, P. L., Bevins, C. L., and Pollack, R. M.,** Affinity alkylation of bacterial Δ⁵-3-ketosteroid isomerase. Identification of the amino acid modified by steroidal 17-β-oxiranes, *J. Biol. Chem.,* 258, 909, 1983.

987. **Bevins, C. L., Bantia, S., Pollack, R. M., Bounds, P. L., and Kayser, R. H.,** Modification of an enzyme carboxylate residue in the inhibition of 3-oxo-Δ⁵-steroid isomerase by (3S)-spiro[5α-androstane-3,2'-oxirane]-17β-ol. Implications for the mechanism of action. *J. Am. Chem. Soc.,* 106, 4957, 1984.

988. **Bantia, S., Bevins, C. L., and Pollack, R. M.,** Mechanism of inactivation of 3-oxosteroid Δ⁵-isomerase by 17β-oxiranes, *Biochemistry,* 24, 2606, 1985.

989. **Penning, T. M. and Talalay, P.,** Linkage of a acetylenic secosteroid suicide substrate to the active site of Δ^5-3-ketosteroid isomerase. Isolation and characterization of a tetrapeptide, *J. Biol. Chem.,* 256, 6851, 1981.

990. **Penning, T. M., Covey, D. F., and Talalay, P.,** Irreversible inactivation of Δ^5-3-ketosteroid isomerase of *Pseudomonas testosteroni* of acetylenic suicide substrates. Mechanism of formation and properties of the steroid-enzyme adduct, *J. Biol. Chem.,* 256, 6842, 1981.

991. **Penning, T. M. and Covey, D. F.,** Inactivation of Δ^5-3-ketosteroid isomerase(s) from beef adrenal cortex by acetylenic ketosteroids, *J. Steroid Biochem.,* 16, 691, 1982.

992. **Martyr, R. J. and Benisek, W. F.,** Affinity labeling of the active sites of Δ^5-ketosteroid isomerase using photoexcited natural ligands, *Biochemistry,* 12, 2172, 1973.

993. **Smith, S. B. and Benisek, W. F.,** Active site-directed photoinactivation of Δ^5-3-ketosteroid isomerase from *Pseudomonas putida* dependent on 1,4,6-androstatrien-3-one-17β-ol, *J. Biol. Chem.,* 255, 2690, 1980.

994. **Penning, T. M.,** Irreversible inhibition of Δ^5-3-oxosteroid isomerase by 2-substituted progesterones, *Biochem. J.,* 226, 469, 1985.

995. **Fayat, G., Fromant, M., and Blanquet, S.,** Aminoacyl-tRNA synthetases: affinity labeling of the ATP binding site by 2′,3′-ribose oxidized ATP, *Proc. Natl. Acad. Sci. U.S.A.,* 75, 2088, 1978.

996. **Schoemaker, H. J. P. and Schimmel, P. R.,** Photo-induced joining of a transfer RNA with its cognate aminoacyltransfer RNA synthetase, *J. Mol. Biol.,* 84, 503, 1974.

997. **Hountondji, C., Lederer, F., Dessen, P., and Blanquet, S.,** *Escherichia coli* tyrosyl- and methionyl-tRNA synthetases display sequence similarity at the binding site for the 3′-end of tRNA, *Biochemistry,* 25, 16, 1986.

998. **Ackerman, E. J., Joachimiak, A., Klinghofer, V., and Sigler, P. B.,** Directly photocrosslinked nucleotides joining transfer RNA to aminoacyl-tRNA synthetase in methionine and tyrosine systems, *J. Mol. Biol.,* 181, 93, 1985.

999. **Kovaleva, G. K., Degtyarev, S. Ch., and Favorova, O. O.,** Affinity modification of tryptophanyl-tRNA synthetase by an alkylating L-triptophan analog, *Mol. Biol. (USSR),* 13, 1237, 1979.

1000. **Kovaleva, G. K., Ivanov, L. L., Madoyan, I. A., Favorova, O. O., Severin, E. S., Gulyaev, N. N., Baranova, L. A., Shabarova, Z. A., Sokolova, N. I., and Kisselev, L. L.,** Inhibition of tryptophanyl-tRNA synthetase by modifying ATP analogs, *Biokhimija (USSR),* 43, 525, 1978.

1001. **Favorova, O. O., Madoyan, I. A., and Drutsa, V. L.,** "Half-site" affinity modification of tryptophanyl-tRNA synthetase leads to "freezing" of the free subunit, *FEBS Lett.,* 123, 161, 1981.

1002. **Akhverdyan, V. Z., Kisselev, L. L., Knorre, D. G., Lavrik, O. I., and Nevinsky, G. A.,** Photoaffinity modification of triptophanyl-tRNA synthetase with γ-(n-azidoanylide)-ATP, *Dokl. Akad. Nauk SSSR,* 226, 698, 1976.

1003. **McArdell, J. E. C., Atkinson, T., and Bruton, C. J.,** The interaction of triptophanyl-tRNA synthetase with the triazine dye Brown MX-5BR, *Eur. J. Biochem.,* 125, 361, 1982.

1004. **Krauspe, R., Kovaleva, G. K., Gulyaev, N. N., Baranova, L. A., Agalarova, M. B., Severin, E. S., Sokolova, N. I., Shabarova, Z. A., and Kisselev, L. L.,** Inhibition of leucyl-tRNA synthetases by modifying ATP analogs, *Biokhimija (USSR),* 43, 656, 1978.

1005. **Wiebauer, K., Ogilvie, A., and Kersten, W.,** The molecular basis of leucine auxotrophy of quinone-treated *Escherichia coli.* Active-site directed modification of leycil-tRNA synthetase by G-amino-7-chloro-5,8-dioxoquinoline, *J. Biol. Chem.,* 154, 327, 1979.

1006. **Rainly, P., Noller, E., and Kula, M.-R.,** Labeling of α-isoleucyl-tRNA ligase from *Escherichia coli* with α-isoleucylbromomethyl ketone, *Eur. J. Biochem.,* 63, 419, 1976.

1007. **Mehler, A. H., Kim, J.-J. P., and Olsen, A. A.,** Specificity in the inactivation of enzymes by periodate-oxidized nucleotides, *Arch. Biochem. Biophys.,* 212, 475, 1981.

1008. **Rainly, P., Hammer-Raber, B., Kula, M.-R., and Holler, E.,** Modification of α-isoleucyl-tRNA synthetase with α-isoleucyl-bromomethyl ketone, The effect on the catalytic steps, *Eur. J. Biochem.,* 78, 239, 1977.

1009. **Santi, D. V. and Cunnion, S. O.,** Macromolecular affinity labeling agents. Reaction of N-bromoacetyl-isoleucyl transfer ribonucleic acid with isoleucyl transfer ribonucleic acid synthetase, *Biochemistry,* 13, 481, 1974.

1010. **Yue, V. T. and Schimmel, P. R.,** Direct and specific photochemical cross-linking of adenosine 5′-triphosphate to an aminoacyl-tRNA synthetase, *Biochemistry,* 16, 4678, 1977.

1011. **Silver, J. and Laursen, R. A.,** Inactivation of aminoacyl-tRNA synthetases by amino acid chloromethyl-ketones, *Biochim. Biophys. Acta,* 340, 77, 1974.

1012. **Frolova, N. Yu., Kovaleva, G. K., Agalorova, M. B., and Kisselev, L. L.,** Irreversible inhibition of beef liver valyl-tRNA synthetase by alkylating derivative of L-valine, *FEBS Lett.,* 34, 213, 1973.

1013. **Hountondji, C. and Blanquet, S.,** Methionyl-tRNA synthetase from *Esherichia coli:* primary structure at the binding site for the 3′-end of tRNA$_f^{Met}$, *Biochemistry,* 24, 1175, 1985.

1014. **Wetzel, R. and Söll, D.,** Analogs of methionyl-tRNA synthetase substrates containing photolabile groups, *Nucleic Acids Res.,* 4, 1681, 1977.

1015. **Valenzuela, D. and Schulman, L. H.,** Identification of peptide sequences at the tRNA binding site of *Escherichia coli* methionyl-tRNA synthetase, *J. Am. Chem. Soc.,* 25, 4555, 1986.

1016. **Rosa, J. J., Rosa, D. M., and Sigler, P. B.,** Photocrosslinking analysis at the contact surface of tRNA^{Met} in complexes with *Escherichia coli* methionine:tRNA ligase, *J. Am. Chem. Soc.,* 18, 637, 1979.

1017. **Podust, V. N., Nevinsky, G. A., and Lavrik, O. I.,** Modification of *E. coli* phenylalanyl-tRNA synthetase by 1,N^6-ethenoadenosine-5'-triphosphate, *Biopolim. Kletka (USSR),* 1, 267, 1985.

1018. **Lavrik, O. I. and Khodyreva, S. N.,** Modification of the α-subunit of phenylalanyl-tRNA synthetase from *E. coli* MRE-600 with N-chlorambucilyl-phenylalanyl-tRNA, *Biokhimija (USSR),* 44, 570, 1979.

1019. **Gorshkova, I. I. and Lavrik, O. I.,** Investigation of the kinetics of the affinity modification of phenylalanyl-tRNA synthetase as a method for studying the interaction between the binding sites of tRNA and other substrates, *Mol. Biol. (USSR),* 9, 887, 1975.

1020. **Budker, V. G., Knorre, D. G., Kravchenko, V. V., Lavrik, O. I., Nevinsky, G. A., and Teplova, N. M.,** Photoaffinity reagents for modification of aminoacyl-tRNA synthetases, *FEBS Lett.,* 49, 159, 1974.

1021. **Gorshkova, I. I., Knorre, D. G., Lavrik, O. I., and Nevinsky, G. A.,** Affinity labeling of phenylalanyl-tRNA synthetase from *E. coli* MRE-600 by *E. coli* tRNA^{Phe} containing photoreactive group, *Nucleic Acids Res.,* 3, 1577, 1976.

1022. **Lavrik, O. I., Nevinsky, G. A., and Riazankin, I. A.,** Influence of the photoreactive ATP analog structure on the affinity modification of phenyl-tRNA synthetase. Modification of the enzyme at two types of the nucleotide sites, *Mol. Biol. (USSR),* 13, 1001, 1979.

1023. **Renaud, M., Fasido, F., Baltzinger, M., Boulanger, Y., and Remy, P.,** Affinity labeling of yeast phenylalanyl-tRNA synthetase with a 3'-oxidised tRNA^{Phe}. Isolation and sequence of the labelled peptide, *Eur. J. Biochem.,* 123, 267, 1982.

1024. **Baltzinger, M., Fasido, F., and Remy, P.,** Yeast phenylalanyl-tRNA synthetase. Affinity and photoaffinity labelling of the stereospecific binding sites, *Eur. J. Biochem.,* 97, 481, 1979.

1025. **Owens, M. S., Clements, P. R., Anderson, A. D., and Barden, R. E.,** S-Methanesulfonyl-CoA: a thiol-specific reagent for affinity labeling of short chain acyl-CoA sites, *FEBS Lett.,* 124, 151, 1981.

1026. **Collier, G. E. and Nishimura, J. S.,** Affinity labeling of succinyl-CoA synthetase from porcine heart and *Escherichia coli* with oxidized coenzyme A disulfide, *J. Biol. Chem.,* 253, 4938, 1978.

1027. **Ball, D. J. and Nishimura, J. S.,** Affinity chromatography and affinity labeling of rat liver succinyl-CoA synthetase, *J. Biol. Chem.,* 255, 10805, 1980.

1028. **Nishimura, J. S., Mitchell, T., Collier, G. E., Matula, J. M., and Ball, D. J.,** Affinity labeling of succinyl-CoA synthetase from *Escherichia coli* by the 2',3'-dialdehyde derivatives of adenosine 5'-diphosphate, *Eur. J. Biochem.,* 136, 83, 1983.

1029. **Powers, S. G., Muller, G. W., and Kafka, N.,** Affinity labelling of rat liver carbamyl phosphate synthetase I by 5'-p-fluorosulfonylbenzoyladenosine, *J. Biol. Chem.,* 258, 7545, 1983.

1030. **Marshall, M. and Fahien, L. A.,** Proximate sulfhydryl groups in the acetylglutamate complex of rat carbamylphosphate synthetase I: their reaction with the affinity reagent 5'-p-fluorosulfonylbenzoyladenosine, *Arch. Biochem. Biophys.,* 241, 200, 1985.

1031. **Dawid, I. B., French, T. C., and Buchanan, J. M.,** Azaserine-reactive sulfhydryl group of 2-formamido-N-ribosylacetamide 5'-phosphate: L-glutamine amido-ligase (adenosine diphosphate), *J. Biol. Chem.,* 238, 2178, 1963.

1032. **Bottcher, B. R. and Meister, A.,** Covalent modification of the active site of carbamyl phosphate synthetase by 5'-p-fluorosulfonylbenzoyladenosine. Direct evidence for two functionally different ATP-binding sites, *J. Biol. Chem.,* 255, 7129, 1980.

1033. **Easterbrook-Smith, S. B., Wallace J. C., and Keech, D. B.,** Pyruvate carboxylase: affinity labeling of the magnesium adenosine triphosphate binding site, *Eur. J. Biochem.,* 62, 125, 1976.

1034. **Hudson, P. J., Keech, D. B., and Wallace, J. C.,** Pyruvate carboxylase: affinity labeling of the pyruvate binding site, *Biochem. Biophys. Res. Commun.,* 65, 213, 1975.

1035. **Chen, S.-L. and Kim, K.-H.,** Identification of the cyclic AMP and ATP binding sites of acetyl coenzyme A carboxylase by use of 5'-p-fluorosulfonylbenzoyladenosine, *J. Biol. Chem.,* 257, 9953, 1982.

1036. **Long, C. W., Levitzki, A., and Koshland, D. E., Jr.,** The subunit structure and subunit interactions of cytidine triphosphate synthetase, *J. Biol. Chem.,* 245, 80, 1970.

INDEX

Footprinting, 32—33
Fragmentation of biopolymers, 125
Free radicals, 43, 57, 69—70

G

Gene-directed mutagenesis, 192—193
General direct criteria of affinity modification, 157—159
Genetic code, 7
Glutamate decarboxylase, 189
Glutamate dehydrogenase, 171
Glycogen, 169, 172
Glycogen phosphorylase, 18—19
Glycolysis, 10—12, 19
Guanosine reactivities, 30—31
Guanosine residues, 42

H

Hairpin structures, 8, 38
Half-of-the-sites reactivity, 170
Haloketone groups, 46
Haptens, 20
Heterobifunctional reagents, 35—36, 89, 91—96, 104, 192, 197—200
Heterolytic reactions, 43
Homobifunctional reagents, 35
Homolytic reactions, 43
Hormone regulation, 19
Hydrogen bonding, 1—4, 83
Hydroperoxy group, 70
Hydrophobic compounds, 85—86
Hydrophobic interactions, 2—3, 176
Hydroxy groups, 89
Hypnotic drugs, 58

I

Immune system, recognition in, 19—21
Immunoassay, 21
Immunochromatography, 25
Immunoglobulins, 19—20
Immunotoxins, 193
Inactivation kinetics, 39
Inactivation of biopolymer, 116—118, 131
Inflammatory cascade, 70
Influenza virus, 13
Initiation, 15
Insulin, 189, 191
Intercalating dye, 185
Intramolecular recognition, 6—10
Δ^5-Δ^4-Isomerase, 58, 65

K

Kinetic complications in affinity modification, 139—142
Kinetic cooperativity, 141, 148
Kinetic curves, 136—138
Kinetics of affinity modification, 129—157
cooperative effects in affinity modification, 148—151

general rules for kinetic treatment of complicated affinity modification processes, 151—157
main sources of kinetic complications in affinity modification, 139—142
range of validity, 129—134
simplest scheme of affinity modification, 129—134
simultaneous enzymatic and affinity modification processes, 134—136
simultaneous modification of several independent centers by same reagent, 136—139
suicide inhibition, 145—147
unstable reagents, 142—145

L

Labeling of biopolymers, 25
Lac operator, 105
Lac promoter, 16
Lac repressor, 16, 105
β-Lactam antibiotics, 195—196
β-Lactamases, 71, 195—196
β-Lactam ring, 71—72
Lactate dehydrogenase, 84
Laser irradiation, 60
Leukotrienes, 70
Ligand binding site, 168, 187—188
Liposome techniques, 81, 193
Localization of ligand binding sites, 187—188
Localization of peptides, 123—124
Locally damaged biopolymers, 192—193
Lysosomal proteolysis, 190

M

Macromolecular complex formation, 112
Macromolecular structure, 61
Macrophages, 190
Mapping of secondary and tertiary structure, 28—31
Mapping protein and nucleic acid complexes, 36
Mass conservation law, 131, 150
Maxam-Gilbert DNA sequencing methodology, 26, 32, 123, 126
Mechanism-based enzyme inhibitors, 64
Messenger ribonucleic acids (mRNAs), 7, 196
Metal-containing groups of affinity reagents, 75—80
o-Methyltransferase, 53
Micrococcus luteus, 170
Mitochondrial F_1 ATPase, 168, see also F_1 ATPase
Modification points, determination of, 118—127
Molecular recognition, 1—21
cell surface, 16—19
double-stranded nucleic acids, 4—6, 10—14
effectors, 14—15
enzymes, 4—5, 14—15
immune system, 19—21
intramolecular recognition, 6—10
multipeptide complexes, 10—14
multisubunit structure formation, 10—14
noncovalent interactions, 1—6
nucleoproteins, 10—14
regulatory processes, 15—16
substrates, 14—15
templates, 14—15

7 Days